The Monkey's Voyage

The Monkey's Voyage

How IMPROBABLE JOURNEYS SHAPED *the* HISTORY *of* LIFE

Alan de Queiroz

BASIC BOOKS
New York

Copyright © 2014 by Alan de Queiroz
Published by Basic Books,
A Member of the Perseus Books Group

All rights reserved. Printed in the United States of America. No part of this book may be reproduced in any manner whatsoever without written permission except in the case of brief quotations embodied in critical articles and reviews. For information, address Basic Books, 250 West 57th Street, 15th Floor, New York, NY 10107-1307.

Books published by Basic Books are available at special discounts for bulk purchases in the United States by corporations, institutions, and other organizations. For more information, please contact the Special Markets Department at the Perseus Books Group, 2300 Chestnut Street, Suite 200, Philadelphia, PA 19103, or call (800) 810-4145, ext. 5000, or e-mail special.markets@perseusbooks.com.

Library of Congress Cataloging-in-Publication Data

De Queiroz, Alan.
  The monkey's voyage : how improbable journeys shaped the history of life / Alan de Queiroz.
      p. cm.
  Includes bibliographical references and index.
  ISBN 978-0-465-02051-5 (hardcover) -- ISBN 978-0-465-06976-7 (e-book) 1. Animals--Dispersal. 2. Plants--Dispersal. 3. Biogeography. I. Title.
  QH543.3.D42 2013
  570--dc23
                        2013036248

10 9 8 7 6 5 4 3 2 1

For Tara, Hana, and Eiji

CONTENTS

Introduction: Of Garter Snakes and Gondwana 1

SECTION ONE:
EARTH AND LIFE

Chapter One: From Noah's Ark to New York: The Roots of the Story 23
Chapter Two: The Fragmented World 47
Chapter Three: Over the Edge of Reason 73
Chapter Four: New Zealand Stirrings 95

SECTION TWO:
TREES AND TIME

Chapter Five: The DNA Explosion 115
Chapter Six: Believe the Forest 133

SECTION THREE:
THE IMPROBABLE, THE RARE, THE MYSTERIOUS, AND THE MIRACULOUS

Chapter Seven: The Green Web 151
Chapter Eight: A Frog's Tale 175
Chapter Nine: The Monkey's Voyage 203
Chapter Ten: The Long, Strange History of the Gondwanan Islands 225

SECTION FOUR:
TRANSFORMATIONS

Chapter Eleven: The Structure of Biogeographic "Revolutions" 257
Chapter Twelve: A World Shaped by Miracles 281
Epilogue: The Driftwood Coast 305
Acknowledgments 307
Figure Credits 310
Notes 311
References 329
Index 350

CENOZOIC	Quaternary	Holocene	Present
			0.01 million years ago
		Pleistocene	
			2.6
	Neogene	Pliocene	
			5.3
		Miocene	
			23.0
	Paleogene	Oligocene	
			33.9
		Eocene	
			56.0
		Paleocene	
			66.0
MESOZOIC	Cretaceous	Upper Cretaceous	
			100.5
		Lower Cretaceous	
			145.0
	Jurassic	Upper Jurassic	
			163.5
		Middle Jurassic	
			174.1
		Lower Jurassic	
			201.3
	Triassic	Upper Triassic	
			237
		Middle Triassic	
			247.2
		Lower Triassic	
			252.2

Introduction

OF GARTER SNAKES AND GONDWANA

Science must begin with myths, and with the criticism of myths.

—Karl Popper

I recently put up a large map of the world in our house, ostensibly for our daughter and son, ages five and two, although to this point I'm the only one who's looked at it much. As something of a map hoarder, if not exactly a connoisseur, I appreciate a map made with care and some measure of creativity, like this one. It's a standard Mercator projection (the type of map that makes Greenland appear the size of Africa), but beyond that there is hardly anything conventional about it. The continents show no political boundaries and are colored in pale earth tones that blend into each other, the transitions having only the vaguest correspondence with the boundaries of actual biomes. Glass-like fragments depicting sea ice fill the Arctic region, with the smaller pieces cascading southward as if raining down on the rest of the world. The oceans, so often represented on maps as featureless blue expanses, are here pleasingly filled with the topography of the sea floor—the ridges and valleys, the broad plateaus and deep trenches, the gently sloping continental shelves. These characteristics make the map feel dynamic, chaotic, and alive, complementing its most obvious feature, namely, that it's populated with the painted images of dozens of wild creatures, from iguanas and sperm whales to water buffaloes and birds of paradise.

The map is entitled "The World of Wild Animals," but, more accurately, it should be "The World of Wild Vertebrates," and, even within that restricted scope, the coverage is decidedly mammal-centric. Nonetheless, it can serve as an introductory lesson for the budding biogeographer, for the student of how living things are distributed across the Earth. Perusing the map, a fundamental fact of biogeography immediately jumps out: different regions have distinct faunas. That, in fact, is presumably the main intended message of the map. Lions, a giraffe, and an elephant are stacked in a column in Africa; kangaroos hop toward a duck-billed platypus and a frilled lizard in Australia; a family of tigers and a family of pandas cozy up to each other in Asia; penguins are scattered across Antarctica, while the frozen seas of the far north carry puffins and auks, black-and-white birds that look a bit like penguins but aren't. These sorts of connections between animal and place are known even to small children. (Our five-year-old can recite at least a few of them, even if she can't consistently identify Africa or Australia on a map.) In time, those children (hopefully) will learn that it is evolution, the great overarching theory of biology, that makes sense of these differences between faunas; the sets of animals are distinct because they have evolved in isolation from each other. The separate landmasses are like different worlds, with long (unimaginably long) independent histories of descent with modification.

There are exceptions to this grand pattern, however, and it is a large part of the business of biogeography to explain these anomalies. On the "World of Wild Animals" map, for instance, we find that both northern North America and northern Eurasia have wolves, moose, and elk, among other shared creatures. These facts do not fit the rule of separate landmasses having distinct faunas, but they're exceptions that are easily explained: North America and Eurasia were connected at various times in the recent past (most recently some 10,000 years ago, during the last ice age) via the Bering Land Bridge, so the histories of those regions are not as independent as their current separation would suggest.* Just a moment ago in geologic time, wolves, moose, and elk could pass on solid ground between North America and Asia.

---

* The biotas of the tropical parts of Eurasia and the Americas are much more distinct, at least in part because the recent incarnations of the Bering Land Bridge have been too cold for tropical organisms to pass over by that route.

Our children's map raises other questions that are not so easily answered, however. That's especially true if one focuses on the landmasses of the Southern Hemisphere. For instance, on our map we see four kinds of flightless birds in the group known as the ratites: a rhea in South America and an ostrich in Africa, facing each other across the Atlantic, and, thousands of miles from these, a herd of emu in Australia and a kiwi poking at the dirt in New Zealand. These four species are clearly distinct from each other, yet, in the grand scheme of things, they are fairly closely related, so how did they end up in these far-flung places, separated by wide stretches of ocean? Similarly, on the map we see a mandrill in Central Africa staring across the Atlantic in the direction of another monkey, a South American capuchin. Again, these species are obviously different, but they are also obviously part of a fairly tight evolutionary group. And again, they present the puzzle of how closely related species can end up on landmasses separated by oceans. Furthermore, in both of these cases, the seafloor topography artfully depicted on our map indicates that the landmasses in question are separated not by shallow shelves, but by deep ocean. This fact adds to the mystery, because it means we cannot invoke movement across a Bering-type land bridge to explain these piecemeal distributions.

As it turns out, the ratites and monkeys are just the tip of the iceberg. There are southern beech trees in Australia, New Zealand, New Guinea, and southern South America. There are baobab trees in Madagascar, Africa, and Australia. There are crocodiles in most warm parts of the world, including all the major Southern Hemisphere landmasses. There are hystricognath rodents (a group that includes guinea pigs) in South America and Africa. These and many other similar examples collectively make up one of the great conundrums of biology, a riddle that has intrigued naturalists since Darwin's time (and, in some sense, even before that). What can explain this profusion of far-flung, fragmented distributions? How on earth could a giant flightless bird or a southern beech, with seeds that cannot survive in seawater, cross a wide expanse of ocean?

For most of these cases, the answer, the one that we now find in textbooks, came from geologists more than biologists: the flightless birds and the baobabs, the crocodiles and the beech tree seeds didn't have to cross oceans, because the oceans weren't always there. At one time, all the major southern landmasses were part of the enormous supercontinent of Gondwana. However, about 160 million years ago, rifts began to form in the Gondwanan crust, like cracks in an eggshell. The supercontinent began to break up along these fissures, the pieces drifting apart at far less than glacial speed as magma welled up through the crust and spread out as new ocean floor. The Atlantic Ocean Basin formed, pushing Africa and South America apart. Zealandia, a continent including present-day New Zealand, New Caledonia, and other islands, drifted away from a combined Australia and Antarctica, the latter two continents also eventually going their separate ways. India, once attached to Australia, Antarctica, and Africa, famously wandered north and plowed into Asia, forming the Himalayas in the process. This is all part of the worldview of plate tectonics, a theory that, with a flurry of evidence, was swiftly transformed to fact in the 1960s: the Earth's crust is made of giant plates that carry continents and get pushed around as magma spreads out from rifts in the crust. Continents drift.

The pieces of Gondwana carried with them not just soil and bedrock, but also the animals and plants of the supercontinent—the ratite birds, the crocodiles, the southern beech trees, and countless others. Where once there had been a single, continuous Gondwanan biota, now there were many descendant Gondwanan biotas wandering off to their separate fates. The reality of continental drift means that there is no need to invoke miraculous ocean crossings by flightless birds and southern beech seeds. The plants and animals of the Southern Hemisphere didn't have to move; the continents moved for them.

The landmasses of the Southern Hemisphere have been called "Gondwanan life-rafts," a set of giant Noah's Arks that carry with them to this day the ancient supercontinent's flora and fauna, albeit transformed by millions of years of evolution. This landmasses-as-life-rafts story is the iconic tale of historical biogeography, the study of how the distributions of living things change through time. It's the textbook example of how the creation of physical barriers—in this case, seas and oceans—can fragment the distributions of groups of organisms. It's a story simultaneously so obvious and so elegant that it's barely worth arguing about.

Or is it?

It's June 2000. My girlfriend (now wife), Tara, and I have flown to San José del Cabo, near the southern tip of Baja California, and, instead of heading down the coast to party in Cabo San Lucas (where we would have been in our element about like flounders on a freeway), we've rented a jeep and driven some thirty miles north into a different world altogether. We're in a rocky arroyo that drains the eastern slope of a small mountain range called the Sierra de la Laguna, in the company of a few cows and burros, but no people. It's hot and bright, the forested hillsides brown and bare of leaves in the dry season, the sun glaring off the white boulders and sand of the arroyo.

The two of us are crouching next to a nasty, spiny shrub that someone has sarcastically and misogynistically dubbed a *buena mujer*. Tara, maybe thinking about now that the nightclubs in Cabo don't sound so bad after all, is reluctantly gripping the neck of a very large garter snake while I work my fingers down the snake's body to where it disappears into a hole beneath the shrub. The snake has some kind of purchase underground and I'm pulling her out a fraction of an inch at a time, trying not to wrench her too hard in the process, trying also (and unsuccessfully) to avoid jabbing myself on the buena mujer. The process is exhausting, not because it's physically difficult, but because we're fighting against the will of another being; with each pull I feel the snake resisting and I sense her muscles straining and tearing. For all she knows, this is a life-or-death struggle, and she imparts that sense of urgency to our side of the encounter as well. Tara, who's more afraid of snakes than I am but also feels more empathy for them, is not enjoying this episode.

After ten profanity-filled minutes, we get the snake out. I've been studying garter snakes for years and usually find them subtly beautiful, but even I have to admit that this is not a pretty snake. She's messy looking, mostly black but with ragged, dark brown stripes along her sides, as if someone used the torn edge of a piece of cardboard to draw her pattern. The fact that she's trying to sink her teeth into me as I drop her into a pillowcase doesn't help. What this snake lacks in disposition and looks, though, she makes up for in other ways. For starters, she's one of the biggest garter snakes I've ever seen. Back home, when we measure her, she turns out to be almost three-and-a-half feet long, huge for a garter snake and the largest specimen of her subspecies ever recorded, a bit of trivia worth a paragraph-long note in a herpetological journal. I end up using her, along with other snakes caught on this trip, in experiments showing that members of her species change the way they forage depending on the depth of the water, a shift that may mirror the way their feeding behavior has evolved. This snake also turns out to be pregnant and, two months later, she will give birth in the lab to a dozen tiny black garter snakes, all much prettier than their mother.

My real reason to remember this snake now is not her size or her offspring or her foraging behavior though—it's her location, the fact that she came from southern Baja California. The distribution of her species,

I.1 A garter snake, *Thamnophis validus*, from the Sierra de la Laguna, near the southern tip of Baja California. Photo by Gary Nafis.

Thamnophis validus, is what got me thinking about organisms catching rides on drifting tectonic plates. It's why I began thinking about the fracturing of Gondwana.

Baja California is not one of the Gondwanan fragments, but its geologic history is reminiscent of the breakup of the southern supercontinent. At one time, the peninsula of Baja California was just another part of the mainland. No sea separated Baja California from the rest of Mexico, so many terrestrial species must have inhabited both what is now the southern part of the peninsula and the adjoining part of mainland Mexico; there was nothing to stop a mouse from walking (or a seed from being carried by a mouse) from the one place to the other. However, between 4 and 8 million years ago, a crack in the Earth's crust began to form, a fissure between Baja California and the mainland. This rift is at the same border between tectonic plates as the San Andreas Fault, along which the Pacific Plate moves northwest and the North American Plate slides southeast, generating countless California earthquakes. In Mexico, instead of plates sliding past each other, that rift formed and grew wider and wider until, at some point, the fissure broke through to the Pacific Ocean, and seawater poured into the gap, creating the Sea of Cortés.* In other words, Baja California is part of another "life-raft," although the raft is still moored at its northern end to the continent. Biologists who study this region believe that when the Sea of Cortés formed, many kinds of animals and plants were isolated on the peninsula, creating odd cases in which populations in southern Baja California have their nearest relatives on the other side of the sea. In western Mexico, then, it's as if we are catching the breakup of Gondwana in a very early stage, with Baja California playing the part of one of the smaller continental fragments, like Madagascar or New Zealand.

Our dark garter snake, *T. validus*, is one of those species that occurs both in Baja California and across the Sea of Cortés on the Mexican mainland. These snakes are found in the slow rivers, irrigation canals, and mangrove swamps of the coastal plain along most of the western edge of the mainland, but in Baja California they occur only near the southern tip, mostly in the rocky arroyos of the Sierra de la Laguna. *T. validus* is one of the species that supposedly caught a ride on the peninsula as it drifted away from the continent (see Figure I.2).

This "incipient life-raft" story is a compelling hypothesis for the distribution of *T. validus*, but nobody had ever collected the critical genetic data to test it. Robin Lawson, a fellow herpetologist and evolutionary biologist, and I decided to do just that. Between us we took two more trips to Mexico, and, with the help of Tara, my graduate student Matthew Bealor, and an amateur snake enthusiast named Phil Frank, we collected *T. validus* specimens from sites spanning about eight hundred miles of Mexico's west coast, from Sonora to Michoacán. Then we sequenced some of the genes of these garter snakes along with the ones Tara and I had collected in the Sierra de la Laguna.

The results were clear and striking: the Baja California snakes were genetically almost identical to some of their mainland counterparts. The genes we were looking at—genes in the mitochondria that code for proteins—evolve very quickly. Thus, if the peninsular snakes had been isolated from mainland snakes for several million years, as the landmass-as-life-raft hypothesis required, the genes of the two groups would have become quite different from each other. The fact that they were instead

---

* This description is a simplified view of the origin of the Sea of Cortés; the process probably occurred in several stages and involved not only the Pacific and North American Plates but also smaller tectonic plates in the region.

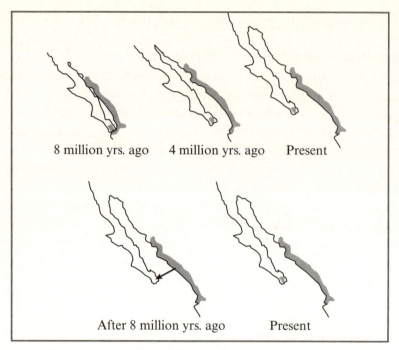

I.2 Two possible explanations for the piecemeal distribution of *Thamnophis validus*. Gray shading shows the range of the species. Upper: fragmentation of the range through the rifting that created the Sea of Cortés. Lower: dispersal across the sea (shown by arrow). Modified from de Queiroz and Lawson (2008).

nearly identical had a clear implication: the life-raft hypothesis, based on the slow movement of tectonic plates, could not explain why *T. validus* is in Baja California.

The best explanation of this extreme genetic similarity is that snakes on the Mexican mainland crossed the 120-mile width of the Sea of Cortés very recently ("very recently" meaning within the past few hundred thousand years) and established a population in southern Baja California (see Figure I.3). They didn't ride with the drifting peninsula, but instead jumped the gap long after the sea had formed. If there was any kind of raft involved, it was probably a literal one, a log or a clump of vegetation driven by an easterly wind and carrying a few snakes (or even just one pregnant female) across the sea.

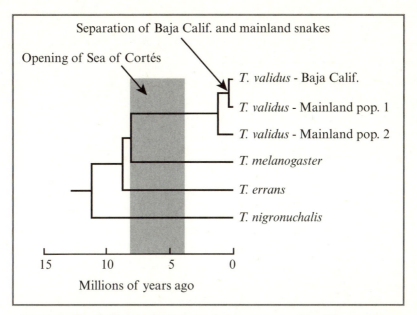

I.3 Part of a DNA-based "timetree" for garter snakes. The tree suggests that Baja California *Thamnophis validus* separated from mainland snakes only within the past few hundred thousand years, much more recently than the physical separation of the peninsula from the mainland (indicated by shading). Modified from de Queiroz and Lawson (2008).

The landmasses-as-life-rafts hypothesis is part of a school of thought with the somewhat imposing name of *vicariance biogeography*. Although I will try to avoid the use of scientific jargon in this book, "vicariance" is a word that I cannot get around and need to define. (For more definitions, see Box.) It refers to the fragmentation of the range of a species or larger group into isolated parts by the formation of some sort of barrier, as with the Sea of Cortés. As another obvious example, consider the effects of rising sea levels after the most recent ice age. At the peak of glaciation, about 18,000 years ago, a vast layer of ice extended from the Arctic past the Great Lakes in North America and as far south as Germany and Poland in Europe. Because so much of the world's water was tied up in this ice, sea levels were much lower than they are today, which meant that many areas that are now underwater were exposed as land. As the ice melted, seas rose by more than 300 feet, and some places that had been parts of continents were transformed into islands, like sand castles surrounded by a rising tide. For example, much of what had been continental Southeast

A FEW THOUGHTS ON BIOGEOGRAPHIC TERMS AND CONCEPTS

The basic notions of long-distance dispersal and vicariance are fairly straightforward, but a few points about these and related concepts may be helpful. This box also serves as a glossary for the very few technical terms commonly used in this book.

Normal dispersal is the expected movement of organisms either within continuous tracts of suitable habitat or between patches of suitable habitat that are close together. Say the climate is warming at the end of an ice age. As the ice retreats and new habitat slowly opens up, beech trees and squirrels on the edge of the area move into the previously ice-covered region. That's normal dispersal for the trees and the squirrels. No improbable jump is required to explain it. *Long-distance dispersal*, in contrast, involves the movement of organisms across an area that is, for those organisms, a substantial barrier to dispersal. Because of the barrier, this kind of movement is both unexpected and unpredictable; long-distance dispersal is thus sometimes referred to as *chance* or *sweepstakes dispersal*. Obvious examples include the movements of nonflying vertebrates from continents to islands many miles offshore or of many kinds of lowland organisms across high mountains. In general, a population founded by long-distance dispersal will be genetically isolated from the source population because movement between them is difficult; thus, populations originating in this way will tend to diverge from the source population. This is why, for instance, native land animals on remote islands are almost always classified as distinct species from related mainland forms. Both normal and long-distance dispersal must be defined in light of an organism's particular dispersal abilities. For example, crossing a mile-wide sea channel would qualify as long-distance dispersal for a frog or a mouse, but would be normal dispersal for many birds.

A *disjunct distribution*, in the simplest terms, is any discontinuous distribution in which some part of the species or larger group is separated from another part. The cases described in this book always involve disjunctions in which the parts are separated by a substantial barrier (or barriers) to dispersal, usually an expanse of ocean. One way to think of these distributions is that movement between the separated parts today would require long-distance dispersal by the organisms in question (if the movement is even possible).

Vicariance is the splitting of the continuous range of a group into two or more parts by the development of some sort of barrier to dispersal. In its strict sense, vicariance refers to the fragmentation of the range of a *species*, and is a mechanism whereby one species becomes two or more species. For example, in the case of the ratite birds, vicariance implies that each geologic fragmentation event—the separation of South America from Africa, India from Madagascar, etc.—divided the range of a ratite species. I follow this strict definition, with a major exception. Specifically, when dealing with molecular clock and other dating studies, I take vicariance to mean the fracturing of the distribution of *any* taxonomic group (whether a species or a higher-level taxon such as a genus or family), a process that might or might not be connected to the birth of new species. As an illustration, suppose that an ancestral ratite species had spread by normal dispersal all over Gondwana, but that, *while the supercontinent was still intact*, this ancestor evolved into distinct species in the areas that would become Africa, South America, and so on. The breakup of Gondwana would then have left ratites on landmasses separated by oceans, as in the strict case, but, in this alternate scenario, new species would have arisen *before* the fragmentation of the supercontinent.

This broader definition has been implicitly adopted in many molecular clock studies, probably because it simplifies distinguishing long-distance dispersal from fragmentation. Specifically, vicariance, broadly defined, subsumes *all* explanations that involve fragmentation of an ancestral range and do *not* require long-distance dispersal. Thus, if we reject vicariance in this sense, we are necessarily also supporting long-distance dispersal. For molecular clock studies, what this means is that results fall into two categories: if a particular evolutionary branching point is estimated to be as old *or older* than the fragmentation event in question, that branching age is deemed consistent with vicariance, while, if the branching point is estimated to be younger than the fragmentation event (as in Figure I.3), then long-distance dispersal is supported. In any case, the general message of the book is not affected by these definitional issues.

> To produce a disjunct distribution, long-distance dispersal has to be followed by the establishment of a permanent population in the new area. In many cases, establishment in a new environment may be more difficult to achieve than long-distance dispersal per se. I will often use "dispersal" to mean "dispersal and establishment"; the meaning in these instances should be obvious from the context.
>
> A *taxon* is a taxonomic group and might refer to a species, a genus, a family, or a group at any other level in the taxonomic hierarchy. *Homo sapiens* is a taxon, as is the genus *Homo*, and the family Hominidae. The plural of "taxon" is *taxa*.
>
> *Sister groups* are lineages that are each other's closest evolutionary relatives. Among living species, for example, the two species of chimpanzees are sister groups to each other, and these two chimp species together form a lineage that is the sister group to humans. The concept can apply to any level in the Tree of Life; marsupial mammals and placental mammals are sister groups, as are green plants and red algae.
>
> A *timetree* is a representation of an evolutionary tree in which the estimated ages of the evolutionary branching points (for example, the split between the human and chimp lineages) are indicated (see Figure I.3).
>
> A *continental island* is one that previously was connected to a continent and became an island, either because of submergence of a land bridge (as was the case for Sumatra, Java, and other islands of the Sunda Shelf), or because of tectonic processes (as was the case for pieces of Gondwana, such as Madagascar and New Zealand). An *oceanic island* is one that emerged *de novo* from the sea and has never been connected to a continent. All of the oceanic islands discussed in this book were created by volcanoes. Hawaii and the Galápagos are classic examples.

Asia was inundated, leaving the higher regions as the islands of Sumatra, Java, and Borneo, among others. With the fragmentation of land areas, terrestrial species that had been spread out across the region during the glacial period inevitably had their ranges broken up as well. Today, populations of the same species of frogs, snakes, monkeys, and other organisms can be found on Sumatra, Java, Borneo, and the Southeast Asian mainland. Many of them probably were in all those places before the rise in sea level; they achieved their piecemeal distributions simply by staying put while the waters rose around them, isolating their populations on the various islands and on the continent. The frogs, snakes, monkeys, and other species experienced a *vicariance event*, a breaking up of their formerly continuous ranges.

The archetypal vicariance event (actually a series of events) is the one I began with, the fragmentation of the distributions of Gondwanan plants and animals through the breakup of the supercontinent. In that case, as in the example of Southeast Asian islands, the newly formed barriers are seas or oceans, but there are many different kinds of barriers, many different ways that members of a group can be cut off from each other. For instance, the onset of a drier climate can turn wooded lowlands into desert, while leaving woodlands intact at higher elevations; the result might be fragmentation of the ranges of woodland species into isolated populations on separated mountain ranges. In effect, the dry climate turns the mountains into habitat islands. Similarly, the formation of a land connection creates a barrier for aquatic organisms, as when the rise of the Isthmus of Panama some 3 million years ago separated populations of fishes, shrimp, and other ocean species in the Pacific and Caribbean. Ultimately, those barriers generate new species because the separated populations no longer exchange genes and eventually evolve in different directions. Many of the sea creatures that had their distributions divided by the Panamanian Isthmus, for example, are now classified as separate species on the Pacific and Caribbean sides.

Vicariance biogeography emphasizes such fragmentation events as explanations for the distributions of species and higher taxa (genera, families, etc.). In particular, when a biogeographer with this mindset comes across a taxonomic group with a distribution made up of disconnected areas—like the flightless ratite birds spread across the southern continents—his first thought is "What external process (say, climate change or continental drift) broke the distribution into pieces?" He may think it's *conceivable* that a piecemeal distribution of that sort could be the result of long-distance ocean crossings, but that possibility will be an afterthought, something almost unworthy of real attention. (In fact, as I will describe later, many biogeographers of this school think that hypotheses invoking long-distance dispersal, whether over land or water, are not only unimportant but unscientific.)

The rise of vicariance biogeography in the 1970s was a big deal within the discipline, to put it mildly. It changed the way biologists thought about the distributions of living things in two fundamental ways. First, as just mentioned, it put the fragmentation of environments at the front of people's minds. Second, because fragmentation affects many groups in the same way—for instance, rising seas will break up the ranges of multiple terrestrial species at once—it made people think about generalities, about patterns of distribution that are shared by different taxonomic groups. Biogeography has a long history of attempts to generalize across such groups, but the emphasis on vicariance made that kind of generalization almost inescapable. In other words, it forced people to consider, not just the geography of their own favorite genus of legless lizards or snapping shrimp, but how the distributions of whole biotas may have been broken up through time. Vicariance biogeography often has been called a scientific revolution: it dramatically changed many biologists' views of the history of life, and the way they approached their science. To teach biogeography today without mentioning vicariance—and tectonic-driven vicariance, in particular—would be like teaching physics without quantum mechanics, or molecular biology without the double helix.

At the time of our snake-collecting trip to Baja California, I knew relatively little about biogeography, and what I did know was mostly filtered through the lens of vicariance. For instance, in teaching an evolution course at the University of Colorado, I had devoted a couple of lectures to biogeography and had used, as my key example, distributions fragmented by the breakup of Gondwana. Thus, when I began reading articles as background for writing the paper on garter snakes crossing the Sea of Cortés, I expected to encounter mostly studies supporting landmass as-life-raft theories, that is, vicariance via continental drift. That is not what I found. Instead, I kept running across recent papers in which the authors *expected* to find evidence for landmasses as life-rafts, but ended up arguing for a very different kind of explanation for disconnected distributions, namely, dispersal of plants and animals across seas and oceans. In other words, lots of biologists were finding just what we had found for the Baja California garter snakes.

Many of these studies were about the southern continents and continental islands, the pieces of ancient Gondwana. The papers arguing for ocean crossings kept piling up on my desk—tortoises from Africa to Madagascar, some two hundred plant species between Tasmania and New Zealand, southern beeches among several Southern Hemisphere landmasses, baobab trees between Australia and Africa, rodents from Africa to South America. At some point in my frenzied reading of all these articles, I went from thinking that there were some really weird cases of oceanic dispersal out there to thinking that the weird cases might actually be the norm. To put it another way, my mind jumped from the iconic view of Gondwanan landmasses as life-rafts to something resembling an airline map, with the route lines tracing countless ocean crossings between the disconnected and now widely separated fragments of the supercontinent.

This epiphany, which I soon learned was happening to other biologists as well, was dramatic. Obviously, the continents had moved—nobody was claiming that the theory of plate tectonics was wrong—and obviously, they had carried species with them, but somehow, these facts did not explain nearly as much about the modern living world as we had thought. Instead, what accounted for many of the most strikingly discontinuous plant and animal distributions was a process that had previously occupied some sleepy backwater in my mind, that is, seemingly implausible, improbable ocean crossings.

The goal of this book is to tell the story of this recent sea change in biogeography, from a view dominated by vicariance to a more balanced outlook recognizing that the natural dispersal of organisms across oceans and other barriers is also hugely important. In a nutshell, the point is to recount how the field of biogeography flipped from landmasses-as-life-rafts and other fragmentation scenarios to something closer to the airline route map, using Gondwana as the geographic focus. Ultimately, I also want to explain what this dramatic shift in thought tells us about both the nature of scientific discovery and the history of life on a grand scale. It may even tell us, on one level, why we are here.

The book is divided into four sections. The first provides the historical background, setting the table for what will follow. This section begins with Charles Darwin and the birth of evolutionary views about the distributions of living things, describes the rise of vicariance biogeography, and ends with inklings of the sea change among New Zealand scientists. The brief second section deals with a critical but controversial source of evidence in biogeography, namely, *molecular clock* analyses, which are used to infer the ages of branching points in evolutionary trees (such as the time at which Old World monkeys and New World monkeys separated from each other). The third section is, in an obvious sense, the "meat" of the book; there I set forth the main examples that have turned biogeography on its head. The four chapters of this section can be seen as successive ratcheting steps in an argument for discarding the extreme vicariance position and replacing it with the view of a living world strongly molded by ocean crossings and other chance dispersal events. Finally, in the fourth section, I present the deep implications—the "big picture" messages—of the new worldview with respect to, first, the way in which science progresses (or fails to progress) and, second, the nature of the long history of life on Earth.

In December 2006, a few years after my garter-snake-induced epiphany, I found myself visiting one of the smaller fragments of ancient Gondwana. Tara, her mother, and our friend Jan—all botanists—had signed up for a field course on the ferns of New Zealand, and it had taken Tara about ten seconds to convince me that I should go too. For a naturalist, New Zealand is one of the wonders of the world; the biologist Jared Diamond has called its flora and fauna "the nearest approach to life on another planet." As pretty as ferns are, I didn't want to spend two weeks fixated on them while crawling on all fours in the mud, but I figured I could go off on my own and try to find some of Diamond's alien life forms, then meet up with the others after their course was over. Perhaps I could see a tuatara, a lizard-like reptile in an order that is thought to have died out everywhere else while dinosaurs still roamed the Earth; or imposing kauri trees, as thick as California's giant sequoias and covered with their own forests of epiphytes; or a Wrybill,* a shorebird with a beak that bends not up or down but sideways (almost always to the right, as it turns out). So, while Tara and the rest of the "ferniacs" left Wellington in their tour bus, I rented a car, headed north for the kauri forest, and eventually ended up traversing most of the length of the country. (While there I was very careful about driving on the left, looking right when crossing streets, and so on, but on returning to the United States, with my brain still reverse-wired, I promptly turned onto the wrong side of a busy boulevard in Las Vegas. Luckily, Tara yelled loudly before we came close to colliding with the oncoming traffic.)

---

* I have capitalized common names of bird species, following the established convention among ornithologists, but I have not capitalized the common names of species in other groups.

As I wandered around New Zealand, from the subtropical forests in the north to the glacial valleys of the south, I was constantly running into signs of Gondwana. Not signs as in biological or geological evidence, but signs as in signage. Almost every nature preserve and national park had signs or pamphlets mentioning the Gondwanan origins of New Zealand's flora and fauna. The country seemed to be part of both the British Commonwealth and an even larger league of nations, the fragments of the former southern supercontinent.

At Nelson Lakes National Park, in the northern part of the South Island, I walked in a mossy forest of the famous Gondwanan trees, southern beeches. Many of the tree trunks were blackened by a fungus that grows on the honeydew that drips out of the rear end of a scale insect, giving the forest a slightly diseased look (although the fungus apparently does no harm to the trees). Still, the trees were beautiful, their foliage delicate and layered, in places making them look like overgrown bonsai. Thumbing through a small field guide to the trees of New Zealand, I picked out, by the size and shape of their leaves, at least three species—red, silver, and mountain beech. *Nothofagus fusca, Nothofagus menziesii, Nothofagus solandri*. According to the landmasses-as-life-rafts story, they are all part of a lineage that has been in New Zealand since the breakup of Gondwana.

For a while the forest trail seemed to wander aimlessly, finding and then losing the course of a small creek, but eventually it gathered purpose on a long set of switchbacks up toward a ridgetop. Reaching the treeline (the *bushline*, to Kiwis) I was startled to find it unlike anything I had experienced in the mountains of North America. As I climbed, the beech trees got smaller and smaller, but the forest didn't thin out gradually, as I had expected. Instead, within just a few steps, the dwarfed but still dense forest disappeared, and I entered a completely treeless alpine zone. It was like walking from woods into a farmer's cleared field. This alpine area turned out to be as strange to me as the abrupt passage into it; in the Rockies or the Sierra Nevada, the vegetation above the treeline is sparse, or very short, or both, but in these New Zealand mountains, much of the treeless alpine area was thickly covered with tall tussock grass.

Just above the treeline I found a flat outcrop that made a good seat, where I caught my breath and admired the view across the deep blue of a large lake, Rotoiti, to the paler blues and greens of the mountains beyond. This seemed to be the place where most people turn around, or, at least, the place where whoever planned the trail *thought* people should turn around, because, above this point, the path narrowed, and the carefully laid switchbacks became a steep beeline to the crest of the ridge.

On this last, gasping scramble to the ridgetop, with my nose almost in the dirt, I gained a greater appreciation for the subtle beauty of the alpine plants. All around me the tussock grass whipped and undulated in the heavy wind. Between the tussocks and beneath my clutching hands were stringy, dark-green plants with scaly leaves like a juniper, shiny yellow buttercups, and mats made up of rosettes of narrow sage-colored leaves. Wandering along the ridgetop I noticed a pale gray mound a couple of feet wide affixed to the flat surface of a rock. Up close, the mound resolved itself into thousands of leaves, each one rolled up into a tiny cylinder, hard to the touch. It was a vegetable sheep, a silly name but an apt one—from a distance, a group of these plants looks like a shepherd's flock. Vegetable sheep are in the sunflower family, but, remembering Jared Diamond's words, I thought of them as sunflowers from another planet.

On the ridgetop, the wind was roaring in my ears and threatening to blow me off my feet. But, to my relief, just a few steps down on the lee side of the ridge it was perfectly calm and quiet, as if someone had flipped off the switch on the wind machine. I had passed a few people on the trail, but now I was alone in the abrupt silence. I sat down, drank some water, and took in the view—the rocky ridge, the washed-out earth tones of the alpine landscape, the dark green of the beech forest below. The place felt untouched and ancient.

If I had visited this spot a few years earlier, I would have thought of the southern beech trees, the vegetable sheep, and the other plants as descendants of the flora that drifted off with New Zealand as it broke away from other parts of Gondwana. No doubt I would have felt the mythic power of that story as I sat in the quiet solitude of the mountains—*Here I am, on an actual piece of Gondwana, surrounded by its ancient flora!* Instead, an entirely different scenario passed through my mind. I imagined a tangle of trees, perhaps blown down by a storm, floating on a wide ocean thousands of miles from land, with fruit still in the trees' branches, and seeds in the dirt stuck to their roots. In the dark recesses of the tangle, I envisioned spiders and crickets and lizards clinging to the branches.

And I thought, "It's time for a new story. It's time to change those signs."

*At 6:00 in the morning on December 14, 2004, an Aldabra giant tortoise (*Dipsochelys dussumieri*), the Indian Ocean's analogue to the oversized tortoises of the Galápagos, ambled out of the sea at Kimbiji, 22 miles south of Dar es Salaam in Tanzania. Inspection of the tortoise's shell showed faint concentric growth rings, indicating that the animal came from the native population on Aldabra, where the high density of tortoises leads to slow growth, rather than from introduced populations elsewhere in the Seychelles or on Changuu Island near Zanzibar. Aldabra also made sense as the point of origin based on the direction of prevailing currents. A trip from Aldabra to Kimbiji would cross 460 miles of ocean waters as the crow flies, and presumably somewhat farther as the tortoise floats.*

The Kimbiji tortoise was emaciated, as one might expect, but even more telling was the fact that its front legs and part of its lower shell were covered with thickets of goose barnacles, like the hull of a boat. Barnacles settle as tiny larvae and, once fixed, do not move. From the size of the largest ones, it was surmised that the tortoise had been in the ocean for at least six weeks.

I.4 The Kimbiji tortoise. Photo by Catharine Joynson-Hicks.

Section One

EARTH *and* LIFE

Chapter One

FROM NOAH'S ARK TO NEW YORK: THE ROOTS OF THE STORY

Prelude: Croizat's Vision

Léon Croizat sat at his desk, writing... and seething. For Croizat, an Italian botanist living in Venezuela, writing seemed as natural as breathing, and nearly as constant—from 1952 to 1962, his especially prolific period, he published four technical biology books totaling close to 6,000 pages—and when he wrote, he was often thinking about Charles Darwin. And when he thought about Darwin, he seethed. It was not about religion—Croizat was as complete an evolutionist as Darwin had been. However, in Croizat's eyes, Darwin had gotten almost everything about evolution wrong. To begin with, Croizat believed that natural selection was a trivial part of evolution, not its main driving force. More than anything, though, he hated Darwin's views of historical biogeography, of the means by which living things had acquired their particular distributions on the Earth.

Croizat had a grand vision, a unified theory of the geography of life. It boiled down to this: the distributions of groups, from orchids to earthworms to armadillos, all reflected the dynamic climatic and geologic history of the planet itself. Sea levels rose to inundate land bridges; ocean basins opened, dividing continents; island arcs plowed into continental margins. These changes in the configurations of landmasses and oceans left an indelible imprint on life. In fact, that imprint was so unmistakable that one could use the distributions of living things to reveal the history of the Earth. Find out where the orchids and the worms and the armadillos live, and the arrangements of the continents through time also would be revealed.

Croizat gave his theory a name befitting its all-encompassing nature, its power to explain the distributions of living things over the entire planet. He called it *panbiogeography*. He also provided a memorable phrase, probably the most memorable one in the history of the discipline, five words that captured the essence of his worldview: "Earth and life evolve together."

Croizat's panbiogeography ran counter to an idea that had a long history among biologists and naturalists, namely, that the discontinuous distributions of species and higher taxa often were the result of chance dispersal, of unpredictable, long-distance jumps. To the extent that such dispersal was common, it meant that distributions did *not* reflect Earth history. Terrestrial organisms, for instance, could move even among landmasses that were widely separated. To Croizat, this was lunacy, mere storytelling founded on absurdly improbable events. Beyond that, it robbed biogeography of any kind of generality, because a different story might apply to every taxonomic group. Perhaps snails had reached the Hawaiian Islands attached to the feathers of a bird, spiders by using long silk strands to float on storm winds, and bean trees as seeds embedded in a raft of vegetation. And perhaps ants and termites and bumblebees had not reached those islands simply because, well, because they had not. This view of biogeographic history was pure chaos, the antithesis of unification. And where did this pabulum come from? It came from Charles Darwin. To most biologists, Darwin was like a secular saint, even a deity, but to Croizat he was a fool and worse—he was the unthinking dilettante who had come up with an unsupportable view of the geographic history of life and somehow convinced almost everyone that he was right. A hundred years after publication of *The Origin of Species*, in which Darwin had presented his ideas on chance dispersal, the field of biogeography was still laboring under the delusions of the "master."

Croizat thought it was time for this long, anti-intellectual chapter to come to an end, and that he, of course, would be the one to end it.

Beginnings

At Down House, his country home in Kent, Charles Darwin worried about the implausibility of long-distance dispersal, especially dispersal over water. He thought it was a problem for his theory of evolution. How could the same species, or two species that were closely related by descent, turn up in regions separated by seas or oceans? For that matter, how did many species find their way to oceanic islands, which were separated from *everywhere* by ocean barriers? Darwin had all kinds of reasons to believe that species were connected by descent, but he thought this problem—the problem of related groups living in areas divided by large bodies of water—could be a sticking point for skeptical readers. In these cases, it almost seemed as if creation were a better explanation than evolution. Wasn't it easier to imagine that God had created the same or related species in these widely separated places than to envision all manner of animals and plants making absurdly long ocean voyages? Could iguanas really have rafted from South America to the Galápagos on their own? Could beech-tree seeds have floated from Australia to New Zealand?

Darwin was aware of the other natural (as opposed to divine) explanation for such distributions, that is, the existence of former land connections. However, over time he had come to view the easy use of such explanations as little better than invoking the supernatural. It seemed like cheating, pulling something out of thin air, or, more precisely, conjuring up land out of the deep, unfathomable ocean. He and his close friend, the botanist Joseph Hooker, had a running argument about the subject. Like Darwin, Hooker had taken a long ocean voyage, as a naturalist aboard the HMS *Erebus* and the HMS *Terror*, and, seeing obvious similarities among the floras of various southern lands, he had suggested that plants had moved across now-sunken land bridges. Darwin did believe that lands had risen and fallen—he had seen evidence of rising land in the Chilean Andes and of subsidence in the coral islands of the Pacific—but he didn't like using land-bridge explanations in specific cases when there was no geological evidence to back them up. In an 1855 letter to Hooker, he wrote, "It shocks my philosophy to create land, without some other & independent evidence" (that is, other than distributions of organisms). Hooker, for his part, was equally skeptical about some of Darwin's ideas on the dispersal of plants and animals across water. He especially didn't like Darwin's penchant for suggesting transport on icebergs. The *Erebus* and the *Terror* had journeyed to Antarctica, crashing their way through ice floes, and Hooker had seen his share of icebergs. He had the impression that not many living things caught rides on them.

The issue of oceanic dispersal was important enough to Darwin that, from 1854 to 1856—while he was still waffling over how to present his evolutionary ideas publicly—he conducted a whole series of experiments at Down House to figure out whether plant seeds and other propagules could possibly cross large water barriers. He put the seeds of eighty-seven kinds of plants in bottles filled with salt water for weeks and months, then planted the seeds to see if they were still viable. He dangled the disembodied feet of a duck in an aquarium to see if hatchling freshwater snails would cling to them. Knowing that some fish would eat plant seeds, he forced seeds into the stomachs of fish, fed the fish to eagles, storks, and pelicans, and then tried to germinate the seeds he retrieved from the birds' droppings.

The experiments convinced him that long-distance oceanic dispersal was a lot more likely than one might think. Many kinds of seeds survived after being immersed in salt water for 28 days, and a few survived for 137 days. The young snails did climb up onto the duck's feet, suggesting they could hitch a ride to wherever a duck might fly (although the distance would be limited to the time it takes a tiny snail to dry up). Some of the seeds from the eagle, stork, and pelican droppings germinated, indicating another possible means of transport by birds. Careful, as always—Darwin was nothing if not a careful thinker—he reasoned that seeds on their own wouldn't make it very far because they would sink. So he also collected dry branches with fruits attached and dropped these into salt water to see how long they could remain afloat. Combining the results of these floating-branch experiments with the seed-viability numbers and estimates of the speed of ocean currents, he calculated that seeds of 14 percent of plant species could travel at least 924 miles and still germinate at the end of the trip.

Darwin wrote quite a few letters to Hooker describing these results and, like most experimentalists, he seemed to take pleasure in conveying the difficulties of the work. "It is quite surprising that the Radishes sh^d [should] have grown, for the salt-water was putrid to an extent, which I c^d [could] not have thought credible had I not smelt it myself," he wrote in one letter. He also enjoyed what sounds like a self-effacing, Victorian version of trash-talking at Hooker's expense: "When I wrote last, I was going to triumph over you, for my experiment had in a slight degree succeeded, but this with infinite baseness I did not tell in hopes that you would say that you would eat all the plants, which I could raise after immersion." Eventually, he changed Hooker's mind on the subject. At one point, Hooker even conceded that "I am more reconciled to Iceberg transport than I was." Darwin had won a round for dispersal explanations.

It was not as if Darwin were the first person to think about oceanic dispersal. In fact, some 250 years earlier, in the late 1500s, there had been a surge of interest in both overwater dispersal and land bridges. What brought on this early attention to the geography of living things was, oddly enough, a shift from allegorical to literal interpretations of the Bible. In particular, taking the story of Noah's Ark at face value meant that all the animals in the world, two by two, must have ended up in a crowd on the top of Mount Ararat after the Flood. This meant that somehow animals had repopulated the world from that single spot, which in turn required them to cross oceans. How had they done it? One theory was that transoceanic journeys had been made in stepping-stone fashion, with the animals swimming from island to island. Another was that animals had traveled as cargo on boats (the same boats with which people had repopulated the world). A third had animals crossing from the Old World to the New World on the lost continent of Atlantis.

An English historian of science named Janet Browne has argued persuasively that these biblically motivated ideas about the colonization of the world by animals (plants weren't part of the Ark story) mark the beginnings of scientific thinking about the distributions of living things on the Earth. They also may represent early inklings of the dispersal-vicariance debate: the stepping-stone and cargo ideas are obviously about long-distance dispersal, and the notion of Atlantis as a land bridge looks like a vicariance hypothesis, with the continuous ranges of animal "kinds" being split into Old and New World portions by the drowning of the lost continent.

Still, if these ideas represented the beginnings of biogeography, it was a case of rational or semirational thinking being piled on a foundation of myth. To my mind, *modern* biogeography—that is, a science that would be instantly recognizable to, say, a grad student poring over evolutionary trees generated from monkey DNA sequences—began with Darwin and his putrid seed bottles, disembodied duck's feet, and eagle droppings. To be precise, it began with two assumptions about the history of life that led Darwin to perform those experiments.

I alluded to one of these assumptions above, the obvious one, the idea of evolution itself, which Darwin had come to accept as fact in 1837, not long after returning from his voyage on HMS *Beagle*. More specifically, the notion that each species originated in a single place, having evolved from some other species, and the related premise that similar species had evolved from a common and localized ancestor, meant that disjunct distributions had to be explained by natural movements of organisms. If a species originated in one place but ended up in a second place, across a sea or ocean, some kind of explanation was needed. In essence, Darwin was faced with the problem of Mount Ararat all over again, only this time without ancient people hauling animals all over the world on boats. He needed natural explanations for such distributions, and that meant either land bridges or his favored mechanism, oceanic dispersal.

Darwin's second key assumption was that the Earth was enormously old, an idea he may have picked up from reading Charles Lyell's *Principles of Geology*. Lyell, following the eighteenth-century Scottish geologist James Hutton, had argued for geological uniformitarianism, the theory that the features of the Earth had been generated by processes people could still observe, such as erosion and vulcanism, acting at relatively constant rates. From uniformitarianism, it followed that some features required an awfully long time to reach their present form: the Grand Canyon, for instance, must have been created over eons, as the Colorado River carved its way down into the rock, inch by inch, year by year. This in turn meant that the Earth itself must be exceedingly old. Exactly how old was a matter of much questionable conjecture, but the planet was clearly many millions of years old, not just a few thousand, as biblical literalists believed. The acceptance of this incomprehensibly long history—what the writer John McPhee would later call "deep time"—meant that the processes and events that influenced the distributions of living things had had a very long time to operate. Of course, the age of a single species or group of closely related species did not extend all the way back to the origin of the Earth, but such groups might still be many thousands or even millions of years old. This realization was critical to Darwin's belief in the importance of dispersal, because, although he had shown that long-distance colonization over water was possible, he was not arguing that it happened frequently. The dispersal of seeds or birds or insects across the Atlantic or to the Galápagos or Hawaii would be rare, at best, so long stretches of time were required to account for the observed distributions.

In short, what Darwin had begun, as an outgrowth of trying to prove the truth of evolution, was the new science of historical biogeography. Soon, he would have company in this new field.

In 1855, while Darwin was fiddling with his saltwater seed bottles and duck's feet at his home in Kent, Alfred Russel Wallace, who was then thirty-two (fourteen years younger than Darwin), was collecting natural history specimens, especially beetles, in the independent kingdom of Sarawak on the island of Borneo. Unlike Darwin the gentleman, Wallace was from a working-class family and had toiled at various other jobs before deciding to earn a living as a collector. He had already spent four years in the Amazon—a trip that ended with the loss of most of his specimens and notes in a shipboard fire and a subsequent week and a half spent aboard a lifeboat in the Atlantic—and now he was on what would ultimately be an eight-year sojourn in the Malay Archipelago.

Wallace is sometimes remembered as just that guy who pushed Darwin to publish his theory of natural selection by coming up with the same idea years after Darwin did. But Wallace was a thinker of great scope and depth in his own right. Like Darwin, he seems to have been an honest and generous man, but he may have been more ambitious than his older colleague, or, at least, had ambitions less tempered by caution. In his mid-twenties, while planning his Amazon trip, he was already hoping to gather facts "towards solving the problem of the origin of species." It was in Sarawak that he made his first big step toward that goal. With the rainy season holding up his collecting, he had some time on his hands and made good use of it, writing a theoretical paper called, somewhat cryptically, "On the Law Which Has Regulated the Introduction of New Species." He wrote the paper in February 1855 and shipped it off to England, where it was published later that year in *Annals & Magazine of Natural History*.

The key observation in Wallace's paper was that close taxonomic connection went hand in hand with close geographic association. For instance, within a widespread taxonomic family, species in the same genus tended to be found in the same geographic area, or at least near each other, whereas species in different genera often were not geographically close to each other. To take a nonrandom example, the garter snakes I study make up the genus *Thamnophis*, a group confined to North America, but there are other genera within the same snake family on every continent except Antarctica. Something analogous could be seen in the fossil record: within a family, for example, genera from the same time period tended to be more alike than those from different periods.

Such observations did not originate with Wallace, but the conclusion—the law—he drew from them was radical. "*Every species*," he wrote, "*has come into existence coincident both in space and time with a pre-existing closely allied species*" (italics in original). What he was implying was that species evolved from other species; similar species were associated in space and time because they arose from a common ancestral species that lived in that same area. For emphasis, he included the statement of his "law," italicized, at both the beginning and the end of the paper, but unfortunately the message was still a bit cryptic. He never quite came out and said species A gave rise to species B. Some readers got the evolutionary message; others didn't. Charles Lyell was so impressed by Wallace's arguments that he started thinking much more seriously about whether species evolved from other species (although it wasn't until ten years after publication of *The Origin of Species* that he finally conceded that they did). Weirdly, Darwin read Wallace's paper and, at least initially, didn't see it as either interesting or evolutionary. In the margins of his copy, he wrote, "nothing very new," and, "It all seems creation with him." His misreading of the "Sarawak paper," as it came to be known, is especially odd, since he was already making virtually the same arguments for evolution based on the geographic proximity of similar species. It is hard not to think that Darwin, worrying that someone would scoop him, subconsciously distorted Wallace's paper into something that didn't overlap much with his own thinking and, therefore, didn't threaten him.

Three years later, the parallel thinking of Darwin and Wallace would become unmistakable, and part of the lore of scientific history when Wallace, holed up with a malarial fever on the island of Ternate in the Moluccas (Maluku Islands), flashed upon the survival of the fittest as the mechanism for evolution. Within a few days, he had written a paper on the subject and, in what must be one of the most bizarre coincidences in the history of science, sent the manuscript to just one person, a man he barely knew, Charles Darwin. This time, Darwin got the point and nearly had a conniption. Had Wallace instead submitted the manuscript to a journal, we might now talk of Wallace's theory of natural selection. Instead, after some behind-the-scenes machinations by Darwin's friends Lyell and Hooker, papers by both Darwin and Wallace were read at a meeting of the Linnean Society on July 1, 1858. Darwin, finally spurred to action twenty years after he first thought of natural selection, quickly wrote *On the Origin of Species by Means of Natural Selection*, the "abstract" of a much longer planned work that was never finished. And the rest is the Darwinian Revolution.

But that is part of another (and frequently told) story. The point I want to make here is that, even before Wallace's fevered "Eureka!" moment about the survival of the fittest, he and Darwin already shared a common intellectual path. Both men had a profound interest in geographic distributions, and that interest had been critical to both of them in recognizing that species evolve from other species. Both had accepted the notion that the Earth and the life upon it have a history extending many millions of years into the past. For biogeography, what all of this meant was that these two men were trying to explain the distributions of living things within a new framework, a new set of assumptions about the nature of the world. It was the framework of descent through deep time.

1.1 Great minds think alike: Alfred Russel Wallace (left) and Charles Darwin independently saw that the distributions of plants and animals are evidence of evolution. Both also came up with the theory of natural selection.

In this new framework, some answers to biogeographic questions were no longer legitimate. Take the work of Edward Forbes, a contemporary of Darwin's, who, during the 1840s and 1850s (he died in 1854) studied geographic distributions of European species, both terrestrial and marine. Forbes was, in Thomas Henry Huxley's words, "an acute and subtle thinker," and, like Darwin and Wallace, he was interested in general explanations for the similarities of species found in different geographic areas. But he was not an evolutionist, and that made all the difference. Finding molluscs in the Aegean Sea that were similar but not identical to those off the coast of Scotland, he thought there must have been separate centers of creation in these regions. According to Forbes, God had seen fit to create nearly identical shelled creatures in two places because the environments were nearly identical. This idea of separate creations of similar (or even the same) species in different regions was popular at the time, but it obviously wasn't the way Darwin and Wallace would have interpreted the same facts. They would have seen the evolutionary connections of these similar species and wondered where their ancestors had lived, and how the descendants had ended up in different areas. On the flip side, in this new worldview, some answers now seemed more reasonable than they had before. Oceanic dispersal events that were exceedingly unlikely over short periods, for example, might become probable given "deep time." Basically, some very wrong assumptions—creation in various forms and the notion of a young Earth—had been replaced with the right ones. In short, before they became linked as the independent discoverers of natural selection, Darwin and Wallace had already become the first modern historical biogeographers.

In terms of published work, the landmark for this new biogeography was *The Origin of Species* and, in particular, its two chapters on geographic distribution. Like so much in *The Origin*, reading those chapters is almost a jaw-dropping experience; even now, more than 150 years later, they could serve as a useful introduction to biogeography (although, as we will see, some modern researchers, following Croizat, view those same chapters as worse than useless). It must have been a revelation for naturalists reading these arguments for the first time, the jumble of disconnected facts of distribution suddenly all making sense, as if the discordant notes of an orchestra tuning its instruments had coalesced all at once into a symphony. It was all about species arising from other species and then moving about the Earth, limited by their powers of dispersal. Suddenly it was clear why island species are usually similar to those on nearby continents; why animals that cannot easily cross large ocean barriers, such as frogs and mammals, are missing from remote islands; why regions with distinct floras and faunas are separated by barriers to dispersal such as deep waters or deserts; why similar environments in widely separated parts of the world are populated by taxonomically distant species. And, in case people doubted the possibility of some of the ocean journeys that must have taken place, Darwin included a discussion of means of dispersal. The seed experiments were in there, and the duck's feet and the pelican droppings. The idea of transport on icebergs got more than its fair share of space.

I think of the publication of *The Origin* as the death knell for the era of simply making stuff up about how God had ordered the distributions of living things. Now a legitimate study could not rest on one person's idiosyncratic view of the Divine. Now things had to make sense in materialistic terms. It was, among many other things, the beginning of thinking about distributions broken up by oceans in modern terms.

Incidentally, Wallace would ultimately spend much more time than Darwin did studying geographic distributions. Wallace would write a two-volume work on animal biogeography and another book on island life, and he became known for delineating faunal regions with borders representing barriers to dispersal. Fittingly, the most famous of these borders, deep water running between various islands in the Malay Archipelago and separating the Asian and Australian regions, has become known as Wallace's Line. Throughout his life, Wallace was considered a kind of poor man's Darwin, both literally and figuratively, but in this one arena he didn't have to play second fiddle: he ended up being dubbed the "father of biogeography."

The Revolution That Wasn't

Darwin and Wallace had another belief in common, namely, that the continents had remained more or less fixed through deep time.* This assumption influenced biogeographic explanations. If, for instance, Africa and South America had always been where they are now, then taxonomic groups found in both places, such as monkeys and ratite birds, must have either crossed the Atlantic Ocean, traveling over the water or on a now-sunken land bridge, or taken the long way round through the northern continents. This belief in the permanence of the continents was by no means universal; for instance, one popular theory in the late nineteenth and early twentieth centuries held that the Earth's crust cycled through massive changes, with the continents of one era sinking to become the ocean basins of the next.** However, there was almost complete agreement on one point: the continents, whether permanent or transient, did not move sideways to any great extent.

---

* For a time, Wallace believed in the horizontal movement of continents (that is, continental drift), but he changed his mind in the early 1860s to a belief in fixed continental positions (Parenti and Ebach 2009).

** This was the Austrian geologist Eduard Suess's theory, which held that the Earth was continuously shrinking as it cooled, producing high and low regions of the crust, like wrinkles in the skin of a drying apple. Suess believed that South America, Africa, and India once had been connected by land, but subsidence had resulted in oceans flooding parts of the giant, conglomerated landmass, thus separating it into pieces. Suess's grand theory has been thoroughly rejected, but his name for the southern supercontinent caught on (although its boundaries became modified). He had called it Gondwanaland after the region of India where sedimentary rocks from the ancient landmass were found (Oreskes 1988).

This is not to say that the notion of continents moving horizontally hadn't been raised. In fact, by the beginning of the twentieth century, the idea of continental drift had a long, albeit mostly obscure, history. Way back in 1596, the Flemish cartographer Abraham Ortelius, seeing the jigsaw puzzle fit between the continents on opposite sides of the Atlantic, had suggested that those landmasses had once been joined and had drifted apart. Nothing much came of Ortelius's idea, but in 1858, the same year that Wallace sent his paper on natural selection to Darwin, a French geographer named Antonio Snider-Pelligrini came up with a new version of continental drift. Anticipating later thought, Snider-Pelligrini suggested that all the continents had been joined together during the Carboniferous period, basing this inference on identical plant fossils found in Europe and North America as well as the long-recognized South America–Africa fit. He also suggested that what had driven the continents apart was material erupting from the Earth's interior. Again, not many people took notice, but from about that time on, this apparently batty idea of continental drift kept cropping up, even if hardly anyone believed it. In an 1898 book and a 1910 paper, Frank Bursley Taylor, an amateur American geologist, presented a more complex scenario for continental movement that included the separation of South America and Africa along the Mid-Atlantic Ridge and the creation of mountains at the forward edges of moving landmasses. Taylor also proposed a mechanism for drift, a combination of tidal forces and a speeding up of the Earth's rotation caused by the capture of a comet during the Cretaceous. That comet, according to Taylor, had become the moon. Taylor's ideas didn't make much of an impact either, doomed perhaps by a lack of evidence and the catastrophic nature of his moon-capture scenario, which didn't sit well with the uniformitarian views of most geologists. Other continental drift proposals met with a similar lack of interest or worse.

Much of what Snider-Pelligrini, Taylor, and others said would turn out to be true. Fossil similarities and the fit of South America to Africa really were evidence of the former attachment of continents, the mechanism for continental drift did involve material erupting from the interior of the Earth, and the Mid-Atlantic Ridge really was the line along which the Atlantic had opened up. But none of these early advocates of continental movement had presented much reason to believe their proposals. It would take a scientist very focused on the task to make a strong case for continental drift . . . and get shot down as well.

That scientist was a quiet, intense man named Alfred Wegener, the son of a minister, born in Berlin in 1880. Wegener's doctoral dissertation

1.2 Alfred Wegener in 1910, the year he started thinking about continental drift.

was in astronomy, but he was an eclectic researcher and, before his work on continental drift, he was best known as a meteorologist. He was clearly a bold character, both intellectually and physically, with, as one colleague noted, "fine features and penetrating blue-gray eyes." He seemed almost a stereotype of a scientist, albeit a positive one—focused, serious, and uncompromising.

Wegener's life reads like an adventure story. In his mid-twenties, he and his brother set a world record by floating in a hot-air balloon for fifty-two hours. Several years later, on a meteorological expedition, he and another scientist made a seven-hundred-mile trek on foot across the Greenland icecap. Running low on food, they had just killed their dog and were about to eat it when a group of Inuits arrived on the scene and helped them find their way to a settlement. Even the writing of his book on continental drift was connected to a physical ordeal: he finished it while recovering from a bullet wound in the neck suffered in battle during the First World War.

Wegener began thinking about continental movement in 1910. He and a friend were browsing through a new atlas when Wegener noticed what Ortelius, Snider-Pelligrini, and others had before. "Please look at a map of the world!" he wrote to his fiancée. "Does not the east coast of South America fit exactly with the west coast of Africa as if they had formerly been joined?"* In 1912, he published two papers on his new continental drift theory. When these papers were heavily criticized, Wegener's father-in-law, a climatologist, warned him against jumping into a new field. By then, however, Wegener was committed to his theory, and, in any case, he wasn't one to let the usual boundaries of academic disciplines hem him in. He knew that the theory was revolutionary and he felt an urgency to pursue it much further. "If it turns out that sense and meaning are now becoming evident in the whole history of the Earth's development," he wrote to his father-in-law, "why should we hesitate to toss the old views overboard? Why should this idea be held back for ten or even thirty years?" He would find out why, although the knowledge undoubtedly wouldn't have stopped him.

An obvious initial goal was to figure out how the continents had once been arranged. Gravity measurements and other evidence showed that continental crust was less dense than the crust of the ocean floor, which indicated to Wegener (and to many geologists of the time) that the rock underlying continents was a distinct and permanent feature of the Earth. If this was true—if the extent of continental crust had remained largely intact through time—it meant that landmasses that once had been broadly attached to each other should fit together like pieces of a puzzle; the solution could be found because all the pieces were still around. At first, Wegener simply used the coastlines as they are, but later he matched the outlines of the continental shelves, which more accurately reflect the borders of continental crust. What he found, with some fudging here and there, was an exceptionally good fit. It looked as though all the continents had once been part of a single supercontinent. Wegener called this enormous landmass the *Urkontinent* (the "original continent"), but it soon

---

* Some have said that Wegener actually got the idea of continental movement from reading Taylor's paper on the subject, but this claim is at best uncertain (McCoy 2006). Wegener reported that he thought of the idea independently. In any case, since the basic idea had been around for several hundred years, and Wegener's legacy is based on the volume and quality of evidence he presented, it does not seem to matter how he first came to think of the phenomenon.

became known by a different name, one that we still use today—Pangea, Greek for "All-Earth."

If Pangea had really existed, then there should be evidence of it in the rocks. In particular, there should be geological features and fossils that were continuous across the various, now separated continents. If, for example, Africa and South America had been connected, one should find some of the same rock formations on both continents, and, when the two puzzle pieces were set against each other in their original positions, those formations should line up. Wegener likened the continents—the fragments of Pangea—to a torn newspaper; the lines of print would run evenly across the page if one put the pieces back together in their proper arrangement.

Some of Wegener's most compelling evidence came from matching up such "lines of print." The lines came in many different forms, including mountain ranges, coal beds, sedimentary and volcanic rock formations, glacial deposits, and fossil occurrences. There are folded mountains in Scotland and Ireland that continue in Newfoundland. There are coal fields in Belgium and the British Isles aligned with coal fields in the Appalachians. There are matching volcanic kimberlite pipes containing white diamonds in Africa and South America. There are fossils of the so-called *Glossopteris* flora on all the southern continents, including Antarctica, and those of the freshwater reptile *Mesosaurus* in southern Africa and southern South America. There are glacial erratics "of a peculiar quartzite grit with banded jasper pebbles" that seem to have arisen in ancient mountains of Griqualand in southern Africa, but are also found in Brazil. Wegener wrote that finding so many matching "lines of print" argued a million to one in favor of continental movement. The million-to-one odds he pulled out of a hat, but the general argument was sound; if you arrange the pieces of a puzzle and twenty swaths of color run through it, all aligned, you've done the puzzle right.

The theory also explained some strange observations about ancient climates, a particular interest of Wegener's from his work as a meteorologist. Wegener focused on the Carboniferous and Permian periods, when parts of South America, India, Africa, and Australia show abundant evidence of an ice age while vast areas of North America, Europe, and Asia were covered by warm, wet forests. (The name "Carboniferous" comes from the coal beds that formed from the remains of the forest plants.) It seemed that parts of the world that are now warm had been cold, while, at the same time, parts that are now cold had been warm. Others had tried to explain this conundrum by moving, not the positions of the continents, but the locations of the poles. However, Wegener saw that there was no polar location that by itself adequately explained the climate pattern. One always ended up with anomalies, such as big glaciers near the Equator. Instead, he showed that the paradox could be explained by continental drift: with all the continents conglomerated into Pangea, and the South Pole positioned near the southern end of the supercontinent, the pattern of glaciation in the south and warm forests in the north made perfect sense.

Wegener found himself more or less in the position of Darwin and Wallace when they had convinced themselves that evolution happened, but they hadn't yet come up with natural selection as the mechanism. Wegener knew that continental drift occurred, but he didn't know what caused it. In his book *Die Entstehung der Kontinente und Ozeane* (*The Origin of Continents and Oceans*), he threw in a couple of ideas, probably knowing that they weren't right or, at least, weren't enough. One was a centrifugal force generated by the spinning Earth that, by acting differently on continental crust and ocean crust, would supposedly push the continents toward the Equator. The second was the tidal force of the sun and moon pulling on the continents and sending them westward with respect to the ocean floor. In both cases, Wegener envisioned the continents pushing through the ocean crust, rock plowing through rock. That would turn out to be a big problem.

His book appeared in German in 1915 (although it was not until the translation of a third edition in 1924 that it became available in English). The first edition was only ninety-four pages long, more like a novella than a novel, but it was monumental in scope. Later editions were even more impressive, as Wegener essentially rewrote the book for each edition, continuing to add new evidence. Reading it is a bit like reading *The Origin of Species*, in that one is constantly struck by how prescient and modern it is. Wegener talks of rift valleys as incipient oceans (which is what they are); of the clockwise rotation of landmasses bordering the Pacific (which is why Los Angeles and San Francisco are heading in opposite directions); and of connections between continental drift, faulting, earthquakes, and volcanoes (which are now all understood as linked phenomena).* From the beginning, many scientists seemed to realize that this was a serious and potentially revolutionary piece of work, and the book was widely read and widely discussed. Conferences were held about it. Wegener was no

---

* There are some conspicuous gaffes in Wegener's book as well. For instance, Wegener thought that the rates of continental movement were sometimes orders of magnitude greater than we now know them to be.

Gregor Mendel, planting his peas in obscurity. The book almost immediately made him well known.

You will often read that the ultimate outcome of all this attention was the rejection and ridicule of Wegener's theory. Damning judgments from geologists and other scientists did come thick and fast. "Wegener's hypothesis in general is of the foot-loose type, in that it takes considerable liberty with our globe, and is less bound by restrictions or tied down by awkward, ugly facts than most of its rival theories," wrote one geologist. Another called Wegener's theory "a beautiful dream, the dream of a poet. One tries to embrace it, and finds that he has in his arms but a little vapor or smoke." A third was even more pointed, describing Wegener's argument as "ending in a state of auto-intoxication in which the subjective idea comes to be considered an objective fact." Jokes about continental drift also circulated in university classes. There was the one, for instance, about half of a fossil specimen from Europe matching perfectly with another half dug up in North America, like an amulet in a fairy tale.

However, the notion that the reaction to Wegener's views was almost totally negative is an oversimplification. It turns out that the response to the drift theory was very inconsistent from place to place. In Britain and, especially, continental Europe, many scientists saw the merit in at least some of Wegener's arguments. Quite a few of them had seen the matching strata, landforms, or fossils on opposite sides of the Atlantic for themselves, and viewed those "lines of print" as strong evidence that the continents had been joined. Although relatively few European scientists became wholehearted supporters of continental drift, the seeds that Wegener planted there were not simply eradicated. The English geologists Fred Vine and Drummond Matthews, for example, both recalled being receptive to the drift theory well before they made their own discoveries about seafloor magnetic anomalies, findings that helped vindicate Wegener's ideas.**

It was in the United States that the reaction to Wegener closely matched the widely held story of rejection and ridicule. It was there that the drift theory was commonly seen as "foot-loose" and Wegener as

---

** Matthews, on a trip to the Falkland Islands, asked a colleague what he should read about the geology of the area. The reply, according to Matthews, was, "Oh well, you've got du Toit [Alexander du Toit's book on the geology of South Africa], . . . if you don't believe in continental drift just take out a tape measure and measure the Devonian sections in the Falkland Islands." Matthews did this and found "they were very much impressively the same as the description [given by du Toit for South Africa], . . . inch for inch they measured up" (Oreskes 1988).

"auto-intoxicated" by his own mental machinations. Critics especially attacked Wegener's proposed mechanisms to account for continental movement; it just didn't seem reasonable that continents could plow like giant barges through oceans of solid rock, and, in any case, the centrifugal and tidal forces that Wegener pointed to seemed totally insufficient for the task. These critics felt that, without a reasonable mechanism, the whole edifice of drift theory was fundamentally unsound. By proposing implausible (and, as it turns out, completely incorrect) mechanisms for continental movement, Wegener left himself open to attack. And it seems that many scientists threw out the baby—the fact of drift, regardless of mechanism—with the bathwater.

However, this account fails to explain why Americans were so much more opposed to the theory than were their European (and European colonial) counterparts. Perhaps it would be more accurate to say that the lack of a reasonable mechanism was exploited by scientists who were strongly biased in the first place to reject Wegener's arguments. Why Americans were especially biased is not an easy thing to answer. However, the historian of science Naomi Oreskes has suggested that American geologists were particularly enamored of new kinds of instruments and the hard numbers that came out of them (a kind of intoxication with technology that has bedeviled scientists of many sorts), and, consequently, that they tended to give little weight to what they viewed as old-fashioned, "subjective" evidence. In this view, the problem that Americans had with Wegener was that his evidence was almost all of this subjective type—things like the identity of landforms and strata in South America and Africa that were invariably based on the opinions of a few geologists.* The Americans wanted numbers, and, as it turned out, those numbers would be a long time in coming.

Despite these complications, it is reasonable to conclude that Wegener's theory was not generally accepted, even in Europe. In hindsight, it does seem like an odd, if not completely inexplicable, episode in the history of science. One wonders, in particular, what was going on in the minds of the many scientists who took the trouble to read Wegener's book, with its piles of evidence, and yet still thought he was utterly wrong. One

---

* Oreskes has also pointed out that the lack of an acceptable mechanism does not generally keep scientists from believing in the reality of a phenomenon. For instance, many people were convinced by Darwin that evolution was a reality without being convinced that natural selection was a plausible mechanism. Similarly, scientists accepted the reality of the ice ages without being convinced of a cause for them (Oreskes 1988).

also wonders how history might have been altered if only he had managed to convert a few more prominent geologists. But it didn't happen. Perhaps, in the most general sense, it's just that scientists have a hard time giving up their entrenched views. In most instances, that kind of conservatism probably makes sense—it keeps people from wasting their time chasing "vapor or smoke." Occasionally, however, scientists have to break out of that conservatism, lest they be left in the dark.

At the meeting of the Geological Society of America in 1922, a geologist named R. Thomas Chamberlin, harshly criticizing the drift theory, said, "If we are to believe Wegener's hypothesis, we must forget everything which has been learned in the last 70 years and start all over again."

That was exactly right.

In the spring of 1930, Wegener embarked on his fourth trip to Greenland, leading an expedition to make meteorological and ice measurements. From the start, the expedition was plagued by delays and equipment problems. In the fall, when two men stationed at an outpost in the middle of the icecap began to run out of supplies, Wegener decided to lead a group out from a camp on Greenland's west coast to reprovision them. Often traveling through deep, fresh snow with their heavy dogsleds, it took the party forty days to reach the outpost, three times as long as it might have under ideal conditions. After resting for just two days at the outpost, Wegener and a companion, an Inuit named Rasmus Villumsen, headed back toward the west coast. The date was November 1, Wegener's fiftieth birthday. A few things are known or can be inferred about their return journey. The extreme cold and howling winds must have slowed them down. They eventually abandoned one of the two dogsleds, and from that point Wegener traveled on skis while Villumsen rode the remaining sled.

Six months later, a search party found Wegener buried near his upright skis, in a grave that Villumsen apparently had dug. His body was lying on a reindeer skin and a sleeping bag and was sewn up in two sleeping bag covers. There is a thought that he died of heart failure. Villumsen made it at least twelve miles farther, but then all trace of him disappeared. His body was never found.

To some, Wegener's death in Greenland might seem like an all-too-fitting end to the even greater tragedy of his life, defined by the fact that he had proposed a great scientific theory, but was maligned for it and never saw his idea vindicated. However, that view assumes that he was almost universally considered a crackpot, which, as described above, wasn't actually the case. In fact, upon his death, the prominent scientific journal *Nature* ran a full-page obituary, calling his passing "a great loss to geophysical science." In any case, the man who led the Greenland expedition does not come across as pitiful in any way. If he was weighed down by anything, it was worry over the logistics of a complex operation at the mercy of the weather, and concern for the safety of his men. The accounts of other expedition members make him out to be an intensely respected, if taciturn, leader, thoroughly focused on the job at hand.

When Wegener's body was found, his nose and hands were marked by frostbite, but his eyes were open, and, according to one member of the search party, "the expression on his face was calm and peaceful, almost smiling." Perhaps there's a personal metaphor in the image of Wegener in death: he was scarred by life, but remained unbroken.

New York Before the Storm

Wegener's theory did not die with him, but at the time of his death it looked to be somewhat moribund. The drift theory did have a substantial minority of followers in continental Europe, but, in Britain, very few scientists thought Wegener was right, and in the United States, virtually none. His book had caused a great stir, but not the worldwide advance in geology—and biogeography—that it could have and should have created. A few scientists, notably the South African geologist Alexander du Toit, adamantly used Wegener's theory to explain plant and animal distributions broken up by oceans, but they were in a small minority. In the terminology of the philosopher Thomas Kuhn, the "paradigm shift" didn't happen, and "normal science" largely went on as before.

Oddly enough, however, in 1915, the same year that Wegener's book was published, a book-length paper came out that laid the foundation for a substantial change in biogeographic thinking that had nothing to do with continental drift. At that time, the land-bridge builders, intellectual descendants of Joseph Hooker (before Darwin converted him), had the upper hand over Darwinian dispersalists. The outlines of now-sunken connections were being drawn on maps willy-nilly wherever closely related species were found on both sides of a sea or ocean. There were, supposedly, bridges between South America and Africa, South America and Australia, Madagascar and India, Europe and North America, Samoa and Hawaii, and on and on (see Figure 1.3). Sometimes, as for the link between South America

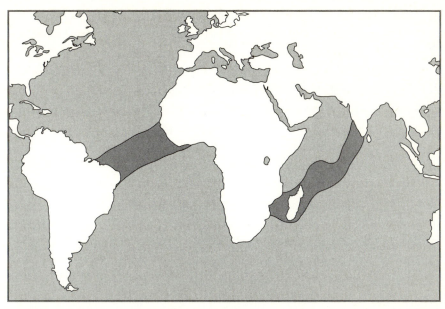

1.3 Two of the many land bridges conjured up to explain plant and animal distributions. Redrawn and modified from Hallam (1994).

and Africa, several alternative bridges were hypothesized, none of them based on any reasonable geological evidence. In retrospect, this episode of biogeographic history seems laughable, but there was a certain logic to it. Despite Darwin's experiments, land-bridge enthusiasts simply could not believe that oceanic dispersal of terrestrial organisms was important, and, like most scientists at the time, they didn't believe in continental drift. Former land bridges were a way out of the dilemma.

The 1915 paper that began to swing the pendulum back was entitled "Climate and Evolution."* Its author was William Diller Matthew, a thin, bespectacled, professorial-looking fellow who worked at the American Museum of Natural History in New York City as a curator with a specialty in fossil mammals. When it came to explaining distributions broken up by oceans, Matthew was very much like Darwin. Both of them believed in the fixed positions of continents, and both were leery about invoking former land bridges without geological evidence. This meant that Matthew, like Darwin, was a dispersalist. Matthew didn't think as much

---

* The title of the paper refers to Matthew's notion that very long-term climate cycles were critical in shaping evolution and dispersal.

as Darwin did about seeds in seawater or snails clinging to duck feet, but that was because he was a mammalogist; most of his study organisms needed natural rafts to disperse across wide expanses of water.

When it came to the frequency of dispersal events, Matthew's argument was all about those rafts. For mammals colonizing large oceanic islands, it went something like this: Take the small number of natural rafts, about 10, that have been seen far out at sea in the past three hundred years or so. Multiply that by 100 to get an estimate of the actual number of such rafts—1,000—in that stretch of time. Assume that the Cenozoic Era is 60 million years long, and you get 200 million rafts during the Cenozoic. Of these 200 million, say that only 2 million have had living mammals on them. Of these 2 million, only 200,000 will have reached land, and of these 200,000, only 200 will have resulted in species establishing themselves in the new area. We only know of two dozen or so cases of mammals reaching large, oceanic islands on their own, so our calculation of 200 is more than enough to take care of all the known cases. And this is for mammals. It's much easier for lizards or tortoises or sunflowers to get to such places.

I know it sounds suspicious. Where do all the numbers come from? Basically, out of William Diller Matthew's ear. But the message is not the exact numbers—Matthew admitted that—it's just Darwin's old argument that given a very long time (60 million years in this case) a lot of things that seem very, very unlikely will happen. Rats can get to the Galápagos. Monkeys can reach Sulawesi. Pygmy hippos (now extinct) can make it to Madagascar. Actually, according to Matthew, the pygmy hippos didn't even need a raft. They could have just swum the 300 miles across the Mozambique Strait.

Like many biogeographers, including Darwin and Wallace, Matthew also argued that the nature of an island's fauna can tell you if the place was ever connected to a continent. He used the mammals of Madagascar as an example. If the island had been connected to Africa at any time, you'd expect to find a large complement of African mammals there. Instead what you see are just a few major groups—lemurs, carnivores, tenrecs, and rodents, plus a shrew that might be introduced, and bats, which can fly there. Also, those mammal groups do not seem to have gotten to Madagascar all at once, but rather, apparently colonized the island one by one at different times. (Matthew probably based this inference on the connections of these groups to African relatives and the fossil records of those relatives.) This pattern of a few colonizations spread out in time is exactly what you'd expect if each group reached the island by chance, overwater dispersal.

Writing before Wegener's book had been translated into English, Matthew was arguing against the idea of a now-sunken land bridge to Madagascar, not against the island as part of the continental drift story. But the same argument could be made against Madagascar as part of Pangea or Gondwana, and, when Wegener's theory became widely known, Matthew argued against it, and not just for Madagascar, but in general. Matthew, like Darwin and Wallace, did not completely discount former land connections, but he thought that most modern groups were too young to have been affected by ancient land bridges (to the extent that those even existed). His emphasis was clearly on long-distance dispersal.

To make a long story short, Matthew had a major impact on biogeography through "Climate and Evolution" and other works and through a wide network of students and colleagues. It would be hard to overestimate the networking aspect in Matthew's case. He became something more than just a respected scientific mentor; one follower, for instance, referred to himself and others as Matthew's "disciples" and called "Climate and Evolution" "a kind of Holy Writ." Some of the scientists within his sphere of influence were George Gaylord Simpson, a paleontologist who replaced Matthew at the American Museum of Natural History when Matthew moved to Berkeley in 1927; Ernst Mayr, an ornithology curator at the museum from 1931 to 1953, and later a professor at Harvard; and Philip J. Darlington, also at Harvard. These three were all prominent scientists: Darlington was a well-known beetle expert and the author of widely read books on biogeography, and Simpson and Mayr were arguably the most significant paleontologist and evolutionary biologist, respectively, of the twentieth century. Influence begat influence, and by the 1940s, dispersalism, at least in the United States, was at its height. Hooker's land bridges had taken precedence for a time, but now Darwin's ocean crossings again held sway.

Because this dispersalist mode of thinking had originated at the American Museum of Natural History and also had a strong following at Columbia University, it was called by some the "New York School of Zoogeography." That moniker was coined by none other than Léon Croizat, who seems to have used the phrase as his way of identifying the enemy. Croizat was the most headstrong and divisive character in a field that would soon have no shortage of such people, and what was about to happen in biogeography would not at all resemble the friendly jousting between Darwin and Hooker.

"On several occasions, when the vessel has been within the mouth of the Plata, the rigging has been coated with the web of the Gossamer Spider. One day (November 1st, 1832) I paid particular attention to the phenomenon. The weather had been fine and clear, and in the morning the air was full of patches of the flocculent web, as on an autumnal day in England. The ship was sixty miles distant from the land, in the direction of a steady though light breeze. Vast numbers of a small spider, about one-tenth of an inch in length, and of a dusky red colour were attached to the webs. There must have been, I should suppose, some thousands on the ship. . . . While watching some that were suspended by a single thread, I several times observed that the slightest breath of air bore them away out of sight, in a horizontal line. On another occasion (25th) under similar circumstances, I repeatedly observed the same kind of small spider, either when placed, or having crawled, on some little eminence, elevate its abdomen, send forth a thread, and then sail away in a lateral course, but with a rapidity which was quite unaccountable."

—*Charles Darwin*, The Voyage of the Beagle

Chapter Two

THE FRAGMENTED WORLD

The Ichthyologist Who Played with Fire

In his book *The Tipping Point*, Malcolm Gladwell likens the spread of products, behaviors, and ideas to epidemics, claiming that, in all these cases, small things, including single individuals, can have unexpectedly large effects. Gladwell describes three sorts of people that often have such disproportionate influence in pushing a trend past a tipping point. First, there are *mavens*, great collectors of information. A maven might be, for instance, the acquaintance who keeps up with the latest news on which foods are good for your heart (this week, it's coconut oil), or the friend who's always telling you about a wonderful new hole-in-the-wall restaurant. Second, there are *connectors*, often extreme extroverts, who have an exceptionally large network of friends, acquaintances, and/or colleagues. Connectors are hubs from which information spreads especially quickly. Finally, there are *salespeople*, who excel at the art of persuasion. A salesperson might be an actual professional who talks you into buying an especially quiet dishwasher, but is just as likely to be a friend of a friend who convinces you of the evils of fracking.

For the spread of vicariance biogeography that took place from the 1960s into the 1980s, Gareth Nelson was all three of these things. In the mid-1960s, Nelson, an ichthyologist—"Gary" to his friends and colleagues—might not have seemed a likely candidate to become the center of a major scientific movement. The late philosopher and historian of science David Hull described Nelson, who had then just finished his

PhD, as "a lanky youngster . . . anything but imposing in his white socks, shirt open at the collar, and brown-and-tan-checked sport coat." But that "lanky youngster" had or would acquire some special qualities. He was exceptionally well-read within his field, so that he became privy early on to new ideas that few had heard of (a maven); he was a natural networker and became the editor of a key scientific journal (a connector); and, critically, he had a bold, persuasive personality (a salesperson).

Relevant to the art of persuasion, Nelson had no shortage of confidence, no hesitancy when it came to speaking his mind, and, above all, no reflexive deference to authority. In fact, it often seemed that he was fueled by defiance. There was the time, for example, when he and Donn Rosen, another ichthyologist, were discussing something about taxonomy, and Rosen suggested they should wait to see what Ernst Mayr, at the time probably the most famous and influential evolutionary biologist in the world, had to say about the topic in an upcoming book. Typically, Nelson was having none of that. "I'm not about to let Mayr do my thinking for me!" he barked. Defiance isn't a universally attractive quality, but it seemed to suit the time (the rebellious 1960s) and place (principally New York City, never known as a paragon of politeness). Defiance was part of Nelson's persuasive appeal.

In short, Gary Nelson—maven, connector, and salesperson all rolled into one—was just the person to become the chief architect for the spread of the ideas of vicariance biogeography. He was the straw that stirs the drink, or, maybe, the guy who fans the flames that end up burning down the building. One could make a case that, without Nelson, the vicariance movement would never have gotten off the ground. Certainly, it wouldn't have unfolded anything like the way it did.

Nelson's epiphany came in 1966, not in some splendid isolation in the desert or jungle, but in the library of the Department of Paleozoology at the Swedish Museum of Natural History in Stockholm. Nelson was fresh from completing his doctorate studying fish gill arches at the University of Hawaii and was on a year-long National Science Foundation research fellowship that had him examining fossil fishes in England and Sweden. Browsing the current literature section in the library, he picked up a hefty new publication, a nearly five-hundred-page monograph by a Swedish entomologist named Lars Brundin on the evolution and taxonomy of a group of tiny, swarming flies called chironomid midges.

Brundin had been studying chironomids since the 1940s and through them had become especially interested in taxa that occur on landmasses "separated" by Antarctica. Among other places, chironomids are found in Australia and New Zealand and on the other side of Antarctica, so to speak, in South America. Groups that show such *transantarctic relationships* are obviously on landmasses separated by oceans, so Brundin, like so many before him, was addressing the great question in biogeography, the question of disjunct distributions. At the time, Nelson had no particular interest in either midges or transantarctic relationships, so it's an open question why he bothered to pick up the doorstop-like monograph in the first place. Perhaps that's just what mavens do—they root around to discover things and, sometimes, that act is the spark that sets off the fire.

The bulk of Brundin's monograph didn't look like promising reading. It consisted of dry descriptions of the midges, not only the adults but also the freshwater larval and pupal stages. The pupae turned out to be especially important from a taxonomic point of view, and Brundin described them with what might have seemed like obsessive zeal; this was stuff that only a fly specialist—in fact, maybe only a midge specialist—could really get into. However, bookending those descriptions of the midges were sections in which Brundin set forth his approach to studying evolutionary relationships and biogeography and, from that approach, drew strong conclusions about how the tiny flies had ended up on widely separated landmasses. In effect, Brundin was using midges to illustrate some very general methods and concepts, things that could apply to Nelson's fishes, or any other group, for that matter. The approach was what caught Nelson's eye; in more ways than one, it was like nothing he had seen before.

First, Brundin used methods developed by the German entomologist Willi Hennig to construct evolutionary trees for the midges. Hennig's "cladistic" methods involved explicitly identifying each grouping of species or higher taxa (genera, families, etc.) in an evolutionary tree by recognizing the traits that had apparently evolved on the branch leading to the group in question. These *shared derived traits* (or *synapomorphies*, in the elaborate jargon of cladistics) identified so-called *clades* (or *monophyletic groups*), lineages that traced back to an ancestor not shared by any organisms outside the group. For instance, members of the group Aves have feathers and toothless beaks, shared derived traits that evolved on the branch leading to the common ancestor of all birds. Those traits indicate that Aves is a clade—in this case, the clade that includes the common ancestor of modern birds and all of the descendants of that ancestor. Tangentially, under Hennig's system, many traditional taxonomic groups were rejected because, although they included the common ancestor of all members of the group, they did not include all the descendants of that ancestor. For instance, the class Reptilia, as traditionally defined, is not an acceptable cladistic group because some descendants of the common ancestor of reptiles—namely, birds and mammals—have been artificially removed from the group. Fishes, amphibians (if extinct forms are included), apes (if we humans are not counted among them), monkeys, invertebrates—none of them are clades, either, and so all are invalid in a cladistic classification system. That fact is not really important to our story, but it was one of the main things about cladism that pissed off traditionalists like Ernst Mayr and George Gaylord Simpson. The cladists, by the way, won this battle, to the point that most taxonomists these days only give formal names to groups they believe are clades.

By identifying all the clades within some group, one is also specifying the pattern of evolutionary branching within that group. Thus, the outcome of applying Hennig's methods to a set of traits—the pupal characteristics of midges, for example—is a branching diagram, a *cladogram*, that is a precise illustration of how one thinks taxa are genealogically related to each other. Each branching point in the diagram indicates a *speciation* event, the splitting of one evolutionary lineage into two. In essence, a cladogram is a representation of a piece of the tree of life.

It all sounds rather simple, like something that should have been figured out in the nineteenth century. (Some people have argued that it *was* figured out in the nineteenth century.) But the fact is that, in the 1960s, most evolutionary biologists were not operating in this way. They weren't consistent about identifying shared derived traits (most of them weren't even clear about the idea and had never heard of the term *synapomorphy*), and they often ended up drawing evolutionary trees that looked like a river delta with a thick "ancestral" branch (say, reptiles) out of which other branches (say, birds and mammals) arose in a muddy fashion. From such river delta diagrams you couldn't tell exactly how things were related to each other (see Figure 2.1). In the usual reptile-bird-mammal tree, birds just arose out of the great Mississippi-like swath of reptiles in general (or, if the diagram was relatively precise, out of a somewhat smaller swath called *diapsid reptiles*); it wasn't at all clear that the closest relatives of birds are actually specific kinds of theropod dinosaurs. The muddiness of these evolutionary diagrams didn't just stem from a lack of information, but reflected a muddiness of thought. Reading the evolutionary papers and books of that time, one gets the impression that people weren't often thinking about questions such as exactly which reptiles are the closest

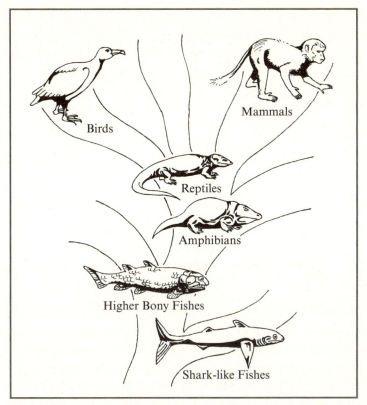

2.1 Muddy diagrams reflecting muddy thinking: a typical depiction of a vertebrate evolutionary tree before cladistics. Redrawn and modified from Romer (1959).

relatives of birds. Hennig's cladograms, in contrast, were perfectly explicit (perfectly Teutonic, it is tempting to say), with a line representing each species lineage or higher taxon intersecting at a particular point with the rest of the lines that made up the tree (see Figure 2.2).*

The logic of cladistics, as expressed by Brundin (who tended to explain Hennig's ideas more clearly than Hennig himself did), immediately

---

* The traditional evolutionary taxonomy of the time also was focused on an often misguided search for ancestors and the notion of a ladder of life that positioned living organisms on a scale of advancement, with humans at the top, of course. Note, for instance, the arrangement of the groups in Figure 2.1, with the main trunk leading from fishes through amphibians and reptiles to the supposed pinnacles of evolution, birds and mammals. Cladists rightfully rejected these notions, replacing them with a more objective view of the diversity of life.

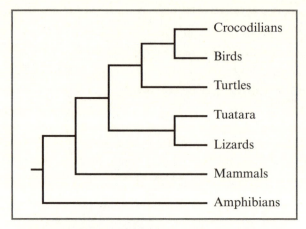

2.2 Evolutionary clarity: a cladogram of tetrapod vertebrates, showing the precise relationships among the different groups.

appealed to Nelson, as it has to many others. (I remember my own "conversion" to cladistics in the early 1980s—under the tutelage of my brother Kevin, who is also an evolutionary biologist and was then a "raving cladist"—as one of the most eye-opening scientific experiences I've had.) More than just presenting the cladistic method of deciphering evolutionary relationships, though, Brundin was also following Hennig in using cladograms to make inferences about biogeographic history. Even before Darwin, people had intuited that "relationships" (whatever that meant in pre-evolutionary terms) were important in explaining geographic distributions. In a sense, one had to know something about relationships to even know that there was a problem to solve. For instance, the fact that species A lives in Hawaii and species B lives in California only implies an interesting question if one senses that A is fairly closely related to B. (The fact that there are bristletail insects of the genus *Neomachilis* in Hawaii and California might mean something. The fact that there are fruit flies in Hawaii and condors in California does not.) With the Darwinian revolution came a clearer recognition of the nature of biogeographic questions. If A is part of the same tight evolutionary group as B, then the question is, "Where did the ancestor of these two species live (California, Hawaii, both places, neither place?), and how, in the course of evolution, did one species end up in one area and the other end up in another?" In many cases, the relationship aspect didn't go much further than that, and biogeographic interpretations were based on other kinds of evidence or assumptions, often debatable ones. In the hypothetical case just given, the conclusion might have been that the group originated in California and later dispersed to Hawaii, based on the not-altogether-airtight assumption that continental species can colonize oceanic islands but not the other way around.

Hennig's insight was realizing that having precise knowledge about relationships—the cladogram with its exact representation of the evolutionary branching pattern—could profoundly affect how one explained geographic distributions. In retrospect, some of his methods were not entirely sound, and Brundin unfortunately followed closely in Hennig's footsteps and made the same mistakes. But the general idea was right, and it remains at the core of understanding the geographic history of living things: to do historical biogeography, you need to know how different species are related to each other, and the more precisely you know it, the better.

That link between cladistics and biogeography was a great insight that Nelson took from Brundin. In Brundin's monograph, Nelson saw cladograms of midges, with lines connecting each taxon to the area (southern

2.3 Two architects of the vicariance biogeography "revolution": Lars Brundin (left) and Gary Nelson in Stockholm in 1988. Nelson was converted upon reading Brundin's monograph on midges. Photo by Christopher J. Humphries.

South America, New Zealand, Australia, and southern Africa) in which that group was found. To the uninitiated, these figures look like some kind of indecipherable electrical wiring diagram, but in them Brundin had found a clear pattern. That pattern had three parts to it. First, every group of midges found in New Zealand was most closely related to a group in South America or in South America plus Australia. Second, Australian groups were typically placed within some larger South American group. The third part is actually a corollary of the first and second, but is worth emphasizing on its own: Australian and New Zealand midges were never each other's closest relatives.

From a traditional "fixed-continent," dispersalist point of view, this pattern didn't make much sense. If chironomid midges had spread to the various areas by dispersing over oceans, one would expect the taxa on New Zealand and Australia—two landmasses quite close to each other—to be, more often than not, each other's closest relatives. Instead, these groups were always connected to South American ones. In other words, the midges really did seem to show transantarctic relationships in a very specific sense. But none of the landmasses inhabited by the midges are connected to Antarctica, so how did this pattern come to be? Was it possible that those areas had all been linked to each other through Antarctica at some time in the past, allowing midges to move easily from, say, southern South America to New Zealand? According to Brundin, not only was that possible, it was exactly what the latest geological evidence showed.

From the Bottom of the Ocean

When we left our discussion of geology, Alfred Wegener lay buried in the Greenland ice (where the ground beneath his frozen remains now lies several feet farther from Europe than when he died) and continental drift was a theory in need of a mechanism. Wegener had suggested centrifugal and tidal forces that had the continents plowing through oceanic rock like giant ice-breaking ships, but almost nobody believed these notions, and with good reason. Even so, there was strong evidence that the continents had moved *somehow*, and, because of that evidence, Wegener's theory continued to have advocates through the 1930s, 1940s, and 1950s.

One of those advocates was Arthur Holmes, a British geologist best known for studies of radiometric dating that indicated the Earth was at least several billion years old, at a time when many geologists thought the correct age was only about 100 million years. Holmes's interest in radioactivity led him to propose that radioactive decay in the Earth's mantle would give rise to convection currents involving mass movements of the liquid magma. If a liquid is heated gently from below, a temperature gradient is set up, but the liquid itself doesn't move en masse. However, if more and more heat is applied, at some point a threshold is passed, and the liquid begins to circulate, hotter liquid rising from the bottom and then falling as it cools. Holmes realized that such convection currents in the mantle, driven by the heat from radioactive decay, could provide a mechanism for continental drift. He envisioned the hottest magma rising underneath a continent and, upon reaching the crust, sliding laterally in opposite directions, pulling the continental crust apart as it did so. The split halves of the continent would then move away from each other, riding on the mantle currents like boxes on a conveyor belt. The problem of continents having to push through the ocean floor was circumvented by descending magma pulling oceanic crust down with it, making room for a continent to move over the top.

If "convection in the mantle" sounds familiar, that's because most geologists today believe Holmes was right about the driving force for continental movement. Other aspects of Holmes's theory are now known to be wrong; for example, he didn't envision magma actually creating new seafloor, as we know it does, and he thought that convection currents would generally rise at the Equator and sink at the poles, which isn't the case. However, given the state of knowledge at the time, his proposed mechanism for continental drift made sense; unlike Wegener's, the mechanism he proposed actually seemed strong enough for the task, and it didn't require having the lighter material of continents plowing through the dense basalt of the ocean floor.

Holmes first published his theory in 1928 and later included it as a small part of his textbook *Principles of Physical Geology*, published in 1944. From his radiometric work, he had become a well-known geologist, and his book became something of a classic. But the theory of mantle convection didn't have much impact. It was plausible, but there was little evidence to suggest that it was right, and why would other geologists believe something new and radical without much to go on? Maybe for the same reason, Holmes had a hard time even convincing himself. In 1953, twenty-five years after his first paper on the convection theory appeared, he admitted, "I have never succeeded in freeing myself from a nagging prejudice against continental drift."

In 1944, when Holmes's textbook was published, Harry Hess was somewhere in the Pacific. Hess, a Princeton geologist, had become a naval reserve officer several years before the war to help obtain gravity recordings in the Lesser Antilles that required the use of a navy submarine. As a result, he had to report for duty the day after the attack on Pearl Harbor, and he eventually found himself on a transport ship, the *Cape Johnson*, making a circuit that would have appealed to Darwin and Wallace in more peaceful times—the Mariana Islands, the Philippines, the Admiralty Islands, Guam, the Carolines.

Even at war, Hess didn't forget about geology. When he had free time ashore, he collected rocks. On board, he was in charge of a fathometer, an echo-sounding device used to help make beach landings, and he kept the thing switched on almost all the time, even when the ship was far out at sea. In this way he was able to gather quite a bit of data on the topography of the ocean floor. From Iwo Jima, where the *Cape Johnson* survived attacks from Japanese planes as well as its own side's "friendly fire," he wrote to one of his former students about the soundings he had collected: "Have been able to get about a dozen or more traverses across deeps and can outline their course pretty well from Iwo to Palau. Have four across the Mindanao deep, too. We filled in a lot of blank spots on the charts."

One thing that stood out in the soundings were many isolated seamounts—extinct, submerged volcanoes that Hess termed *guyots* after a nineteenth-century Princeton geographer, Arnold Guyot. These would turn out to be important, but Hess didn't see why just yet; like Darwin, he came back from his long ocean journey with some interesting observations but no grand theory. But Hess's mind was turned toward the depths now, toward the floor of the ocean.

That was the right place to look and soon it would be the right time to be looking. In the 1950s, there was an explosion of new information about the nature of the ocean floor, not only its topography, but also its magnetic, seismic, and heat flow characteristics. Reams of sounding records eventually showed a giant, more or less continuous network of submerged mountain ranges, nearly 50,000 miles long, running through all the world's oceans. The range that traced a sinuous path down the middle of the Atlantic, from the Arctic Ocean to the latitude of the southern tip of South America, was especially well-studied, and turned out to have some unexpected properties. Running down the center of this Mid-Atlantic Ridge was a valley, suggesting an ocean-long rift in the Earth's surface. That impression was reinforced, to some at least, by seismic recordings indicating that the crust in the area of the ridge was especially thin, along with temperature measurements showing it had an especially high rate of heat flow. It looked like magma was rising to the surface at the ridge. Another odd observation was that nothing in the oceans seemed to be very old; when geologists sampled rocks and associated fossils from the deep sea, from oceanic islands, or from Hess's guyots, they were always of Cretaceous age or younger.

By 1960, Hess had put all the facts together and had written a paper outlining a new theory. That paper, "History of Ocean Basins," was published, after some delays, in 1962.* It built on Holmes's idea of mantle convection and incorporated the new information about the ocean floor to develop a theory of seafloor spreading that students today would easily recognize as part of the modern geological worldview. Hess followed Holmes in suggesting that convection was driving magma on great cyclic paths from deep in the mantle to the crust and back again, but the new observations allowed Hess to flesh out the theory. The mid-oceanic ridges were where "cells" of magma rose up from the mantle. From there, magma would spread out in both directions from the ridge, cooling as it went, creating new ocean floor, and pushing continents apart, as with South America and Africa on opposite sides of the Mid-Atlantic Ridge. Where the new ocean floor met a continent, it would be impelled (by the force of the magma coming up behind it) downward, back into the mantle (in other words, it would be subducted, although Hess didn't use that term).

Hess wasn't a bold, outspoken scientist in the manner of Wegener, and, in "History of Ocean Basins," he was sometimes circumspect in his language. "I shall consider this paper an essay in geopoetry," he wrote, almost disparagingly, in the first paragraph. Still, he knew he was onto something big. In a later section, he starts with a conjecture and turns it, between sentences, into an assertion: "If it [mantle convection] were accepted, a rather reasonable story could be constructed to describe the evolution of ocean basins and the waters within them. Whole realms of previously unrelated facts fall into a regular pattern, which suggests that close approach to a satisfactory theory is being attained." He had constructed that "reasonable story" and it did indeed explain many "previously unrelated facts." Why are there no ancient rocks in the deep ocean? Because the ocean floor is part of a giant cycle of creation and destruction,

---

 * Because of the delays, Hess's paper appeared after one by R. S. Dietz which presented a very similar theory. Dietz himself gave credit to Hess for coming up with the theory first, although he also claimed that he wasn't aware of Hess's work (which had been circulating before it was published) when he developed his ideas. It was Dietz who coined the term "seafloor spreading" (Lawrence 2002).

and all of the really old ocean crust has long since been returned on this "conveyor belt" back to the Earth's mantle. Why are sediments on the ocean floor relatively thin, not at all what one would expect from several billion years of erosion and deposition, and why are they thinnest of all at the mid-oceanic ridges? Because, again, the ocean floor is nowhere very old, and is, geologically speaking, brand new at the spreading centers. Why is there a ring of volcanoes encircling the Pacific? Because ocean crust thrust down into the mantle is melted into magma, which sometimes makes its way back to the surface. Why do ocean ridges often run right down the middle of an ocean? Because the new crust forming at the ridges flows out equally in both directions.

The theory of seafloor spreading also explained Hess's guyots, although one has to read between the lines of his 1962 paper to see this fully. A guyot begins as a volcano formed at a mid-oceanic ridge, with its top above the water. At some point the volcano is no longer active, no longer building itself up, and its emergent peak gets flattened by erosion. From its place of origin, the guyot rides the "conveyor belt" out from the ridge, sinking along the way as the crust cools and subsides. So, when a lot of these volcanoes have formed, the result is a series of flat-topped seamounts, with those farthest from the spreading center also the most deeply submerged. Without the theory of seafloor spreading, guyots are unexplained curiosities. With the theory, they suddenly make sense, another piece of the great puzzle falling into place.

For biogeographic purposes, the critical point is that Hess, building on the work of Holmes and others, had basically solved the problem of how landmasses can drift apart. Although Hess never mentioned Wegener in his paper, it is clear that Wegener's problematic mechanisms for drift—or, perhaps, people's reactions to those mechanisms—were on his mind. From that perspective, the key was what happens at the leading edge of a moving continent: when continental crust meets oceanic crust, the latter, being more dense, is thrust down toward the mantle. When continent meets continent, the crust gets folded up, like two rugs pushed against each other, as when India collided with Asia to form the Himalayas. Like Holmes, Hess didn't have continents plowing implausibly through the dense volcanic rock of the ocean floor. And Hess made sure to mention that fact more than once, distancing himself from Wegener.

Hess's theory of seafloor spreading is the basis of what we now call plate tectonics, each plate being a region of the crust and upper mantle that moves as a unit. For instance, there is a South American Plate bounded in part by the Mid-Atlantic Ridge to the east and the Pacific edge of the continent to the west, where the Nazca Plate is being subducted. The theory of plate tectonics certainly wasn't finished by Hess (nor is it anything like complete today). Still, if Hess didn't erect the whole building, he at least laid the foundation and put up a couple of walls.

One might think that geologists would have rushed to embrace this new theory that finally provided a reasonable mechanism for continental drift and explained many of the odd new facts about the ocean floor. But that isn't what happened. Hess wasn't considered an auto-intoxicated dreamer like Wegener; instead, his theory was mostly just ignored, at least for a time. Still, "History of Ocean Basins" was an indication that the old view of fixed continents and oceans was getting creaky, even in the United States, where that idea was most deeply entrenched. The information that was piling up was about to make the old edifice collapse.

Here is one version, the textbook version, about what happened after Hess's paper was published: In 1962, Fred Vine, a beginning graduate student at Cambridge, took an interest in seafloor spreading—partly because he saw Hess give a talk about it—and began looking for a connection between Hess's theory and some new data on the magnetic properties of the ocean floor. In particular, what Vine and his adviser, Drummond Matthews, were investigating was a peculiar magnetic pattern on opposite sides of oceanic ridges in the Atlantic and Indian Oceans.

As magma cools, iron-containing particles within it align themselves with the Earth's magnetic field. Oddly, in some volcanic rocks, the particles are aligned in the opposite direction of the current magnetic field, an indication that the direction of the field has flipped back and forth through Earth's history. In the recordings that Vine and Matthews were examining, the rocks on the mid-oceanic ridge had the orientation of the current magnetic field, but next to the ridge on either side were areas of rock with the opposite orientation, and then, outside those areas, more rock with the current orientation. Vine and Matthews realized that this pattern was exactly what one would expect if new ocean floor was being formed at the ridge and then flowing outward on either side: the rock at the ridgetop, having just recently solidified, showed the current magnetic field orientation, while the alternating magnetic orientations going outward recorded the formation of older rocks and the flip-flopping of the direction of the field through time. Vine and Matthews published their findings in *Nature* in 1963, their beautiful insight providing a striking demonstration of the reality of seafloor spreading.* Now, finally, geologists were converted en masse to a belief in plate tectonics.

As far as I know, everything about that story is true, except for one thing: the mass conversion didn't occur just yet. Interviewed years later, Vine said of the paper's reception, "It was the classic lead balloon, . . . one can't overemphasize that." He chalked up the negative reaction in part to the fact that the symmetric pattern around the ridges wasn't as clear as it might have been. Vine also realized that he and Matthews were relying on two assertions that at the time were not widely accepted: the reversal of the Earth's magnetic field through time, and the notion that the field's orientation could be read in the rocks of the ocean floor. If people didn't believe those two assertions, they surely weren't going to believe the radical conclusions Vine and Matthews had drawn from them.

Even so, the conversion was not far off. Within a few years, extensive new data from the ocean floor and from lava flows on land validated the idea of reversals in the Earth's magnetic field, one of the underpinnings of Vine and Matthews's work. Far more complete magnetic recordings across several mid-oceanic ridges were published, with stunning "zebrastripe" diagrams showing many bands of rock alternating between the current field orientation and its reverse, the patterns on the two sides of the ridge looking like mirror images. Where the theory predicted that tectonic plates should be sliding against each other (as the Pacific and North American Plates are doing at the San Andreas Fault), seismic studies showed exactly that kind of motion. The formation of chains of oceanic volcanoes, such as the Hawaiian Islands and the seamounts extending northwest from them, was elegantly explained by the movement of a plate over a hotspot, a region where a stationary plume of magma rises up from the mantle.

Wegener, had he survived (he would have been in his mid-eighties at the time), might have been especially pleased to see two new developments. One was a computer-generated fit of the continents bordering the Atlantic Ocean, based on a mathematical theorem for the movement of a rigid plate on the surface of a sphere. The fit was remarkably good and

---

* A Canadian geologist named Lawrence Morley, looking at magnetic evidence from the northeast Pacific, had independently come to the same conclusions as Vine and Matthews. However, Morley's manuscript was rejected by two scientific journals, and he wasn't able to publish his ideas until 1964, the year after the Vine and Matthews paper appeared. Nonetheless, the connection between the magnetic data and seafloor spreading is now usually referred to as the "Vine-Matthews-Morley" hypothesis (Lawrence 2002).

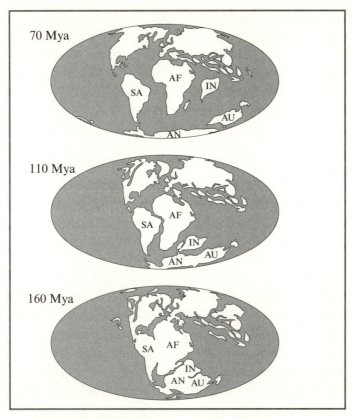

2.4 Wegener vindicated: reconstructions of the globe emphasizing the breakup of Gondwana. AF = Africa, AN = Antarctica, AU = Australia, IN = India, SA = South America. Redrawn and modified from Scotese (2004).

remarkably similar to the reconstructions of Pangea that Wegener had done by eye. The other was a study looking at the continuity of rock formations in North America and Europe, based on similarity in the age of the rocks as measured by radiometric dating, and using the new computer-generated map as a foundation. The result was a striking alignment of the formations on opposite sides of the Atlantic. It was Wegener's matching "lines of print" argument all over again.

In November 1966, a symposium on continental drift was held in New York City, with many of the most influential geologists in North America present. By the end of that two-day meeting, it was clear that the new theory had won over the doubters. When the scientist who was supposed to summarize arguments against seafloor spreading and continental movement was called away on other business, no one bothered to stand up in his place. The weight of evidence—more than fifty years after publication of Wegener's book, almost forty since Holmes's paper on convection in the mantle, and four since Hess's "History of Ocean Basins"—had finally brought down the building and raised a new one.

I have, admittedly, spent a long time telling the story of the validation of plate tectonics and continental drift. Partly that's because I like the story and find Harry Hess, in particular, a sympathetic figure. He seems to have been that rare thing, a brilliant but modest scientist. (Unfortunately, he was also that not-so-rare thing, a heavy drinker and smoker, and died of a heart attack in 1969, when he was sixty-three. Unlike Wegener, however, he at least lived long enough to see his theory widely accepted.) Obviously, plate tectonics is also a vital part of the vicariance revolution, providing a mechanism for fragmentation on a global scale. Beyond both of those points, though, there is a message in the prolonged resistance to the notion of continental drift that will be echoed in other parts of this book. The message is this: Acceptance of theories in science, even theories that we think do a great job of explaining the evidence, is not a straightforward process; the clean, textbook version of how science proceeds is almost invariably misleading. Often, things are said that are ignored or outright rejected, even ridiculed, yet in hindsight turn out to be exactly right (at least to our current understanding). And we're not just talking here about Galileo or Copernicus, but about science in the twentieth and twenty-first centuries. In the case of continental drift, Wegener, Holmes, Hess, Vine, and Matthews all presented theories and/or evidence that we now see as substantially correct and crucial, yet their work was either ignored or, as Vine said about his paper with Matthews, went over like a lead balloon. As we shall see, notions about oceanic dispersal have had a similar history of rejection and ridicule, especially in the past few decades. In fact, if anything, "ridicule" is too soft a word to describe some of the things that have been said (and continue to be said) about long-distance dispersal.

The New New York School

When Lars Brundin was writing his midge monograph, plate tectonics had not yet been widely accepted. Holmes and Hess had published their theories, but the definitive zebra-stripe magnetic orientation diagrams and other evidence were just coming out. Brundin, though, was not a man afraid of new ideas. He saw that continental drift could explain the phylogenetic patterns he was seeing in the midges, saw, in fact, that the explanation of those patterns almost *had* to be drift. It probably also helped that he was from Europe, where the new geological theory was gaining traction faster than in the United States.

In plate tectonic reconstructions of the past locations of landmasses, both New Zealand and Australia were connected in the Mesozoic to southern South America via parts of Antarctica. That was the key to explaining why Brundin's midges showed transantarctic relationships. Antarctica had been at the center of the southern part of Gondwana and thus shared with those other areas a good part of its biota, apparently including the midges. When the land connections were eliminated by continental drift, and Antarctica was transformed into an isolated, largely midge-free block of ice, the tiny flies were still left in New Zealand, Australia, and South America. More than this, the connection between New Zealand and South America, through West Antarctica, had been severed well before the connection between Australia and South America, which had been linked through East Antarctica. The expectation, then, was that in a cladogram that included midges from New Zealand, South America, and Australia, those from New Zealand would branch off first, reflecting the sequence of geological breakup (see Figure 2.5). To put it in cladistic terms, New Zealand midges would be sister to a group that included both South American and Australian midges. This is exactly what Brundin found, and not just once, but over and over, in different subgroups within the Chironomidae. The repetition was important because a pattern seen just once, even if it matched the history of Gondwanan fragmentation, could have been a coincidence, and might still have been explained by long-distance dispersal; the chironomid midges showed the pattern far too frequently and consistently for that explanation to hold water.

What Brundin had done was the first really impressive study in modern vicariance biogeography. The ancient land connections were there to be read, and Brundin, having had the patience to look at the pupal characteristics of hundreds upon hundreds of midges and build cladograms from them, had found the patterns that revealed those connections. He had shown that the transantarctic relationships of midges were explained by the fragmentation of southern Gondwana. And, moreover, he had found "not the slightest evidence of chance dispersal over wide stretches of ocean." Specifically, because New Zealand and Australian taxa were

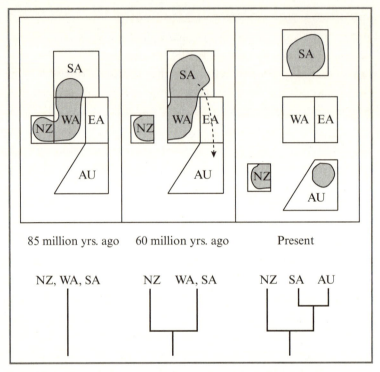

2.5 Vicariance meets cladistics: Lars Brundin's scenario for the geographic history of chironomid midges. Upper: distribution of a group of midges through time in the context of the breakup of New Zealand (NZ), Australia (AU), West Antarctica (WA), East Antarctica (EA), and South America (SA). Initially, the midges occurred widely in New Zealand, West Antarctica, and South America. Subsequently, New Zealand broke away from the other landmasses, and, later still, midges colonized Australia from South America by normal, "garden-variety" dispersal (arrow). In the present, midges are absent from Antarctica because of the severe climate, but occur in the now widely separated landmasses of New Zealand, South America, and Australia. Lower: the evolutionary trees at the same three stages in the group's history. Note that Brundin's scenario does not require any long-distance overwater dispersal, only normal dispersal coupled with the tectonic separation of landmasses.

never sister groups, there was no indication that midges had ever colonized one of these areas from the other by crossing the Tasman Sea that lay between them.

Unlike some of the biogeographers he inspired, Brundin did not see long-distance dispersal as something that either never happens or is impossible to study.* For instance, he wrote, "No one denies . . . that the transporting capacity of gales, hurricanes, and tornados is impressive and that small insects can cross wide expanses of sea with the aid of air-borne dispersal." Nonetheless, he had sharp words for what he saw as the fuzzy and misguided approach of the likes of George Gaylord Simpson, Ernst Mayr, and P. J. Darlington—the dispersalists of the "New York School." He chastised them for their ignorance about phylogenetic relationships, that is, for not understanding how to build cladograms and not realizing that this was the key to understanding biogeographic history. He scolded them for clinging to the assumption that the positions of continents and ocean basins are fixed, or that, if continental drift had occurred, it had happened too long ago to be relevant for modern groups. And he was condescending about their facile reliance on chance dispersal. "Several troubled biogeographers," he wrote dismissively, "have found consolation and relief in the thought of a raft with a sufficient store of food and water put at the disposition of a pregnant female at the right moment." These improbable dispersal stories smacked of *deus ex machina* and seemed to Brundin "comparable with a confession of failure." Brundin's venom, his disdain, poured off the page: "There is something negative, sterile, and superficial involved in the above [dispersalist] approach which offends a critical mind," he wrote. His own emphasis was instead on the general pattern revealed by phylogeny, typically a pattern produced by the fragmentation of areas, that is, by vicariance.

I am trying to imagine what it must have been like for the young Gary Nelson, reading Brundin. The strengths of Brundin's work were many to someone open to new ideas. By the standards of the time, Brundin's Hennigian approach to inferring phylogenetic relationships was a great step forward, and he was clearly right in saying that phylogeny was the key to understanding biogeographic history. The case of the midges, with their repeated phylogenetic pattern matching Gondwanan breakup, also showed the power of the resuscitated theory of continental drift for biogeography. All of that must have impressed Nelson. On top of that, the

---

* In fact, some later vicariance biogeographers criticized Brundin for being too much of a dispersalist.

whole thing was *radical*, a slap in the face of the grand old men of the New York School, who seemed to be doing everything wrong where Brundin was doing it right. Brundin even made that affront an explicit point, writing, "Biogeography has a great future. But the rate of progress is strongly dependent on due distrustfulness of authorities." I imagine that this attitude struck a chord with the rebellious young Nelson. Take up the cause of the fresh and clear-minded, or stick with the muddled establishment. Seems like an easy choice. Hell, I'm neither young nor particularly rebellious, but reading Brundin now I almost wish I had been there too, ready to join the revolution.

As it turns out, Brundin was working in the Swedish Museum of Natural History, where Nelson read the midge paper, so Nelson had a chance to meet him soon afterward. Brundin was then in his late fifties, a thin, somewhat severe-looking man. Nelson found him to be polite, maybe more so than one might have expected from his sharply worded monograph. Oddly enough, according to Nelson, the two of them didn't really talk about science, either then or on the few other occasions when their paths crossed. Nonetheless, Nelson had discovered a great intellectual influence, one that he has carried with him ever since. Not long ago I emailed him and, in a follow-up, he wrote, without any provocation from me, "If you have not read the first 50 pages of Brundin [i.e., the midge monograph], I would suggest you do. He is much clearer than Hennig; and unlike Hennig, he has potent critiques of the dominant paradigms of systematics and biogeography." (There might have been a subtext to that message too, namely, that I was misguided and needed to read Brundin to straighten myself out—Nelson knew that I thought chance dispersal was important.)

Nelson was on a path now, some would say a warpath. As a maven, he had acquired the special information in Brundin's monograph; now he would switch to his roles of connector and salesman. His first stop, while still on his research fellowship, was the British Museum of Natural History in London. During a weeklong visit there, he discussed Brundin's approach with three other ichthyologists: Humphry Greenwood and Colin Patterson, who were the resident curators, and Donn Rosen, who was visiting from the American Museum of Natural History. All of them were smokers, so, to avoid the possibility of setting the alcohol-preserved fish collections on fire, they would meet outside of the building under the colonnade to smoke and chat. At the time, Nelson was the only one who had read Brundin; thus, despite the fact that he was the youngest and only unestablished member of the group, he was in the role of teaching the others about cladistics and its implications for studying biogeography. How far he convinced them during that visit is a little unclear, but even if he only made them look into Brundin's work for themselves, he had planted a seed. Not long afterward, both Patterson and Greenwood became prominent converts, despite the fact that accepting cladistics invalidated much of the research they had already done. Patterson, in particular, who was known for his dramatic talks delivered in an impressive baritone—the "voice of God," as it has been described—would become a fervent advocate of cladistics and vicariance biogeography.

Returning to the United States, Nelson began a job as a curator at the American Museum of Natural History in New York, ironically the very place from which William Diller Matthew and George Gaylord Simpson had spread the dispersalist view against which Brundin was now railing. Nelson wasted no time in continuing his evangelical work. He started by continuing his discussions with Donn Rosen, haranguing his new colleague as they walked the halls of the museum's research area. Other staff members closed their doors to shut out the loud, heated arguments reverberating in the corridors. Like Patterson and Greenwood, Rosen had already published research papers that would be rendered obsolete by cladistics. But no matter. He eventually saw the light and became one of Nelson's closest intellectual allies and a valued confidant. Within the next few years, many others at the museum followed, all of them becoming cladists, and most also adopting the related approach of vicariance biogeography as defined by Hennig and Brundin and in later papers by Nelson, Rosen, and others.

They were definitely onto something, as Brundin's midge work had indicated. The importance of using phylogenetic relationships in an exact way quickly became obvious to most scientists studying biogeography, even to old-school types like Ernst Mayr. (However, Mayr ceded this point only grudgingly, writing, "The Hennigian methodology may be more rigorous than the frequently rather superficial analyses of earlier authors, but the basic approach is the same." In this he exemplified a common reaction to a new theory or approach: claim that your side had it more or less right all along.) Also, by the late 1960s, plate tectonics had become the accepted paradigm in geology, and no one could doubt that continental drift had influenced the distributions of living things, although people continued to argue about just how pervasive that influence had been. Already the dispersalist P. J. Darlington's assessment of continental drift, from his 1965 book *Biogeography of the Southern End of the World*, sounded quaint and ridiculous: "I have therefore become a Wegenerian, but not an extreme one. I doubt the former existence of a Pangaea or Gondwanaland, and

I think that the movements of continents have been simpler and shorter than most Wegenerians suppose." Darlington seemed to imply that the continents had always been apart, as they are today, just not quite so far apart. He was, of course, completely wrong.

One thing vicariance biogeography did not quickly produce was a wealth of clear examples to show that the fragmentation of some area, by plate movements or other processes, explained disjunct distributions. For instance, thumbing through *Systematics and Biogeography*, the thick 1981 book written by Nelson and an American Museum arachnologist, Norm Platnick, one is struck by how few real-world cases of vicariance are included. Through tree diagram after tree diagram, the authors show how to construct cladograms and use them in biogeographic studies, but the book gives little reason to believe that fragmentation actually explains many piecemeal distributions.

There were a few compelling examples, however, to add to Brundin's work on the midges. One came from Nelson's ichthyological buddy Donn Rosen, who had been studying swordtails and other guppy relatives in Middle America since before his conversion to cladism. Looking at two groups of these fishes, Rosen found that the geographic histories implied by the cladograms of the two groups were strikingly similar. Specifically, when one replaced the fish taxon names on the cladograms with the areas in which each taxon was found—producing what are known as *area cladograms*—the patterns were largely the same for the two groups. Rosen then took his analysis a step further by calculating the probability of obtaining the observed level of agreement by chance and found that the probability was very low. The exact probability he came up with, 1 in 105, was later shown to be inaccurate, but the general point he made seemed reasonable to most biogeographers: the similar area cladograms of the two groups suggested a common history of range fragmentation rather than chance dispersal, which would have produced a more haphazard pattern. A hole in Rosen's example was that it wasn't clear what geological or other events had fragmented the fishes' distributions and given rise to the matching histories, but he could at least point to some possible causes, such as the rise, in the Pliocene, of an east-west belt of volcanic mountains in Mexico (the range that includes Mexico's highest peaks, such as the Pico de Orizaba and Popocatépetl). His case was strengthened by the fact that at least parts of his fish area cladograms matched those of other, completely unrelated groups, such as box turtles and red-bellied snakes. The common cause of vicariance should affect many organisms in the same way—the uplift of a new mountain range, for instance, would isolate populations of many species on the two sides—and that was just what Rosen seemed to be seeing.

Another of Nelson's American Museum converts, a young ornithologist named Joel Cracraft, studied the anatomy of an iconic Gondwanan group, the ratites—large, flightless birds such as ostriches, rheas, and kiwis, among others. As with Brundin's midges, Cracraft's cladogram of the ratites seemed to fit the sequence of Gondwanan breakup. Cracraft also combed both the plant and animal literature and came up with other cases showing an apparent match between phylogenetic history and continental separation, some more compelling than others. Among his plant examples was the first application of cladistic vicariance thinking to another iconic Southern Hemisphere group, the southern beech trees of the genus *Nothofagus*. For both the ratites and the southern beeches, one could also make a more traditional kind of argument that favored vicariance; few people thought that large flightless birds or plants, such as *Nothofagus*, with seeds ill-suited for long-distance dispersal, could traverse an ocean on their own. Groups like these had always been conundrums for dispersalists, and now, with the acceptance of plate tectonics, they seemed like obvious examples of Gondwanan vicariance.

2.6 Southern beech trees, genus *Nothofagus*, in Chile. The occurrence of members of this genus in southern South America, Australia, New Guinea, New Zealand, and New Caledonia has been attributed to persistence through the breakup of Gondwana. Photo by the author.

The logic of the cladistic approach, its prospects for producing general explanations on a global scale, and a few nice real-world examples were certainly critical to the rise of vicariance biogeography, but there were other forces at work too. For one thing, there was a kind of scientific seduction involved, embodied by Nelson, in his role as persuasive salesman. Michael Donoghue, a systematist and evolutionary biologist at Yale, who, among many other things, studies the biogeography of Northern Hemisphere plants, was a graduate student in the 1970s, and remembers going to a bar near the American Museum with Nelson. As Donoghue tells it, "We were sitting at a place that had a table that had like a glass top.... And he [Nelson] said, 'How do we know that biogeography isn't like the following: . . . I take a hammer and I hit it right in the middle of the glass and it just splinters the whole thing into different, like, continents.... How do we know it wasn't like that?'" I asked Donoghue what he thought of Nelson's shattered-glass analogy, and he said, "Well, I thought it was bold. I thought at least here's a guy who's thinking outside the box. And that was the attraction of Gary Nelson."

Nelson and other vicariance biogeographers could impress potential converts by being bold and personable, but they could also be bold and bullying. There was definitely a sense of "if you're not with us you're against us, and if you're against us you're a fool." I remember in graduate school seeing a talk by one of the American Museum cladists, and feeling afterward as if I and everyone else in the audience had been berated. The speaker's demeanor during his talk and, especially, in the question-and-answer period that followed, was like a prolonged sneer. That was hardly an isolated case. Both in person and in print, scientists of the cladistic vicariance school often sounded like arrogant, sarcastic (though intellectually rarefied) schoolboys. As another case in point, Nelson began one of his cladistic papers, not with a quote from some scientific luminary like Darwin or Hennig, but with a Stevie Wonder lyric turned into a condescending dig at his opponents: "When you believe in things you don't understand, then you suffer; superstition ain't the way."

That attitude clearly turned off many people, especially the "establishment" scientists who were being sneered at. For instance, Darlington, in a rebuttal to Brundin, was obviously miffed by the cladists, writing that their "central attitude is one of self-conscious superiority." John Briggs, a well-known biogeographer and author of several books on the subject, described with displeasure vicariance scientists "shouting at meetings and interrupting presentations," a common complaint from more traditional scientists offended by the lack of decorum. On the other hand, the sneering, antiestablishment, anything-goes stance may have been like catnip for young scientists—mostly young, male scientists—already leaning in that direction. Although it would be hard to prove it, I have the sense that cladism and cladistic vicariance biogeography were movements heavily fueled not just by intellectual considerations, but also by this radical belligerence that at times seemed like a stratagem. That attitude translated to an effective style of argument based on equal parts shouting and sarcasm, and also may have served as a recruiting tool—*Join the irreverent radicals and be irreverent and radical yourself! You too can use Stevie Wonder lyrics to mock your elders! (And, by the way, if you don't join us, we'll be mocking YOU.)*

The whole cladism/vicariance biogeography camp seemed a bit like an unruly cult, an intellectual gang, complete with passwords ("synapomorphy," "area cladogram," "component analysis"), a deity (Hennig), and gang leaders (Nelson chief among them), but it was a cult that grew quickly. By 1980 the gang had formed an official organization, the Hennig Society, which published the journal *Cladistics* and had board members, an annual meeting, and all the other usual trappings of an academic society. Because, of course, it is not the goal of most rebels to remain outside of the establishment, but, rather, to *become* the establishment.

Michael Donoghue had said that he was attracted to Nelson's boldness, his ability to think outside of the box. Replacing the muddled, river-delta view of phylogeny with an explicit, cladistic approach was a valuable leap away from the old biogeography, as was the incorporation of continental drift. It wasn't all good though. As I will describe in the next chapter, Nelson and many other vicariance biogeographers adopted an approach that, in some ways, was even more constraining than the old New York School dispersalism. On the charitable side, one could say they did it all in the search for a kind of Holy Grail, the ability to make grand generalizations about biogeographic history. But, in the process, they tried to bury the old dispersalism, and in that attempt they ended up arguing themselves into a strange intellectual corner where they envisioned an idealized history of life that never was.

Anoles, the common arboreal lizards of the Neotropics, are found throughout the Caribbean and must have reached many islands by natural overwater dispersal. The brown anole, Norops *(formerly* Anolis*) sagrei, seems to be an especially frequent oceanic voyager, having spread from Cuba to many smaller islands in the vicinity. (This species has also been introduced to the southeastern United States, southern California, Mexico, Hawaii, Taiwan, and Singapore.) Some of these journeys probably required rafting, but two researchers, Amy and Tom Schoener, wondered whether the lizards might have simply floated in the sea in some cases—for example, to move between islands in the Bahamas. To evaluate that possibility, they conducted what amounted to a version of Darwin's experiments on the survival of seeds in salt water, only with lizards as the subjects. They caught brown anoles in the Bahamas and put thirty-nine of them, one at a time, into a saltwater tank in which small waves were generated. Then they recorded the animals' fates.*

All thirty-nine of the anoles were alive and floating after one hour, and ten were still alive and floating after twenty-four hours. (It's a little unclear whether the Schoeners rescued the twenty-nine lizards that were not floating at the end of the twenty-four hours. Incidentally, this study was done in the 1980s, when rules about experimentation on animals, especially non-fuzzy animals like lizards, were a bit looser than they are now. It's doubtful whether this kind of experiment would be approved by an institutional animal-use committee today.) The lizards usually floated with their forelegs extended downward and their heads held well out of the water. These results marked the anoles as champion floaters compared to small mammals such as mice, gophers, and moles, which, as previous experiments had shown, could only float for a few minutes, on average. The Schoeners suggested that brown anoles could indeed have dispersed among islands by floating free in the ocean.

Why these anoles are such good floaters isn't entirely clear. However, it seems to have something to do with surface tension: when the Schoeners added detergent to the water tank, to break surface tension, all the lizards immediately sank.

Chapter Three

OVER THE EDGE OF REASON

Jurassic Hawaii

Twenty miles off the southeastern coast of the Big Island of Hawaii and 3,000 feet below the surface of the ocean, another island is forming. At that spot, lava periodically spews from the summit of an underwater volcano and cools to basaltic rock, building up the mountain in fits and starts. Along with Mauna Loa and Kilauea on the Big Island, this submarine volcano, named Loʻihi, is the active manifestation of the Hawaiian Hotspot, where magma welling up from the Earth's mantle has created a string of islands as the Pacific Plate slides northwest over the stationary plume of magma. Within the next 100,000 years, Loʻihi will likely emerge above the waves, the newest island in the Hawaiian chain and a dramatic embodiment of the geologic origins of the entire archipelago.

That volcanic origin means that the Hawaiian chain is a series of classic oceanic islands, ones that have never been connected by land to a continent. Beyond that, the archipelago is also the remotest of the remote, located some 2,400 miles from North America and even farther from any other continent (unlikely as that may seem when one is lying in the shadow of a giant resort hotel on the crowded beach at Waikiki). Those facts of geologic origin and location suggest an obvious biogeographic conclusion: the rich native biota of Hawaii must be derived from numerous chance, long-distance colonizations. Almost all biogeographers, even many who are otherwise in the vicariance camp, agree that this must be the case.

But not Michael Heads.

Heads is a New Zealand botanist and biogeographer who thinks that successful long-distance dispersal is essentially a mythological process; at best, according to Heads, such events happen so rarely that there's no reason to even consider them as viable explanations for piecemeal distributions. He believes that virtually all successful dispersal is of the "normal" variety, meaning the usual movement of organisms from one piece of suitable habitat to another.

Heads's views remain uncompromising even when it comes to Hawaii, despite the obvious volcanic origins and extreme remoteness of that archipelago. "The taxa that colonized Hawaii," he says, "did so by normal dispersal . . . the same process you see in your garden." He believes that organisms made these predictable, garden-variety jumps to Hawaii, not over vast expanses of the Pacific but, rather, from nearby islands that no longer exist. Superficially, this doesn't seem too farfetched because, in fact, the hotspot that created the current islands also produced a long line of former high islands that now exist as low atolls or submerged seamounts, stretching first northwest from Kauai and then making a bend to head nearly due north all the way to the Aleutians. In theory, many organisms could have made a series of easy jumps as each new volcanic island formed along this line from Alaska to the modern Hawaiian Islands.

Geologically, then, Heads's idea for the origins of the Hawaiian biota sounds possible. However, when we start thinking about the plants and animals themselves, we run into a big problem. Normal dispersal by island-hopping, either from north or east of Hawaii, could not have happened recently, because the relevant islands did not exist. Finches or violets or tarweeds taking short, garden-variety hops from Alaska or California toward Hawaii within the recent past would have ended up afloat in the Pacific.* Thus, if Hawaii *was* colonized via normal dispersal, we should find that the evolutionary connections of island lineages to mainland ones are not recent, but instead are relatively ancient, on the order of 70 million years old or older (the age of the earliest islands in the hotspot chain). The problem is that hardly any Hawaiian lineages seem to be that old.

---

* In a 2011 paper, Heads also mentioned other possible sources of the Hawaiian biota, such as former high islands that now exist as the submerged Musicians Seamounts north of Hawaii and the low-lying Line Islands to the south. The same argument about the age of connections applies to these sources as well. He also mentioned former lands east of Hawaii that are now accreted to or subducted beneath western North America, but these would not have formed closely spaced stepping stones allowing normal dispersal.

Molecular evidence indicates that almost all of the island taxa, including the famous honeycreepers, the Nene Goose, and the silversword plants, separated from continental relatives within the past 20 million years, and most of them did so within the past 5 million years. If those dates are roughly correct, the ancestors of all those organisms must have crossed a wide expanse of the Pacific to reach Hawaii. They could not have colonized the archipelago by normal dispersal, but instead must have done so by more unusual means in each case—birds blown on storm winds or beetle eggs attached to a natural raft of vegetation or something of that sort. In other words, they must have arrived by chance, long-distance dispersal.**

Even if one doesn't believe the molecular dating evidence, there's a similar, commonsense argument against Heads's view. This argument is based on the fact that quite a few taxa in Hawaii are classified as subspecies of more widespread species. For instance, the pueo is considered a subspecies of the Short-Eared Owl, the aeʻo a subspecies of the Black-Necked Stilt, the opeʻapeʻa a subspecies of the hoary bat, the oʻhelo papa a subspecies of the beach strawberry, the kupukupu a subspecies of the Boston swordfern, and the paʻu o Hiʻiaka a subspecies of the oval-leaf clustervine. None of these forms were introduced to Hawaii by humans in recorded history, and prehistoric introduction by Polynesians is extremely implausible for most of them; bats, nondomesticated birds, and plants such as the strawberry and swordfern that are not found in Polynesia were not likely passengers or stowaways on the prehistoric boats that reached Hawaii. Thus, the ancestors of these subspecies apparently arrived on their own and afterward evolved slight differences from their relatives elsewhere. If we make the reasonable assumption that these Hawaiian forms only separated from their close relatives, which are still considered part of the same species, within the past few million years (and probably within the past few hundred thousand years in some cases), then we must also conclude that they could not have come from nearby land, because, again, there was no nearby land. They must have reached the islands by natural long-distance dispersal.

** A recent study suggests that there were no islands in the chain emerging above water between 29 and 33 million years ago. If this is true, the island-hopping route from Alaska could not have populated the current islands (Clague et al. 2010).

Heads is recalcitrant in the face of these well-known observations about the Hawaiian biota. The evidence for long-distance colonization of the archipelago doesn't budge him at all. He doesn't believe the divergence date estimates or the commonsense argument about subspecies. He apparently thinks that most of the Hawaiian biota has been in the mid-Pacific region, on various ancient islands or larger landmasses, since the Jurassic, which would push their origins back to well before the beginnings of the hotspot chain. And his antidispersal ideas don't stop with Hawaii. He believes the same thing about all oceanic islands—the Galápagos, Easter Island and the Marquesas, Ascension, Tristan da Cunha and the Azores, Mauritius, Rodrigues, and Île Amsterdam, to name just a few. Together, the volcanic islands of the world are home to thousands of endemic species, and, according to Heads, the ancestors of all of them reached those islands by normal, everyday dispersal. The same peculiar story must then be repeated over and over again: An island may seem to have been extremely isolated for its entire existence, but that is never really the case. There was always some ancient connection to other land: either a direct bridge or a series of closely spaced, stepping-stone islands over which every instance of colonization could have taken place. There is always reason to reject the evidence for recent evolutionary connections of island species to continental ones and, therefore, the idea of chance, long-distance colonization. Despite what their DNA and morphology indicate, the origins of the island owls, stilts, iguanas, silversword plants, ferns, damselflies, crickets, lava lizards, tortoises, mockingbirds, pigeons, prickly pears, skinks, bristletail insects, and butterflies are all ancient, long predating the current islands. That whipsnake on Isla Clarión, four hundred miles off the west coast of Mexico, may look as if it only branched off from mainland whipsnakes within the past couple of million years, but its origins actually stretch back ten or twenty or thirty times further into the past.

This description may make Heads sound about as legitimate a scientist as a flat-Earth advocate, but that is far from the case. His papers appear in well-respected scientific journals, and he recently wrote a book on the biogeography of the tropics, published by the University of California Press. His antidispersal views are definitely extreme, but, in fact, they're not far out of line with what many biogeographers started thinking in the 1970s and 1980s, when the field, while taking advantage of cladistics and incorporating continental drift, also made a turn down an intellectual cul-de-sac. In Heads's case, at least, it's clear where the strange thoughts came from.

Rebel Among Rebels

Most of the vicariance scientists mentioned in Chapter Two—Lars Brundin, Gary Nelson, Colin Patterson, Donn Rosen, and others—worked within typical scientific institutions, in particular at large natural history museums. They did the things that institutional scientists are supposed to do, such as conducting original research, going to scientific meetings, mentoring graduate students, serving as editors of scientific journals, and publishing peer-reviewed papers. They may have been intellectual rebels, but they were rebels working within the system.

Then there was Léon Croizat, who not only did most of his work outside of the system, but also reveled in his status as a nonconformist, an iconoclast, an individualist rather than a cog in the wheel.

From the start, Croizat's career trajectory didn't resemble that of a typical biologist. He was born to French bourgeoisie parents in Turin, Italy, in 1894, but, by the time he was a young man, the family had fallen on hard times and he was scrambling to make a living. Fearing the fascists, he fled Italy in the early 1920s with his wife and two children and eventually settled in Massachusetts, where he worked at the Arnold Arboretum at Harvard, first mapping the grounds and then as a technical assistant. Although he wasn't formally trained as a scientist, Croizat had been interested in biology from an early age. In the early 1930s, he started publishing botanical papers, at first only horticultural articles on cacti

3.1 Rebel, iconoclast, and opponent of Darwin's ideas: Léon Croizat, the originator of **panbiogeography, in Caracas in 1974.** Photo by Jonathan Baskin.

and other succulents, but later also technical papers on plant taxonomy, especially on the spurge family (Euphorbiaceae). In a couple of those papers he criticized the work of another botanist at the arboretum—someone well above him in the hierarchy—and got himself into hot water. The director of the arboretum, E. D. Merrill, apparently served as Croizat's protector in that case, but following Merrill's dismissal (his "defenestration," in Croizat's words), Croizat also lost his job. That episode may have been the first significant run-in Croizat had with another scientist, but it wouldn't be the last.

Unable to find another job, in 1947 Croizat immigrated to Venezuela, where he and his first wife divorced. He then married a woman who eventually became the head of a successful landscaping business in Caracas. That was a fateful turn of events for Croizat. Through his voracious reading of the scientific literature in the library at Harvard, he had developed radical thoughts on biogeography and evolution, and his second wife's business allowed him to focus full time on expanding and self-publishing these views. Untouched by the usual channels of scientific criticism, and unfettered by editors, he poured out a series of books in the 1950s and 1960s that are impressive for their breadth of knowledge and their sheer bulk; the three-volume *Panbiogeography* alone is more than 2,700 pages long, and all told his self-published books run to more than 5,000 pages.*

These books are, in a word, unconventional, and, in small doses, they provide a refreshing change from the usual dry science texts. Croizat's writing is personal and culturally wide-ranging (he was fluent in more than half a dozen languages), and he constantly challenges authority, from Charles Darwin and Alfred Russel Wallace to George Gaylord Simpson and Ernst Mayr. As a very typical example, consider his sarcastic outburst against William Diller Matthew, the father of the New York School, and his herd-like disciples, from volume 1 of *Panbiogeography*:

> Gone forever are the days of that melanchonious soul, the individual thinker, carrying the weight, and challenge, of individual ideas when shaving in the morning and donning—fancy that—a night-gown at bedtime; the present and the future do belong to "collegiate" undertakings, mass-education, rosy vulgarization, two chickens in every pot and two cars in every garage. Matthew is the man of the times; he sees big, he sees rosy, he sees easy, he does not tire. So structured, and so precisely

---

* The one book he did not publish on his own, *Manual of Phytogeography*, adds another 700 pages to his body of work.

chiselled out in a messianically authoritative handicraft, Matthew's "zoogeography" proves of course impervious to the pinpricks of reason.

On the other hand, Croizat is often maddeningly vague and absurdly repetitive. A passage that begins as refreshingly opinionated frequently deteriorates, within the space of a few pages, into a tedious and venomous rant. For instance, in *Space, Time, Form*, the book that sums up his views (an 881-page summing up), he refers to his nemesis Darwin as "a very unhappy thinker," "*congenitally not a thinker*," "not a clear, cogent thinker," "*essentially not a thinker*," "not *born* a thinker," "anything but a thinker," "definitely poor as a thinker," and "quite limited intellectually" (italics in original), along with many other descriptions to the same effect. Reading Croizat, one starts to think not only that he could have used a good editor, but that he needed a better filter in his brain as well.

Style aside, though, Croizat had some things to say that would eventually permeate the field of historical biogeography. An obvious one was his constant harping on the dispersalists. He was convinced that Darwin, far from having finally put biogeography on the right path, had instead created a science based on a set of unsupported and unsupportable assertions (that was the context for most of those Darwin-as-an-unhappy-thinker quotes). Darwin and Wallace and, eventually, Matthew, Simpson, Mayr, and many others thought that species and higher taxa had "centres of origin" from which they had often spread out, generating wider distributions. For instance, Matthew, probably betraying some form of Eurocentrism, believed that most groups had originated in the Northern Hemisphere and subsequently dispersed to the south. Dispersal, in this view, was often of the long-distance, chance sort, and gave rise to isolated populations that eventually evolved to become distinct species. As we have seen, oceanic islands like Hawaii were key examples for the dispersalists; since these islands had originated and persisted as remote and isolated bits of land, native lineages, such as the honeycreepers and the silverswords, must have reached them by long-distance, overwater dispersal. Not surprisingly, this school of thought, beginning with Darwin and his experiments, emphasized "means of dispersal," which dictated how easily different kinds of organisms could move about the globe and surmount barriers such as oceans. According to the dispersalists, there was a good reason why South America and Africa share quite a few genera and even some species of plants, but almost no land vertebrates: plants, with seeds that can float or be blown by the wind or survive a long journey on a raft, have a much easier time crossing the Atlantic than do lizards or frogs or rats.

Croizat wasn't buying any of this. "Centres of origin," "chance dispersal," and "means of dispersal" were all dirty phrases for him, reflecting how Darwin and his congenitally limited intellect had initiated the biogeographic equivalent of the Dark Ages. In Croizat's view, Darwin had gone about things entirely backwards, coming up with a dispersalist theory of biogeography and then looking for facts that supported it. Croizat seemed to hold to the Baconian view that knowledge about the workings of nature would emerge naturally, without the bias of already having a pet theory in mind, if one simply compiled enough facts (assuming, of course, that one was "born a thinker" and could decipher the meaning of those facts). Croizat claimed that his views on the history of life had come about in just this way.

The facts in this case were the facts of distribution. More specifically, Croizat refined a method that involved placing points on a map to indicate each region where members of a group—a genus of sunflowers or a family of beetles, for instance—were found, and then connecting the dots to form a distributional line that he called a *track* (see Figure 3.2). Examining many such cases, he found that the tracks for unrelated groups often overlap; for example, the tracks for baobab trees and ratite birds,

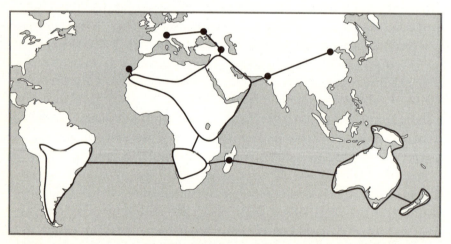

3.2 Croizatian tracks for the ratite birds (ostriches, rheas, kiwis, and relatives). Enclosed areas represent the group's modern (including recent historic) distribution. Circles are fossil localities. Lines connecting areas are panbiogeographic tracks, interpreted as reflecting an ancestral Gondwanan distribution later fractured by continental drift. One can picture the Atlantic and Indian Ocean tracks getting shorter and shorter as one goes back in time (and the ocean basins contract) until finally the tracks are reduced to nothing and the ancestral, largely continuous distribution is seen. Redrawn and modified from Craw et al. (1999).

among others, cross the Indian Ocean, connecting Africa and Australia. The line representing the overlap between multiple tracks Croizat called a *fundamental*, or *generalized track*. This approach of examining individual tracks, and from them identifying fundamental tracks, was the essence of Croizat's panbiogeography, and the overall pattern of fundamental tracks was the key to his view of biogeographic history. He claimed that the "patterns of geographic distributions of plants and animals—whatever their 'means' [of dispersal]—are absolutely congruent, as a fact of nature, within a minimum of fundamental tracks and centers."

To Croizat, the observation that there are relatively few fundamental tracks led to an obvious conclusion: the processes that have given rise to distribution patterns must be general ones that have affected diverse organisms, with completely different "means of dispersal," simultaneously. Long-distance dispersal is, by definition, a random process, so it couldn't possibly account for these repeated geographic patterns. Lizards on rafts, seeds blown by hurricanes, snails on birds' feet—these things would have produced a chaotic cobweb of individual tracks, not a small number of shared, fundamental tracks. That unhappy thinker Darwin may have believed that chance dispersal was a key to biogeography, but the facts clearly showed otherwise.

In place of the notion of groups spreading out from centers of origin via long-distance dispersal, Croizat saw vicariance almost everywhere. In his view, the history of a group typically involved a phase of spreading out by means of normal, garden-variety dispersal, followed by a phase in which the group's geographic range remained more or less static. *Mobilism* and *immobilism*, he called the two phases. During the immobilist phase, some external change, such as tectonic movement or a rise in sea level, often broke the overall range of the group into fragments. Populations in the isolated fragments would then be free to evolve into distinct forms. In other words, diversity was generated by the splitting of an ancestral geographic range, that is, by vicariance. Fundamental tracks crossing oceans reflected the shared ancestral ranges of many groups and their later sundering by a common process of fragmentation, tectonic or otherwise. For instance, a fundamental track crossing the Indian Ocean showed that various groups, such as the baobabs and the ratites, had once lived on a contiguous African-Australian landmass that had been broken apart by the opening of the ocean.

Vicariance, of course, was not a new idea in the 1950s, going back as it did to the biblical literalists and perhaps even to the ancient Greeks, not to mention Darwin, Wallace, and the other "dispersalists," who all believed to some degree in the importance of range fragmentation. What distinguished Croizat was what one might charitably call the "purity" of his vision: he saw vicariance as all-important, and long-distance dispersal as insignificant. As a consequence, "means of dispersal" were not even worth studying. Darwin's experiments with floating branches and seeds in bird guts, his calculations based on the speed of ocean currents, his conjectures about transport on icebergs, they were all a colossal waste of time. For Croizat, the diversification of living things was driven by geologic and climatic processes that fragmented ancestral ranges. "Earth and life evolve together," he famously wrote. That's pretty much all there is.

Even at the time, there were some serious weaknesses in Croizat's argument. Two things, in particular, come to mind. First, his claim that there are very few fundamental tracks was misleading. The tracks of individual lineages on New Zealand, for instance, run all over the place—to New Guinea, New Caledonia, South America, Australia, Tasmania, and Southeast Asia, among other places. Certainly, if all New Zealand lineages had tracks connecting up to, say, southern South America plus Australia, that would argue strongly for origins by vicariance (via Gondwanan breakup) rather than oceanic dispersal, but that is simply not the case. Similarly, tracks for the Hawaiian Islands run to various parts of the Americas, Asia, Australia, New Zealand, and other Pacific islands. And the same pattern holds for virtually any sizable area on the planet. Croizat and his followers raised the reasonable point that a diversity of tracks is actually expected because landmasses are often made up of pieces with different geologic origins; for example, New Caledonia is an amalgamation of several different island arcs, and thus should contain lineages with different tracks. However, if any number of fundamental tracks can be accommodated under Croizat's worldview, it becomes unclear whether the number of such tracks is a valid argument in his favor and against long-distance dispersal. If the signatures left by both vicariance and chance dispersal are a spider's web of tracks, how do we choose one explanation over the other?

Second, the fact that unrelated organisms with different means of dispersal share the same track doesn't necessarily argue against long-distance dispersal; it is possible for rare, chance dispersal events to collectively produce a clear pattern. As a case in point, an island close to a continent would likely be colonized by many groups from that continent. Similarly, if a strong, directional ocean current flows between two landmasses, as is often the case, dispersal might give rise to many tracks connecting those two areas. Every one of these dispersal events could be something rare and unpredictable—rats on a tangle of vegetation or seeds inside the floating carcass of a bird—but, having followed the same path across the sea, the lineages involved would share the same fundamental track. What is haphazard at one level can be ordered at another: throw a dandelion seed, a beetle, and a monkey into the air in random directions during a hurricane, and they'll all end up sailing away from you with the wind.

In short, the pattern of fundamental tracks doesn't necessarily have an obvious interpretation. The "facts" of distribution are not as easy to read as Croizat made them out to be.

Until the mid-1970s, Croizat's work was somewhat widely known among botanists, but had hardly been cited at all by zoologists. This lack of attention was significant because most of the influential biogeographers of the time—including dispersalists like George Gaylord Simpson, Ernst Mayr, and P. J. Darlington—were zoologists. Croizat and his defenders have argued that he was the victim of a "conspiracy of silence," citing the fact that Simpson, Mayr, and others had read parts of his books (and, in Simpson's case, had even corresponded with Croizat), but did not refer to him in their own publications. The counterargument is that these dispersalists had such a low opinion of Croizat's work that they simply didn't think it was worth bringing up. Mayr wrote privately that Croizat had a "totally unscientific style and methodology," and Simpson went even further, dubbing Croizat "a member of the lunatic fringe."

In any case, in the 1970s, Croizat's panbiogeography did become more widely discussed, thanks to a zoologist who was much more receptive to his ideas than the dispersalists were. That zoologist was none other than Gary Nelson, the American Museum ichthyologist and hub of the cladistic vicariance movement.

Nelson first read Croizat in the 1960s while in graduate school, where, in his words, he "pondered for a time" the three volumes of *Panbiogeography*. However, that initial rumination doesn't seem to have led anywhere. It was perhaps only after reading Lars Brundin's monograph on the chironomid midges that Nelson began thinking about Croizat more deeply. Brundin, while noting (without explanation) that Croizat's approach was "not wholly sound," was nonetheless impressed by his "blazing sermon," a sermon that emphasized range fragmentation and blasted the dispersalists. Given Nelson's respect for Brundin, those words may have spurred him to take another look at Croizat.

By the early 1970s, Nelson had soaked up Croizat's message and had become convinced that the man was something of an overlooked visionary. In a 1973 paper, he wrote enthusiastically of Croizat's work and indicated how it might dovetail with the emerging field of cladistic vicariance biogeography. He was especially impressed with Croizat's single-minded search for general patterns, reflected in the discovery of fundamental tracks shared by diverse kinds of organisms, but thought that the methods of panbiogeography could be improved by taking into account the evolutionary relationships within each group. Instead of just connecting the dots of distribution, as Croizat had done, one could, in effect, analyze sets of evolutionary trees (actually, to be technically accurate, cladograms) of many different groups superimposed on distribution maps; using trees would provide more exact information about how those groups might have become fragmented in the course of their evolution. Basically, Nelson envisioned adding another piece to the vicariance revolution by merging Croizat's search for generality with the explicit cladistic methods of Hennig and Brundin. The vicariance school would then contain both the big-picture approach of Croizat and the precision of cladistics.

At about this time, Nelson took advantage of an odd opportunity to push Croizat's work. In his role as editor of the journal *Systematic Zoology*, Nelson had received a manuscript from Croizat criticizing dispersalism and defending panbiogeography. Like almost everything written by Croizat, the manuscript was extreme by academic standards; it was personal, caustic, and blatantly one-sided. Most manuscripts submitted to journals get sent out to a few experts for peer review, but Nelson ended up sending Croizat's paper to nineteen reviewers, an almost unheard of number. Not surprisingly, most of the reviewers thought the manuscript wasn't fit for publication and, even after Croizat revised it, the reviews (now reduced to "only" nine) were split between acceptance and rejection. At this point, Nelson made another move that, according to the usual standards for journal editors, was both unusual and bold: he suggested to Croizat that the two of them, along with Nelson's American Museum buddy Donn Rosen, should collaborate on a paper, using Croizat's manuscript as a starting point. Nelson likely saw this as a chance to further Croizat's cause in a way that would end up being more palatable to academic biologists than if Croizat had simply been left to his own unfiltered devices.

Croizat agreed to the collaboration. The resulting paper, published in 1974 and innocuously entitled "Centers of Origin and Related Concepts," is something of a mishmash of ideas, but is significant for several reasons. First, it helped introduce Croizat's use of individual and fundamental tracks to a wider audience. Second, it set forth, in terms no one could miss, the idea that vicariance is the dominant process in generating biogeographic patterns. For instance, the authors concluded that historical biogeography "is to be understood first in terms of the general patterns of vicariance displayed by the world biota." Finally, the paper brought Croizat's over-the-top criticism of Darwin and dispersalism into the mainstream of biogeography. "Having failed to dissect these concepts (center of origin, vicariance) to their core," Croizat et al. wrote, "contemporary zoogeographers founder in a self-created morass of chance hops; great capacities for, or mysterious means of, dispersal; rare accidents of over-sea transportation; small probabilities that with time become certainties; and other pseudo-explanations." Like Brundin's work, it was a call to overturn the status quo, but with Croizat's characteristic lack of restraint. The paper described Darwinian biogeography as "a world of make-believe and pretense" and ended with a final jab at the New York School: "No one well informed of the zoogeography of our times can have an illusion about its manifest disreputability." Croizat's vision of a befuddled Darwin and his equally benighted minions was now out there, in a well-known journal, with the names of two established scientists, Nelson and Rosen, to validate it.

Disconnection

In his midge monograph and other papers, Lars Brundin had presented vicariance as the most significant reason for disjunct distributions, and he had bashed the dispersalists for their uncritical reliance on rare, chance events. Nonetheless, Brundin still believed that long-distance dispersal was significant. However, the Croizat, Nelson, and Rosen paper embodied a point of view altogether more extreme, in which vicariance was thoroughly dominant and long-distance dispersal was barely worth thinking about. To borrow a term used by Stephen Jay Gould in describing the spread of adaptationism,* the Croizat et al. paper reflected a "hardening" of vicariance biogeography into a purer form, a form disinclined to

---

* More specifically, Gould argued that the modern evolutionary synthesis (i.e., the integration of evolutionary theory with the genetics of populations) "hardened" from a relatively pluralistic form into one that exaggerated the scope of natural selection.

tolerate any remnant of Darwinian dispersalism. A few years later, in that same spirit, Nelson came up with a notorious phrase for dispersalism, deriding it as "a science of the improbable, the rare, the mysterious, and the miraculous."

This pro-vicariance, antidispersal view was quickly propped up by the spread of two ideas that, depending on one's point of view, have served ever since as either important assumptions or unfortunate stumbling blocks in biogeography. One was the notion that long-distance dispersal hypotheses cannot be refuted and are therefore unscientific. Consider, for instance, a group (say lungfish, ratite birds, or southern beech trees) found on several Southern Hemisphere continents. The hypothesis that such a distribution was the result of Gondwanan fragmentation would have to be rejected if the evolutionary relationships within the group failed to match the sequence of continental breakup. In contrast, there is no pattern of evolutionary relationships that can definitively refute long-distance dispersal for a particular group. If relationships for a group match Gondwanan breakup, this could just be a coincidence; the distribution might still be explained by a dispersal event that happened to occur in a way that mimics vicariance. More generally, there is no distribution pattern, no matter how scattered, that cannot, in theory, be explained by a set of long-distance jumps. And finally, failure to ever observe chance, long-distance dispersal by a particular group of organisms does not disprove dispersal explanations for that group; indeed, the very nature of such events makes it highly unlikely that they *will* be observed. In short, it seemed there was no type of evidence that could really rule out dispersal.

Right from the start, some people recognized that there was something odd about this notion of dispersal hypotheses as unscientific and therefore not worth considering. For instance, in the mid-1970s, a young ichthyologist named Robert McDowall (who will enter our story again as an older man) pointed out that we know long-distance dispersal happens, so it's very strange to construct a whole approach to biogeography that ignores dispersal on the grounds that it's untestable. As an obvious example, McDowall pointed out that White-Faced Herons had colonized his native New Zealand from Australia by long-distance, overwater dispersal in the twentieth century (see Figure 3.3). People knew this because they had recorded the first instances of these birds breeding in New Zealand; the herons had not been established, and then they were—there was nothing hypothetical about it. What good is a "rigorous" science of biogeography, McDowall asked, if it forces us to ignore such events that obviously affect distributions? Nonetheless, the dispersal-is-unscientific argument

3.3 Rare and mysterious but true: White-Faced Herons (*Egretta novaehollandiae*) dispersed naturally from Australia and established a population in New Zealand in the 1940s. Photo by Glen Fergus.

took root, no doubt in part because Popperianism, which identifies falsification as the hallmark of a scientific hypothesis, was (and remains) the philosophy of choice for many biologists. In fact, in their 1981 book *Systematics and Biogeography*, Gary Nelson and Norm Platnick claimed that their approach was the union not just of Hennigian cladistics and Croizatian panbiogeography, but also of Karl Popper's philosophy of science. The implication was that, to be a true scientist, you had to be a falsificationist, and therefore, you couldn't be a dispersalist.

The second prop for the pro-vicariance, antidispersal view was the idea that almost all evidence about the ages of taxa was so unreliable as to be essentially worthless. According to the dispersalists of the New York School, many groups were simply too young to have been influenced by continental drift. If these groups showed disjunct distributions that seemed to reflect ancient continental breakup, that was simply an illusion. For instance, Ernst Mayr claimed that most living groups of mammals, birds, and flowering plants had originated long after the separation of South America and Africa, and therefore could not have been affected by that event. But if vicariance was as dominant as Croizat, Nelson, and their followers believed, this argument couldn't possibly be right. These groups had to be older than they appeared to be.

The dispersalists based their age estimates in part on the direct evidence of the fossil record, with the earliest fossil specimen of a group taken as a rough indication of the group's time of origin. Of course, everyone knew that the fossil record was incomplete—Darwin himself had made that point forcefully in *The Origin of Species* to explain why we don't have fossils of all the intermediate forms that his theory implies must have existed—but many biologists nonetheless believed that the record was good enough to give ballpark estimates of the ages of many taxa. For instance, the fact that the oldest fossils of most orders of mammals date to within the past 50 million years was used to claim that most of these groups originated after the Mesozoic, which ended 66 million years ago. Vicariance scientists argued that these claims were completely unjustified. The oldest fossils of a taxon certainly set a *minimum* age for the group, but the actual age of origin will almost always be older, and one never knows just how much older. A group with a fossil record going back 49 million years might be 50 million years old or it might be 150 million years old.

By the 1970s, a new method of estimating ages had emerged, and it too became a target for the vicariance scientists. Specifically, biologists were using genetic differences between species as a way of estimating the time since the lineages in question had split from each other. These early studies assumed a strict molecular clock, that is, they assumed that changes in genes occur at a constant rate. Under this assumption, once the clock's rate of ticking was calibrated using fossil evidence for any one evolutionary split—for instance, between house mouse and Norway rat—genetic differences between other groups could be easily translated into the time of separation of lineages. Look at the differences in genes (or, back in the 1970s, the proteins encoded by genes) between a human, a chimp, and a gibbon, and you can see that human and chimp separated about 7 million years ago, while the human plus chimp lineage separated from the gibbon about 20 million years ago.

There were early indications, which have been abundantly verified, that the molecular-clock assumption was wrong, that the clock of genetic changes doesn't actually tick at a constant rate. Vicariance scientists used this evidence of a faulty timepiece to completely discount any estimation of the age of a group using genetic differences. In fact, if anything, ages based on the molecular clock were considered even more suspect than those based directly on the fossil record, because the clock had to be calibrated using fossils! You had to set the rate of ticking based on a fossil age estimate that could be way off, and then assume that the rate never changed, an assumption that was demonstrably false. Two wrongs couldn't make a right.

The skepticism about estimating the ages of groups based on the fossil record or molecular dating analyses is a fundamental part of the extreme vicariance worldview. To a degree, this skepticism is healthy; the fossil record really is incomplete, often incredibly so, and molecular dating, even in its later, more sophisticated incarnations, is based on assumptions that are hard to validate and are often simply wrong. (Whether the ages estimated from such analyses could be as consistently wrong as some vicariance scientists believe is something I will take up in Chapter Six.)

Nonetheless, there were signs that vicariance biogeography had developed an acute disconnection from reality related to this skepticism about the estimated ages of groups. Often, those signs can be traced back to Croizat and his extreme worldview. I started this chapter with one such disconnect, Michael Heads's claim that not one of the thousands of native lineages of the world's oceanic islands are derived from long-distance colonization. Heads is what one might call a panbiogeographic fundamentalist, and his claim about Hawaii and other oceanic islands echoes Croizat's views, although Heads, in this case, might be even more extreme than Croizat himself. As mentioned earlier, if most Hawaiian lineages are as young as current evidence indicates, then they reached the islands when the configuration of land and ocean was more or less as it is today, which means that they must have traveled vast distances over water to get there. Thus, Heads's claim against long-distance dispersal can only be true if island lineages are much older than the evidence suggests.

Another sign of this separation from reality comes at the end of Nelson and Platnick's book *Systematics and Biogeography*, and, if anything, is even more bizarre than Heads's notion about oceanic islands. Nelson and Platnick, in keeping with their Croizatian search for general patterns, perform a thought experiment about the evolutionary history of *Homo sapiens*: What if the branching order in the phylogenetic tree of humankind matched the history of continental breakup? If that were the case, wouldn't we conclude that the geographic history of humans was part of the same general process of tectonically driven vicariance that we see in so many other lineages, from chironomid midges to ratite birds to southern beech trees? Wouldn't it, therefore, also mean that humans are far more ancient than anyone imagines?

It's a grand vision for humanity, placing us within the great story of the fragmentation of the world's biotas through continental drift. It's an epitome of the Croizatian vision that "Earth and life evolve together." It's also completely looney. It reminds me of old science-fiction movies that had cavemen fighting dinosaurs, the movies that well-informed preteens would point to and say, "That's dumb, people never lived at the same time as dinosaurs." What Nelson and Platnick were saying was that, if the evolutionary and tectonic patterns matched, one could infer that the human lineage—meaning people as people, not as proto-humans or tree-living apes or any more distant ancestor—extended back 66 million years and more to the Mesozoic, which was . . . the Age of Dinosaurs. Never mind that the intensively studied human fossil record indicates that the genus *Homo* is only a few million years old. Never mind that the genes of humans and chimps are so similar that they suggest that the common ancestor of these two lineages (which was certainly not a person in the usual sense) existed only 7 million years or so ago. Never mind a whole host of fossils showing that a succession of progressively deeper human ancestors—the first hominids, the first apes, the first monkeys—were not around in the Mesozoic. All that evidence is worthless. Molecular clocks don't work. The fossil record is hopelessly incomplete. People might have lived with dinosaurs.

As with Heads's view of oceanic islands, I find it incredible that Nelson and Platnick actually believed what they said about humans in the Mesozoic. I'm wondering if, in the final stages of writing their book, they were on some kind of "vicariance high" and had a temporary loss of all perspective. To reassure myself that this wasn't the case, I emailed Nelson and asked him about that section of the book. Would a match between the evolutionary tree of humans and the sequence of Mesozoic seafloor spreading events still make him think that continental breakup might explain human geography? He answered with one word: "Yep."

Black and white answers attract us. We want our politicians to tell us they're 100 percent behind the middle class and small business and our troops overseas, 100 percent against giant multinational corporations and lobbyists and the enemies of democracy. We want to think that we'll be healthy if we just take that cholesterol drug, happy if we just listen to each other. We experience a record heat wave and want to put it all down to human-caused global warming, or, if we are of another ilk, to blame it on sunspots. Sometimes, things actually are simple and straightforward, but, more often, we're just filtering out the real complexity. I can't say I have any real handle on why we do this. Maybe sorting through the complexities makes a person too indecisive and being indecisive gets you selected out of the gene pool. Maybe seeing things in black and white helps us make valuable bonds with other people who share our unequivocal beliefs, often at the expense of some other group: "We are Christians and they are infidels," or "We are capitalists and they are communists," or, most generally, "We are good and they are evil."

In any case, I have a strong impression that what led vicariance scientists to their extreme views was, at least in part, this attraction to the unambiguous. Many of the tenets of vicariance biogeography can be seen in that light. For instance, we can believe that distributions broken up by oceans reflect a complicated mixture of vicariance and dispersal, or that vicariance is dominant, so let's believe vicariance is dominant. We can believe that evolutionary relationships and molecular clock analyses and the fossil record should all be used to interpret biogeographic history, or that only evolutionary relationships are critical, so let's believe that only evolutionary relationships are critical. We can take an eclectic approach and follow the evidence in whatever form it takes, or be strict Popperians who think that only vicariance is falsifiable, so let's be strict Popperians. Basically, vicariance scientists chose a simple, straightforward approach—using cladograms or, if they were panbiogeographers, tracks—to uncover the workings of a single, general process, namely, vicariance, rather than dealing with messy methods and messy interpretations. They chose black and white over countless shades of gray.

In doing so, they seemed to think they were finally making a real science out of biogeography. There was Brundin (actually a moderate, as vicariance biogeographers go) saying that the dispersalist approach, which he was trying to replace, "offends a critical mind." There was Croizat harping on how biogeography had followed the misguided teachings of Darwin for a hundred years, and how Croizat's own panbiogeography, grounded in incontrovertible facts, "will have powerful influence upon the whole of biological thinking." There were Croizat, Nelson, and Rosen proclaiming that the way out of the Darwinian morass was, among other things, "to formulate explicit methods of statistical analysis (based on the concept of generalized tracks) that yield unambiguous and repeatable results." One gets the sense that some vicariance scientists thought they were transforming the mushy science of biogeography into something more like the successful hard sciences of physics and chemistry. It may not be a coincidence that two biologists of the cladistic vicariance school, Dan Brooks and Ed Wiley, actually came up with something called the "non-equilibrium thermodynamics theory" (sometimes called "evolution as entropy"), which placed all of evolution within a framework borrowed from Newtonian physics.

Unambiguous, general, statistical, and repeatable doesn't sound all that bad, of course. However, in my experience, evolutionary biology (of which historical biogeography is a part) is not a science that lends itself to the kind of narrow methodological approach or, especially, the extreme generalization that came out of the vicariance movement. The impulse to make grand generalizations has a long and fairly unhappy history in this field. For instance, some evolutionists have pushed the view that almost all features of organisms are adaptive (Wallace, among others, was an early proponent of such "panselectionism"), while others, in complete opposition, have claimed that virtually all features arise automatically because of the way development happens (and has to happen), with little input from selection. Similarly, some have consistently attributed the success of especially large evolutionary groups, such as insects, to special traits ("key innovations"), while others see the success of such groups entirely as a product of chance. None of these extreme and extremely simplistic views stand up to scrutiny; the complexity, the plurality of nature always intrudes. I'm reminded of something a friend of mine, an evolutionary biologist named John Gatesy, likes to say when he senses that some biologist is missing the trees for the forest: "The only good generalization in biology," says John, "is that there are no good generalizations in biology."

Of course, it could be that vicariance scientists have hit upon the exception to the rule that there are no rules. In theory, the living world could be as they imagine, with just about all disjunct distributions caused by the fragmentation of ancestral ranges. In that world, Croizat's phrase "Earth and life evolve together" would be an accurate general description of biogeographic history. It could all be as Gary Nelson envisioned when he talked to Michael Donoghue in that bar in New York, the hammer smashing the glass table into pieces.

It could be like that, but it isn't.

*In the early 1900s, professor G. E. Beyer of Tulane University in New Orleans collected many Upland Sandpipers (*Bartramia longicauda*) on the Gulf Coast as the birds passed through during their spring migration. (More than likely, he killed them with a shotgun, a standard method for collecting birds in those days.) These birds invariably had small freshwater snails of the genus* Physa *attached to the feathers on the undersides of their wings. Beyer wrote: "I used to count the number of snails regularly; at one time I found as many as forty-one, often between twenty and thirty, never less than ten or twelve." He thought the sandpipers might have placed the snails in their wing feathers on purpose as a source of food, as if they were carrying provisions in a backpack.*

Although Beyer couldn't tell whether the snails were the same as the local Physa *species, he noted that, at the time of year the birds were collected, the local* Physa *snails were not often seen—they only became abundant later in the spring. Furthermore, only sandpipers collected soon after their arrival from the south carried snails. These observations indicated that the birds were not finding the snails locally, but instead had carried them on their journey across the Gulf of Mexico, either from islands in the Caribbean or from points farther south.*

Chapter Four

NEW ZEALAND STIRRINGS

Henry of Gondwana

On that trip to New Zealand in the winter of 2006–2007, I traveled to Invercargill, one of the southernmost cities in the world, to rendezvous with Tara, her mother, and our friend Jan after their fern class. I arrived a day before they did, so I had some time to wander, more or less aimlessly, around the city. It was early summer in the Southern Hemisphere, but in this sub-Antarctic location, a cold sea wind was blowing, funneled through streets lined with ornate Victorian and Art Deco buildings dulled by an overcast sky.

From the expansive gardens of Queens Park—the Central Park of Invercargill—I ducked out of the weather into the Southland Museum to find an odd and oddly entrancing mix of displays: a film on the stark and perpetually windswept islands south of New Zealand (I asked at the desk about the next showing and, instead of giving the time, a staff member just walked me over to the auditorium and turned on the projector); an exhibit telling the story of the wreck of the *General Grant* in 1866 and how nine men and a woman survived on one of those desolate archipelagoes for over a year, eating seals and feral pigs; intricate Maori wood carvings and household artifacts; the leg bones of one of the giant, prehistoric, flightless moa birds (sort of like an emu or a cassowary, but much bigger). Oddest of all was the zoo section of the museum, a zoo dedicated entirely to one kind of animal, the lizard-like tuatara, one of the many species peculiar to New Zealand.

Through a glass partition, I watched a few tuatara doing what tuatara do most of the time, namely, not much. One of them was larger and—was it possible?—more stately looking than the others. He was an attractive pale green shading toward olive, maybe a foot and a half long counting his tail, and, with elongated scales running along his spine, he somewhat resembled an iguana, only stouter. Rows of bumps ran along his sides and tail, like lines of hills in miniature. He looked like a survivor.

That tuatara's name is Henry, and people who should know think he must be more than 110 years old, which is old for a tuatara, but not record-setting. When I saw him, he had shown no interest in sex since being captured almost fifty years earlier, but a couple of years later he gained some notoriety by suddenly turning amorous and mating with a female tuatara named Mildred, who then laid a clutch of eggs that hatched out into eleven tiny tuatara. Assuming his estimated age is correct, when Henry himself hatched out, New Zealand was a colony of Great Britain under Queen Victoria, the Wright brothers were making plans to build an airplane, and Alfred Russel Wallace was in his seventies and still actively writing, although he was by then thinking about the evils of vaccination more than biogeography. You could think of Henry as a relict from a different age.

According to the usual story about the biogeography of New Zealand, Henry is also a relict from the much more distant past—from the Mesozoic Era, when what is now New Zealand was part of Gondwana.

4.1 A relict from the Victorian era and perhaps from Gondwana: Henry the tuatara in the Southland Museum, Invercargill, New Zealand. Photo by the author.

He is a member of a group called the *sphenodontids*, which contains just the one species of tuatara, *Sphenodon punctatus*, found naturally only on thirty-two small islands off of New Zealand. The closest living relatives of tuatara are lizards (including snakes, which, cladistically speaking, are a group of lizards), but the two lineages are not very close, having split from each other some 250 million years ago, somewhat before the origin of the dinosaurs. (For comparison, a chicken and a turtle show about the same degree of separation.) Tuatara, despite their lizard-like appearance, are anatomically distinct; for instance, unlike lizards, they have chisel-like downgrowths at the front of the upper jaw that look like teeth but are actually part of the skull, and male *Sphenodon* do not have a penis (lizards have two). As Henry has demonstrated, tuatara also can live a long time, longer than any lizard, and they're active at colder temperatures than any lizard. At least some of these differences are an indication of the deep evolutionary separation of the tuatara and lizard branches.

The sphenodontids were never a large group—even at their peak they didn't approach the diversity of living lizards—but in the Mesozoic there were several genera, and they crop up in widely separated regions, such as Mexico, England, Argentina, South Africa, Morocco, India, and China. In one sense, then, it has been clear for a long time that living tuatara are relicts, a single species confined to a minuscule fraction of the globe, the remnants of a larger and far more widespread group. However, with the emergence of vicariance biogeography and plate tectonics, tuatara have become known not just as taxonomic relicts but also as relicts of the Gondwanan biota.

The story of the tuatara—and, I should say from the start, it may even be the correct one—goes something like this: By about 83 million years ago, in the Late Cretaceous, a rift in the crust of eastern Gondwana had become a spreading ridge, like the one that made and is widening the Red Sea, like the one that formed earlier in the Cretaceous in western Gondwana and became the Atlantic Ocean. The magma rising up at the ridge spread out to form the floor of a sea, the Tasman Sea. On the western side of the sea lay Antarctica and Australia, which were still joined, and on the eastern side a continent about the size of India that most of us would not recognize. That continent encompassed what we now call New Zealand, New Caledonia, Campbell Island, the Chatham Islands, Lord Howe Island, and a few smaller bits of land. Geologists have dubbed it Zealandia.* On a map that shows ocean depths, one can see an area of relatively

---

* The name "Tasmantis" for this continent apparently has precedence over "Zealandia." However, Zealandia is now the more widely used name.

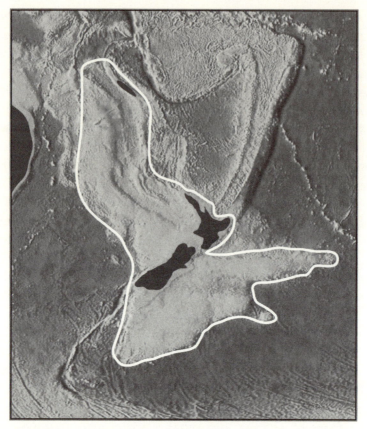

4.2 The now mostly submerged continent of Zealandia (roughly outlined in white) indicated on a relief map of the ocean floor. New Zealand is shown in black near the center of the continent. Modified from an image from the National Oceanic and Atmospheric Administration.

shallow water that looks vaguely like the head of an antelope, with its downward-pointing snout several hundred miles south of New Zealand and two horns stretching northward, one of them containing New Caledonia near its tip (see Figure 4.2). To a reasonable approximation, that antelope head is the old continent of Zealandia.

As Zealandia moved eastward, it slowly sank, probably because the crust it was made of was becoming both cooler and thinner. By the Late Oligocene, some 25 million years ago, it may have sunken far enough that almost all of the original continent was underwater; most geologists think that, at that time, the part that we now think of as New Zealand had an area considerably smaller than it has today. In any case, the current extent and mountainous landscape of New Zealand is a product of more recent collisions along a plate boundary that formed within Zealandia, the boundary between the Australian and Pacific Plates.

The ancestors of the tuatara, so the story goes, were living on the far eastern side of Gondwana when the spreading ridge began forming in the Cretaceous. Thus, when Zealandia separated from Antarctica/Australia, it carried tuatara ancestors with it. It is usually assumed that the Antarctic side also held some of these animals for a time, although there isn't any clear evidence of that; no sphenodontid fossils have ever been found in Antarctica or Australia, but, then again, sphenodontid fossil sites in the Southern Hemisphere are few and far between in general. In any case, the tuatara ancestors persisted on Zealandia as it drifted east and subsided, and they survived the time of greatest marine transgression in the Late Oligocene and Early Miocene, when the land area might have been quite small. With the uplift that created modern New Zealand, tuatara populations perhaps had something of a renaissance; their remains have been found in widely scattered locations on both the North and South Islands. Then, some seven hundred years ago, the ancestors of the Maori reached New Zealand, bringing Pacific rats (kiore) as stowaways, and the rats ate tuatara eggs and probably devastated their populations. When Europeans arrived, they brought cats, pigs, and both Norway and ship rats, and those animals, particularly the rats, likely finished off the tuatara on the two main islands, leaving them only on much smaller offshore islands that remained rat-free. Those small islands had been connected to the main islands of New Zealand during the most recent ice age and were probably part of the contiguous range of tuatara at that time, before the climate warmed, melted much of the ice, raised the sea, and left some *Sphenodon* isolated on the remaining bits of land. Living tuatara are thus relicts twice over, having first survived on Zealandia, while sphenodontids everywhere else went extinct, and then persisting on those offshore islands while the plague of rats wiped out their brethren on the North and South Islands. Today they pass a rather strange existence, by reptilian standards, often sharing burrows with seabirds, and coming out mostly in the cool, maritime night. On Stephens Island, in the strait that separates the North and South Islands, if one walks in the forest at night, one can hear crunching noises in the darkness, the sounds of tuatara eating the giant, cricket-like weta. If the biogeographic story is right, they are the sounds of one Gondwanan relict feasting on another.

What the Fossil Plants Said

"With regard to general problems of biogeography," Gary Nelson wrote in 1975, "the biota of New Zealand has been, perhaps, the most important of any in the world. It has figured prominently in all discussions of austral biogeography, and notable authorities have felt obliged to explain its history: explain New Zealand and the world falls into place around it." In fact, New Zealand had been a focal point of biogeography since before *The Origin of Species*. Darwin's friend Joseph Hooker placed it among the group of southern lands that he believed must all have been connected by land bridges, because of the many plants they had in common. Not surprisingly, Darwin and later dispersalists, such as Matthew and Darlington, had argued instead that New Zealand was colonized by chance ocean crossings; that would explain, among other things, why its biota didn't match expectations for a continent, why, for instance, the place didn't have any snakes or land mammals. Now, as the vicariance revolution took root, Nelson was trying to turn New Zealand into an archetype of the new worldview. New Zealand's isolation from the other major landmasses of Gondwana since the Late Cretaceous meant that it could not have been colonized by normal, overland dispersal in the recent past. That, in turn, meant that New Zealand provided an ideal arena for judging the importance of long-distance dispersal versus vicariance. Could it be that the ancient connection to Gondwana explained the existence of virtually all New Zealand taxa? To Nelson, the answer seemed clear. Waxing poetic, he saw "stranded upon the shores of New Zealand not the waif beginnings of its modern biota, but only bits and pieces of the center of origin/dispersal paradigm." New Zealand, in Nelson's view, held a relict biota, the product not of chance dispersal but of continental drift, and the sooner people accepted that notion, the better.

The vicariance worldview, it must be said, never came to thoroughly dominate historical biogeography in the way that, say, plate tectonics took over geology. However, in New Zealand, it came fairly close, not in the consistent use of cladograms or Croizat's tracks, but in the belief that an ancient vicariance event was the key to understanding the biota. New Zealand had all these weird, seemingly relict creatures—the tuatara and the kiwi and the nocturnal, flightless Kakapo Parrot, to name a few—and it had been part of Gondwana, thus part of the iconic story in vicariance biogeography (see Figure 4.3). The pull to fit New Zealand into this new worldview, maybe even make it the centerpiece example of that view, was apparently strong. *Explain New Zealand and the world falls into place*

4.3 Another possible Gondwanan relict in New Zealand: the Kakapo (*Strigops habroptilus*), a nocturnal, ground-dwelling, flightless parrot. Painting by John Gerrard Keulemans.

around it. I asked several New Zealand biologists what people there were thinking at the time, roughly from the mid-1970s to the early 1990s, and they all had the same memory: vicariance had taken over as the paradigm for biogeography. In 1982, two New Zealand biologists wrote, typically for the times, "By the middle of the Cretaceous period the character of New Zealand's modern forests was ensured." In other words, it was all about Gondwana. The story of the tuatara was also the story of the moa, the kiwi, and the Kakapo, the geckos and frogs, the southern beeches, the massive kauri trees, and countless other lineages.

On the face of it, the fossil record of the weird vertebrate lineages seems to present a problem for this Gondwanan relict idea. That hypothesis requires the continuous presence of those lineages in New Zealand and Zealandia for the past 75 million years or so (by which time Zealandia was fully isolated by the Tasman Sea), but the fossil record doesn't show anything like that. For instance, fossils indicate that the various species of moa birds were dominant herbivores in the forests and shrublands of New Zealand before human settlement, but the moa record peters out quickly as one goes back in time, with the earliest remains only about 16 to 19 million years old. The record for kiwi is even shallower, with the oldest definite fossils less than a million years old, although there are 10-million-year-old footprints that might be a kiwi's. Tuatara, perhaps the most iconic of all the supposed relict lineages, first show up in the same Miocene fossil beds as the earliest moa, not even close to the 75-million-year mark. The same goes for all the other possible living vertebrate relicts of New Zealand; none of them have fossil histories that suggest they were on Zealandia when it drifted off on its own.

The problem with this argument against the Gondwanan relict idea is that there aren't a whole lot of vertebrate fossils of any kind on New Zealand from near the time of its separation from Antarctica/Australia: a handful of dinosaurs, a pterosaur, a bird, a turtle, and that's it.* No amphibians, lizards, or mammals have been found. Yet it is almost certain that many vertebrates were present, just as they were on every other sizable chunk of Gondwana. In other words, the fossil record of New Zealand vertebrates for that time period is just the tip of what must have been a substantial iceberg. There's an old cliché in science that applies here—"Absence of evidence is not evidence of absence"—and that is especially true when the nature of the observations makes it highly unlikely that one will see what is there (or what *was* there, in this case). The absence of moa, kiwi, tuatara, and other vertebrate lineages from the record for early Zealandia doesn't really mean much; it's like not finding the needle in the haystack.

However, the Gondwanan relict idea was not just about vertebrates or animals in general, but about the entire New Zealand biota. For instance, the authors of one early, widely cited paper that used plate tectonics to explain plant biogeography, made the claim that "much of the present lowland flora of New Zealand is similar to that of temperate Gondwanaland 80 million years ago." Indeed, some of the classic examples of supposed Gondwanan holdovers are plants, including the giant kauri trees and the southern beeches. So how good was the plant fossil record of New Zealand, and what did it have to say?

---

* The list here is of the number of different taxa, not individuals, found in the Late Cretaceous fossil record of New Zealand (roughly 66 to 75 million years old). If the Paleocene, the first epoch of the Cenozoic, is added, the list includes two more kinds of birds, both penguins (Tennyson 2010).

The answer to the first question was, "A lot better than the vertebrate record." That was true even if one just considered so-called *macrofossils*—leaves, stems, flowers, and seeds—and was more obviously the case if one included the plant *microfossil* record, the remains of tiny, often microscopic, plant parts, especially fossil pollen.

Pollen is tough stuff. The outer surface of a pollen grain is made largely out of something called *sporopollenin*, a substance so resistant to degradation that its chemical composition isn't completely known, because it has never been fully broken down into its constituent parts. In addition, the sporopollenin covering is made up of two layers, with rods running between them that act as struts, strengthening the entire structure. It's all apparently "designed" by natural selection to protect what's inside, namely, the genetic material in the plant's sperm. Pollen gets out in the environment—on the wind, on the ground, on bees, beetles, and butterflies—and it needs to be constructed in a way that keeps the sperm from getting crushed by other objects or by the pollen walls drying up and shrinking in upon themselves.

Couple the toughness of pollen with the astronomical quantities of it that plants produce, and the end result is a whole lot of fossilized pollen. Add to this the fact that the distinctive form of the pollen of many plant taxa can be used for identification—one group might have grains with spines and large, round pores, another a scalloped surface with slit-like openings and no spines—and you have an incredibly useful source of information about the past. There are many places in the world where you can take a sediment core and, from the pollen remains alone, get a reasonable idea of changes in the vegetation through time. Of course, there are also unique problems in studying fossil pollen, such as the fact that the stuff can get blown in from far away, and you can't always be sure about matching up a type of pollen with the plant it came from, especially for ancient samples. Nonetheless, fossil pollen has been critical in the reconstruction of past floras, including those of New Zealand and Zealandia.

What the plant record—much of it a pollen record—seemed to show was not the dogged persistence of an ancient Gondwanan flora, but a history of taxa in constant flux. Lineages were evolving, of course, as they do all the time everywhere, but new ones were appearing and others disappearing at a rate that, if one believed in the relict story, was somewhat alarming. For instance, a 2001 survey of the fossil records of major flowering plant groups of modern New Zealand (which doesn't differ much from what was known by about 1980) shows that, going forward in time, three groups (including the southern beeches) first appear in the Late

Cretaceous, three in the Paleocene, seventeen in the Eocene, twelve in the Oligocene, nine in the Miocene, three in the Pliocene, and two in the Pleistocene. Meanwhile, many lineages blinked out, never to return (at least until people brought some of them back), including a slew of tropical or subtropical groups such as acacias, *Eucalyptus*, and some palms. If one takes this record as roughly correct, what it shows is that very few modern flowering plant lineages have a contiguous history on New Zealand dating back to its separation from Antarctica/Australia. Instead, most of the original lineages—the "passengers" on the old continent of Zealandia—seem to have gone extinct, and many others have arrived, presumably by long-distance, oceanic dispersal.

To get a more personal, tangible sense of the history of the Zealandian flora, I thought back to a place that Tara and I had visited in the Southern Alps, the range that runs most of the length of the South Island of New Zealand. Near Arthur's Pass, in the heart of the mountains, we stopped to walk a small nature loop by the highway and then hiked up a classic U-shaped valley, carved by Pleistocene glaciers. We were in the subalpine, in an area almost devoid of trees, but with stunted beech forest still above us. Tara is a botanist and has a special affinity for the treeless higher reaches of mountains—her dissertation was on seedlings in the alpine tundra of the Rockies—so she was stopping everywhere on the nature trail, taking close-up photos and looking through *A Field Guide to the Alpine Plants of New Zealand*. She got slightly more mobile on the hike up the valley, but still only made it about half a mile from the trailhead before the allure of the plants stopped her progress for good. I continued up the valley, taking in the scenery, which was spectacular, but earthier than one might expect in New Zealand—dingy gray scree slopes, crumbling mountain heights dotted with snow, reddish-brown shrubs set against straw-colored tussock grass, a cold stream churning over dark rock. At some point I tried not to intellectualize what I was seeing, tried to free my mind for a while from plant taxonomy and deep history, but with limited success; in New Zealand, biogeography was always entering my thoughts. I gave up on my Zen interlude. Instead, as I walked back down the trail, I tried to burn the images of certain plants into my brain for later identification.

Much later, from our photos and from memory, I made a partial list of what we had seen that day (see Figure 4.4). The list was short but taxonomically varied: a shrub of the plantain family in the genus *Hebe*, with

4.4 The world's largest buttercup, *Ranunculus lyallii* (left), and one of the many species of *Celmisia* daisies, *C. semicordata*, from Arthur's Pass on the South Island of New Zealand. The fossil records of these and all other plant lineages that the author could identify at this site suggest that their ancestors arrived by oceanic dispersal. Photos by the author.

scale-like leaves that make it look like a juniper; *Celmisia semicordata*, a daisy with large, yellow-centered, white flowers (for some reason an inordinate number of New Zealand plants have white flowers) and long, pointed leaves like a lily; a speargrass (genus *Aciphylla*), another non-lily with lily-like leaves, this one in the carrot family; *Ranunculus lyallii*, also white-flowered, and botanically famous for being the world's largest buttercup; the ubiquitous tussock grass (*Chionochloa pallens*); yet another white-petaled species, a small orchid (possibly *Aporostylis bifolia*), with its single flower on a tall, hairy stem; and, in a miniature bog, two miniature plants—a carnivorous sundew (*Drosera*) and a creeping plant in the genus *Coprosma*, in the coffee family, with long finger-like yellow pistils sticking up out of the female flower cup.

I worked through this list, finding out from the literature on fossil plants when each of these lineages first appeared in New Zealand. This was neither a large nor a random sample of the flora, but it was still striking what the fossil record of these plants suggested about their origins. None of them had New Zealand histories that stretched back before or even close to the time of Zealandia's separation from Antarctica/Australia, 75 million years ago or so: *Coprosma* was the oldest, appearing in New Zealand in the Eocene, perhaps 50 million years ago, while all the other lineages only turned up within the past 35 million years. Taken at face value, the fossil record indicates that, of this haphazard (but common) set of plants, not a single one is a Gondwanan holdover. They all seem to have arrived by crossing the sea.

A paleontologist named Charles Fleming had been pushing this primarily dispersalist view of the New Zealand flora (and biota as a whole) since before the validation of continental drift, basing his conclusions in large part on fossils. It was Fleming who had first suggested that the area of New Zealand had been substantially reduced in the Late Oligocene and Early Miocene, leading to the extinction of some plant groups. It was also Fleming whom Gary Nelson was singling out when he wrote of the "bits and pieces of the center of origin/dispersal paradigm" being stranded on the shores of New Zealand.

Even in the 1980s, at the height of the vicariance movement, there were a few New Zealand scientists, in addition to Fleming, who didn't buy the Gondwanan relict story for New Zealand. Dallas Mildenhall, a colleague of Fleming's at the New Zealand Geological Survey, remembers that "there was a disconnect between paleontologists and biologists looking at the modern biota," with the paleontologists, a fairly small group, seeing a history of changing lineages that required overwater dispersal, while almost everyone else believed the Gondwanan relict story. Mildenhall is a pollen expert and thinks the fossil pollen record, even with its flaws, clearly indicates the late arrival of many New Zealand plant lineages. In 1980, he published a paper making that point and even suggesting that the history of *Nothofagus*, the southern beeches, might require some long-distance dispersal. By that time, *Nothofagus* had become one of the prime examples of Gondwanan vicariance, so Mildenhall was potentially striking at the heart of the vicariance movement.

My sense, however, is that Mildenhall's paper didn't have much of an immediate impact on the field of biogeography. Through the 1980s, his paper was cited fairly often, but mostly by other paleobotanists, and even then it was rarely referred to for its general message about the dispersal origins of the New Zealand flora. For New Zealand neontologists—biologists who study living organisms rather than fossils—it was still all about Gondwana.

"Do Araucarias have double trunks?" That was the kind of question Dallas Mildenhall would get in letters from a New Zealand high-school student named Mike Pole. This was in the mid- to late 1970s. Mildenhall would reply in detail, not wanting to dampen the kid's botanical and paleontological enthusiasm. There was probably no danger of that though.

Mike Pole had collected his first plant fossil when he was nine and, as a teenager, he was a paleo nut, swapping specimens with his mates and precociously realizing that there were big knowledge gaps to be filled in the study of New Zealand fossils. While still in high school, he read Charles Fleming's 1962 paper "New Zealand Biogeography—A Paleontologist's Approach," along with papers by other scholars with such titles as "Reconstructing Triassic Vegetation of Eastern Australasia"—not exactly the usual teenage reading material. Even then, Pole knew he wanted to have a career working with plant fossils.

He also knew that it was the New Zealand flora, in particular, that he wanted to study, so he ended up getting his undergraduate degree at the University of Otago in Dunedin, about 70 miles from his hometown of Alexandra. In Dunedin, he and his adviser, a paleontologist named Doug Campbell, would sometimes walk from campus over to Campbell's place for lunch and chat about biogeography. A botanist who had recently gotten his PhD at Otago would join them when he was home visiting from New Guinea, where he was teaching at a university. New Zealand is a small academic world, and that botanist was Michael Heads, who would later argue that long-distance dispersal has no importance, that everything is "garden-variety" dispersal followed by vicariance. Pole credits Campbell and Heads for helping him to start thinking critically about biogeography, but Heads's Croizatian views clearly did not rub off on him.

Pole continued into the PhD program in Dunedin, and, while preparing an identification key to the modern forest trees of New Zealand, it dawned on him that the fossils that were the subject of his dissertation were neither ancestors nor even close relatives of the living groups. He had been at least slightly brainwashed by the Gondwanan relict idea, but now he was discovering for himself what Fleming and Mildenhall had seen before, namely, that the fossil record suggests a massive turnover of New Zealand plant lineages in the past 60 million years or so.

In his dissertation, Pole compared the past floras of Australia and New Zealand (using leaf fossils rather than pollen), emphasizing how dynamic they were and how, in both places, the plant lineages that were present reflected the climate at the time. Perhaps most striking was the fact that in the Early to Middle Miocene (roughly 10 to 20 million years ago), the New Zealand flora in some places came to appear very "Australian," with flowering plants, such as acacias, *Eucalyptus*, and the conifer-like *Allocasuarina*, that are adapted to forests that regularly burn. Pole suggested that the similarity between the Australian and New Zealand floras at this and other times was a result of relatively easy movement of plants across the Tasman Sea; the sea was obviously a barrier, but it was a barrier that many plants could overcome. Plants such as *Eucalyptus* had presumably crossed the Tasman at all times, but it was only when the climate of New Zealand became sufficiently "Australian" that they were able to survive there. And when New Zealand's climate shifted again, making the forests less prone to burn, the fire-adapted Australian plants disappeared. In short, the character of New Zealand's flora was dictated not by the set of original, Gondwanan inhabitants, but by long-distance dispersal and climate.

In a subsequent paper, published in 1994, Pole fleshed out the case for the importance of the long-distance colonization of New Zealand. He spent a good deal of time developing an argument reminiscent of Darwin, noting that several oceanic islands have native plant lineages that on New Zealand are considered to be Gondwanan relicts rather than overseas colonists. A conspicuous case is the Norfolk Island pine, *Araucaria heterophylla*, a tree with an attractively symmetrical, triangular shape and dark green foliage, well known as a house plant. *A. heterophylla* is in the conifer family Araucariaceae, a name that for anyone with much knowledge of worldwide biogeography immediately conjures up thoughts of Gondwana. Araucariaceae species have no special means for dispersing their seeds and so are thought to be poor long-distance colonists; thus, their presence in areas as far-flung as southern South America, Australia, New Zealand, and New Caledonia has typically been explained by Gondwanan fragmentation. Pole, however, pointed out that Norfolk Island is a volcano that emerged from the sea within the past 3 million years. He also noted that the island is 435 miles from New Caledonia, the closest likely source for the tree's ancestors. Theoretically, the Norfolk Island pine shouldn't be on Norfolk Island, but there it is, and its ancestors must have arrived by oceanic dispersal.

Pole listed many other examples of this kind—plants that had been categorized as "non-oceanic" because they supposedly couldn't get to oceanic islands, and yet were found on Norfolk, Lord Howe, Campbell, and other islands in the New Zealand region. These islands are all situated on ocean rises or plateaus that had once been part of Zealandia, but there is no geological evidence to suggest they have long histories as land; they may be perched on continental crust, but they are all apparently recent islands, most of them having emerged from the sea as volcanoes within the past few million years. Pole's point was that, if "non-oceanic" plants had reached these oceanic islands, then they and other such lineages could have made it over water to New Zealand as well. For instance, who was to say that the ancestors of New Zealand's single living species of Araucariaceae, the giant kauri tree, which was almost universally considered a Gondwanan holdover, couldn't have arrived there as seeds or trees on a natural raft? In essence, Pole was repeating Darwin's argument about the effectiveness of long-distance oceanic dispersal, but using it to shoot down the Gondwanan relict idea, rather than to argue against the idea of separate creations on different landmasses, as Darwin had done.

Like Mildenhall, Pole also questioned the ancient persistence story for that archetypal Gondwanan lineage, the southern beeches. In particular, he observed that there are three groups of *Nothofagus*, distinguishable by their pollen, but that only one of them, the *fusca* group, has a fossil record in New Zealand going back to the time of the isolation of Zealandia in the Late Cretaceous. The other two, the *menziesii* and *brassii* groups, are known from fossil pollen at that time, but from South America and Australia, not New Zealand.* The implication was that these latter two, although living on Gondwana, had missed the Zealandian boat as it left the harbor and had colonized the area only *after* its separation from Antarctica/Australia. Pole argued that, although the *menziesii* and *brassii* groups, along with many other plant taxa, might be Gondwanan in the sense of having come from some landmass that had once been part of the southern supercontinent, they were *not* relicts with an uninterrupted presence in New Zealand dating from the time of continental breakup. They were not Gondwanan in the sense implied by the vicariance worldview. In pushing his dispersalist view, Pole went farther than even Fleming or Mildenhall had: he suggested that perhaps the *entire* New Zealand flora was descended from overwater colonists.

These days, Mike Pole spends much of his time in Mongolia (where he often stays in a traditional, tent-like *ger*) and Kalimantan, the Indonesian part of the island of Borneo. He lives an adventurous life, dealing on a somewhat regular basis with floods, blizzards, and corruption; wading up streams in the Indonesian rainforest; and sleeping out under the stars in

---

* The assignment of *Nothofagus* fossil pollen to particular groups is questionable, and it is actually possible that none of the three groups mentioned existed at the time of the birth of Zealandia (Cook and Crisp 2005).

the Gobi Desert. Sarawak, where Alfred Russel Wallace wrote his paper "On the Law Which Has Regulated the Introduction of New Species," the one in which he used geographic distributions of related species to argue for evolution, is also part of Borneo, but Pole isn't there to follow in Wallace's footsteps. Instead, as he says, he's switched over to the "dark side" and now works as an exploration geologist, looking for coal. I was a little disappointed to hear this, not so much because of thoughts that coal is dirty energy, but because Pole wrote engaging, thought-provoking biogeographic papers; his 1994 paper on the New Zealand flora, in particular, influenced my own thinking about why things are found where they are. I thought he might have given up on that part of his career.

However, it turns out that Pole still lives part of the time in New Zealand, still finds the time to do research on fossil plants, and still thinks about the origins of the New Zealand flora. Like Gary Nelson, he believes that explaining the origins of New Zealand's biota is a key to understanding the biogeographic history of the entire world. Of course, he also thinks that Nelson got the answer completely backwards; in Pole's view, New Zealand points to the global importance of chance, overwater colonization, not ancient vicariance.

In his 1994 paper, Pole had made the strongest and most pointed case up to that time for the origins of New Zealand's flora by oceanic dispersal. Nonetheless, a skeptic with a commitment to the vicariance worldview could have found holes in his argument. Pole, like Fleming and Mildenhall before him, had relied heavily on fossil occurrences, yet it is well known that the fossil record is incomplete. In particular, the apparent massive turnover in New Zealand's flora since the birth of Zealandia could, in theory, be an artifact of a fragmentary record. Everyone agrees that New Zealand's plant fossil record is far better than its vertebrate fossil record, but that doesn't necessarily mean the plant record is reliable in an absolute sense. One possibility, for instance, is that many lineages persisted during certain periods only in small refugia where they were unlikely to leave any trace in the fossil record. Such groups could be Gondwanan relicts that Pole interpreted as later arrivals or as taxa that disappeared from and later recolonized New Zealand.

Similarly, Pole's examples of "non-oceanic" Gondwanan plants that, against expectations, had colonized volcanic islands such as Norfolk and Lord Howe don't indicate anything definite about the origins of New Zealand's flora. The fact that oceanic dispersal *can* occur doesn't mean that one should believe such a chance explanation when a more straightforward one—namely, persistence through Gondwanan breakup—would do just as well. If vicariance was the default explanation, as many biogeographers believed, then more direct evidence—something that would actually force one to reject the fragmentation hypothesis—was needed.

In that 1994 paper, Pole mentioned two sets of studies that were clearly outside of his own area of expertise, but supported his case for long-distance dispersal. These studies had to do with two iconic Gondwanan groups, the southern beeches and the ratite birds, and suggested relatively recent, overwater colonizations of New Zealand by at least some members of both groups. The authors of the ratite studies, for instance, claimed that the kiwi lineage had reached New Zealand from Australia within the past 45 million years, requiring a crossing of the Tasman Sea by these flightless birds.

In describing these two studies, Pole didn't quite sound as if he fully endorsed them. His skepticism might have had something to do with the general approach the authors employed: they had assigned ages of colonization using a "molecular clock," and the problems with that approach were well known. Problematic or not, however, those studies of southern beeches and ratite birds were heralds of what was to come. They were like the first drops of rain before the downpour.

In July 1892, a natural floating island was spotted off the northeastern US coast, at about the latitude of Philadelphia and some 300 miles from the nearest land. The island was roughly 9,000 square feet in area, contained living trees 30 feet tall, and is said to have been visible from 7 miles away. The same island was again seen in September, by which time the Gulf Stream had pushed it more than 1,200 miles northeast of its previous position.

Section Two

TREES *and* TIME

Chapter Five

THE DNA EXPLOSION

A Question of Timing

One of the stereotypically tedious parts of school history classes is committing to memory parts of the historical chronicle, that is, lists of events and their associated dates, all placed in chronological order. The Battle of Hastings in 1066, Columbus's discovery of the New World in 1492, the signing of the Declaration of Independence on July 4, 1776—these are all familiar entries in the chronicle. Although memorizing such dates may be tiresome, knowledge of the chronicle is clearly vital to making sense of history, to understanding how and why things have happened. Just consider the sorts of historical connections we might contemplate if we had no idea about the proper chronological order of events. We might imagine that Magellan used Captain Cook's maps to chart a route across the Pacific, or wonder if the financial meltdown of 2008 helped promote the rise of Hitler and the Nazi Party.

For human history, especially recent human history, such examples sound silly; the chronicle is typically so well established that we don't waste time considering connections that violate the sequence. However, if history is considered in the most general sense—the history that includes cosmology, geology, evolutionary biology, linguistics, and other areas—there are many cases in which the absolute and relative timing of events is not an obvious collection of facts, but instead is something quite difficult to establish. For instance, starting with the first strong evidence for the Big Bang Theory, in the 1920s, estimates of the age of the universe have ballooned in a series of steps, increasing from about 2 billion years to 13.8 billion years.* Similarly, although we can be sure that human language arose after our evolutionary separation from the chimp lineage, support for a more precise age for this critical event has been elusive. Obviously, there's no written record of the Big Bang, or the origin of human language, and the evidence that does exist for dating these events can be hard to interpret. As a result, the possible chains of cause and effect are also often unclear. It has been suggested, for example, that the origin of language gave rise to selection for increased brain size, but that connection remains speculative, in part because of the uncertainty about exactly when language (or, more precisely, certain steps in the evolution of language) arose. Continuing our Hitler analogy, it's as if we are often operating without clear knowledge of whether the 2008 financial crisis came before or after the Third Reich.**

Historical biogeography, like all historical disciplines, would benefit greatly from having an established chronicle of relevant events. More specifically, the problem of explaining piecemeal distributions fairly screams out for such timing information on the age of evolutionary branching points. In many of these cases, the competing explanations involve processes taking place at different periods; typically, one is an ancient vicariance event, such as the opening of the Atlantic Ocean, and another involves the more recent dispersal of organisms across an ocean or some other barrier. So, if you knew when two lineages split—say, a group of rodents living in South America and another in Africa—there's a good chance you would be able to either reject or support the ancient vicariance explanation; the split could be too recent to be explained by vicariance, or, alternatively, it could be old enough to fit that hypothesis. Biogeographers of all persuasions agree that having accurate ages for branching points in the tree of life would be enormously useful; all agree that knowing *when* would go a long way toward figuring out *how*. What they have

---

* At one time, the age of the universe estimated from its apparent rate of expansion was considerably younger than the estimated age of the Earth, indicating that something was seriously amiss with at least one of these estimates. Present knowledge suggests that, in fact, both estimates were too young at the time, but the estimate of the age of the universe was far too young.

** The chronicle tends to be more difficult to reconstruct as one delves into the more distant past. However, problems of an uncertain chronicle also can arise for the very recent past, as in a criminal case in which the whereabouts of a person at a particular time are critical, but hard to establish. That sort of example illustrates that what is often important is the accurate placement of an event in time relative to other events.

conspicuously failed to agree upon is the practical role of such timing information in real studies of biogeography.

In the early 1990s, when Mike Pole was forming his ideas about the origins of the New Zealand flora, historical biogeography was deeply divided over this issue. It was like a nation polarized into two warring political parties, along with a large number of people of undecided allegiance. On one side were the hard-core vicariance scientists, including Gary Nelson and other cladists, along with panbiogeographers like Michael Heads, whose focus was on cladograms (without connected age information) or tracks as *the* fundamental kinds of evidence. They were notably disinterested in using fossils to place ages on evolutionary groups, a disinterest stemming from their belief that the fossil record is too incomplete to provide useful information for that purpose. In the extreme, these were the scientists who were entertaining ancient vicariance explanations even for such things as the origins of the Hawaiian biota or the distribution of *Homo sapiens*. They typically believed that the only good way to infer the age of an evolutionary branching point was to connect it to some fragmentation event, tectonic or otherwise. One could "know," for instance, that Australian and New Zealand southern beeches had split from each other roughly 80 million years ago if, by looking at cladograms, it was established (and I use that term loosely) that their separation had been caused by Gondwanan breakup. In this view, time—the age of a branching point—was never used to discriminate between dispersal and vicariance, but instead was an *outcome* of already "knowing" that vicariance was the explanation.

On the other side were those who thought they had a rough handle on when many groups first appeared on Earth (and in particular places) and were willing to use that information to interpret biogeographic history. Think of those classroom posters that show the geologic time periods with the history of life superimposed on them—the first insects (crawling next to the word "Silurian"), the first mammals (Triassic), the first birds (Jurassic), and so on. This second group of scientists basically was made up of believers in such timelines, at least as approximations. Not surprisingly, most of them either were paleontologists or had a strong interest in the fossil record. They were people like Mike Pole and Dallas Mildenhall, immersed in the paleobotanical record of Zealandia; Anthony Hallam, a paleontologist who studied molluscs and other shelled invertebrates; and John Briggs, who did research on both living and fossil fishes. They weren't ignorant about the effects of continental drift; in fact, both Hallam and Briggs wrote books about plate tectonics and its revolutionary impact on biogeography. However, all of them were arguing, as New York School dispersalists like William Diller Matthew, George Gaylord Simpson, and Ernst Mayr had decades before, that the fossil record tells us that some groups—a lot of groups, actually—are probably too young to have been split up by ancient fragmentation events. For example, Briggs, in his 1987 book *Biogeography and Plate Tectonics*, suggested that a group of freshwater killifishes found in Africa and South America might have come into existence long after the opening of the Atlantic. If that was the case, at least one of these fishes must have crossed the ocean, somehow tolerating the salty waters. Like others in this school of thought, Briggs had no particular preference for vicariance or dispersal as explanations; he would follow the evidence, and for him, the evidence included information about the ages of groups.

At the same time, many biologists who had or might have had some interest in biogeography were not in either camp. Some of these people weren't ready to accept the extreme views of the vicariance side, yet were also leery of putting too much faith in the fossil record. Michael Donoghue, the botanist who had been attracted to Gary Nelson's intellectual boldness, was one of those. Despite being "raised" as a cladist, he was turned off by the endless, inconclusive cladograms in Nelson and Platnick's vicariance tome, *Systematics and Biogeography*—"chicken scratchings," Donoghue called them—yet he wasn't ready to believe, à la Pole and Mildenhall, that long-distance dispersal was commonplace. Up to the early 1990s, Donoghue had mostly set aside an early interest in biogeography and was pursuing other things. Others had never really been drawn in to begin with, perhaps because the field, after the burst of enthusiasm following the revelation of continental drift, seemed a bit stagnant. However, I suspect that most of these "undecideds," when they were thinking about biogeography at all, had leanings toward vicariance rather than dispersal, because vicariance seemed like the more global and cutting-edge idea. As I mentioned in the Introduction, I knew little about biogeography through the 1990s, but when I had to lecture about the subject in an evolution course, I chose to talk mostly about Gondwanan breakup, not ocean crossings. Global fragmentation just seemed like the "cooler" thing to focus on. In short, the undecided vote seemed, if anything, poised to tip toward the vicariance side.

What shifted this balance, particularly for the undecideds, was the use of molecular data, especially DNA sequences, to put ages on evolutionary branching points. Molecular dating had actually begun long before this, in the early 1960s,* but such studies became much more widespread in the 1990s, and that upward trend has continued to the present. For many people, this approach suddenly made establishing the evolutionary chronicle a reality. It was like finally being able to show, after years of ignorance, that Hitler really had come to power decades before the financial crisis of 2008.

Given the importance of molecular dating for biogeography, a huge question that we have to deal with is whether the ages estimated in this way can really be trusted. Some evolutionary biologists, including, not surprisingly, some of the hard-core vicariance crowd, continue to think that molecular dating is basically worthless, and therefore, that any conclusions that depend on it are equally worthless. In a 2005 paper, Michael Heads wrote that "degree of divergence is a guide neither to the time involved in evolution, nor the age of that evolutionary [splitting] event," and that the molecular clock approach "does not solve biogeographical problems but simply leads into a morass of mysteries and paradoxes." Similarly, Gary Nelson, now retired but still quick with a witty phrase, has ridiculed the approach and its use in biogeography as a futile "molecular dating game."

In Chapter Six, I will get into the thorny but critical issue of whether we should trust molecular dating studies. First, however, I want to address, somewhat idiosyncratically, the question of why this approach took off when it did. In a sense, there can never be a complete answer to a question like that; one can always delve deeper into the long sequence of historical cause and effect, or flesh out in greater detail what happened at key points. In the case of the molecular dating explosion, one could argue for the importance of things like the discovery of the structure of DNA, and, later, of the enzymes that make strands of DNA replicate themselves. The idea of the molecular clock itself, first proposed by an Austrian biochemist named Emile Zuckerkandl and the Nobel Prize–winning chemist Linus Pauling in 1962, and the invention in the 1970s of methods for obtaining long sequences of DNA, were also critical. However, I take all that as background and instead focus on a critical insight that a particular scientist had in the early 1980s. In doing so, I'm not subscribing to a "great man/woman" view of history. Rather, I'm simply emphasizing an event that clearly had a rapid and far-reaching effect. This event may also qualify as a potential "point of no return," that is, an occurrence that set off

---

* The early molecular clock studies were based on amino-acid sequences in proteins rather than on base-pair sequences in DNA.

an unavoidable cascade of effects. Maybe I'm also biased by the fact that I experienced part of the effect of the event in question firsthand, within a few years of when it happened.

Such scientific turning points are not always memorably discrete, even to the people making them. For instance, although there may have been a particular moment when Darwin became a confirmed believer in evolution, his thoughts on the subject had been percolating for years; when he finally converted, it was like fitting a few pieces into a puzzle whose basic form he had already seen. However, if we can take the word of its architect, the turning point I'm about to describe did indeed come in a lightning-like epiphany. That epiphany, that flash point, can be located very precisely in time and space, to a night in May 1983, on Highway 128 in the Coast Range north of San Francisco, at mile marker 46.58.

The Chain Reaction Begins

It was an unseasonably warm night, and the air was thick with the sweet scent of blooming California buckeyes. Kary Mullis was at the wheel of his silver Honda Civic, driving north from Berkeley toward his cabin in Anderson Valley, his girlfriend asleep in the passenger seat. He was thinking about DNA replication.

Mullis and his girlfriend were both chemists working for Cetus, a Bay Area biotech company that, among other things, developed cancer therapies and ways to diagnose genetic diseases such as sickle-cell anemia. Mullis was hardly your stereotyped dull corporate scientist in a white lab coat, however. In fact, he was (and is) a risk-taker and close to a certifiable nut. He had experimented with LSD and other hallucinogenic drugs, even brewing up new compounds and trying them out on himself. (A Berkeley professor in whose laboratory Mullis worked while getting his doctorate had once suggested, with surprising restraint, that Mullis might want to clear the psychoactive substances out of the lab freezer, in case the cops came around.) In Aspen, Colorado, he once skiied down the middle of an icy road with cars whizzing by on both sides, apparently unconcerned because he had it in his head that he would eventually be killed by crashing into a redwood tree, and there weren't any redwoods in Aspen. Weirder still, he had once passed out while getting high on laughing gas, with the tube from the tank stuck in his mouth, and claimed that he was saved by a stranger, a woman who noticed his prostrate form as she floated by on the astral plane. Somehow, bodiless, she managed to get the freezing tube out of Mullis's mouth. Parts of his lips and tongue were frostbitten, but he survived the incident and, years later, met the corporeal version of his savior in a bakery, as if by destiny.

Whatever else he was, though, Mullis was a thinker and a problem-solver. The drive from Berkeley to Anderson Valley took two and a half hours, and it was a time for him to use that gift, to focus on difficult research problems, free from distractions. On this night, he was trying to come up with a quicker genetic test to tell if someone has a disease, such as sickle-cell, tied to a single base pair in their DNA. Somehow, he got to thinking about short sequences of bases—things that Cetus scientists had gotten very good at synthesizing in the lab—and how you could use them to latch onto a part of someone's DNA that had the complementary string of bases, a GTTCCC in the synthetic sequence, for instance, matching up with a CAAGGG in the person's genome. From the place of attachment, you could then get the DNA to start replicating itself, creating a new piece of DNA, in the same way that DNA replicates itself when a cell divides. These were known facts, things that could be done.

Like anyone who worked with DNA, Mullis knew that getting a large enough amount of any particular stretch of the genetic material to enable one to sequence it—to read the order of As, Gs, Cs, and Ts—wasn't easy. The standard procedure was to insert the target DNA into bacteria and then get the bacteria to multiply on petri dishes, replicating the alien DNA along with their own. It was a messy process and far more time-consuming than the sequencing itself. In other words, generating a sufficient quantity of the targeted DNA was the rate-limiting step. Mullis knew that if you could solve this problem by coming up with a simple, fast way to create lots of DNA, it would be a huge methodological breakthrough. It could make DNA sequencing easy.

Suddenly, driving along Highway 128, with the long, pale, flowering heads of the buckeyes drooping down into the beams of his headlights, his mind full of synthetic bits of DNA, he realized how it could be done. It was a classic Eureka! moment, like Wallace, in a malarial fever, flashing on the mechanism of natural selection. Maybe it was Mullis's tendency to think outside of the box at work. He has said that his use of LSD some years before probably helped, opening up new pathways in his mind. In any case, he could hardly believe his thoughts. He pulled off the highway, grabbed a pencil and an envelope out of the glove compartment, and scribbled down a few notes. A large buckeye loomed over the little Honda. His girlfriend stirred in her sleep. There was a small white highway marker where he had pulled out, mile marker 46.58.

What Mullis had realized was that, by using two different synthetic DNA sequences that would attach to areas that weren't too far apart on a DNA strand, you could start in motion a process that would duplicate the target DNA—the stretch between the two attachment sites—over and over again. The key was that a new strand that had begun replicating from one of the bits of synthetic DNA would, in the next round of duplication, provide an attachment site for the other bit of synthetic DNA, which would then initiate replication in the other direction. Run the process long enough, round after round of duplication, and you'd end up with astronomical numbers of the target sequence. The first thing Mullis scribbled down was just the progression of increasing amounts of target DNA with each round of duplication. Starting from one molecule, you'd have two molecules after one round, four after another round, then eight, sixteen, and thirty two. Ten rounds would give you 1,024 molecules, and up and up exponentially. It would work, he thought. By the time he had driven another mile down the highway and pulled over again, he was already thinking about his Nobel Prize.

Mullis called his new method the *polymerase chain reaction*—PCR for short—after the enzyme, DNA polymerase, that catalyzes replication. When he described the new technique to his coworkers at Cetus, including his girlfriend, almost none of them saw that he was onto something groundbreaking. Maybe it was partly because he was known for being a bit flaky. (Did they know about his astral guardian angel?) Mullis himself has suggested that it was also the usual resistance to new ideas, even relatively straightforward ones, that kept them from recognizing the importance of his proposed method. In any case, it was some seven months later, and only after he had mucked around in the lab and generated some promising preliminary results, that many other scientists at Cetus got excited. With others on board, some of them much more careful experimental scientists than Mullis, it was soon apparent that PCR was a viable technique.

Mullis had another big breakthrough when he realized that the process could be streamlined by using a version of the DNA polymerase, the replication enzyme, from a bacterium called *Thermus aquaticus* that lives in the hot springs of Yellowstone National Park. In PCR, after a short period of DNA replication, the newly formed double strands have to be heated up to break them apart again to allow the next round of replication to take place; the polymerase from *T. aquaticus* (or *Taq*, for short), unlike the kind found in most organisms, stays intact when heated, and therefore does not have to be replenished in the test tube after each heating step. Using *Taq* and a heating block called a *thermal cycler*, programmed to

5.1 Kary Mullis invented the polymerase chain reaction (PCR), and thus changed the science of biogeography, among other things. Photo by Erik Charlton.

heat up to break the DNA strands apart and then cool down to begin each period of replication, you could just toss in your ingredients and let the thing run, and after a few hours, a handful of copies of the targeted piece of DNA would turn into millions.

Mullis hadn't been suffering from delusions of grandeur when he dreamed of a Nobel Prize that night under the flowering buckeyes. He shared the Nobel in Chemistry in 1993. While he was in Stockholm for the awards ceremony, the Swedish police came to his hotel room, responding to reports of a red laser beam, like those used on rifle sights, coming from his window. Sweden has a low crime rate, but someone had been murdered in Stockholm a year or so earlier by a sniper using one of those laser sights, so that red beam was making people nervous. When the police questioned him, Mullis had to confess: he had been playing with a new laser pointer, shining it near passersby on the street below to see how they'd react. It was just Mullis being Mullis.

The transformation that PCR brought about took place very quickly and happened to coincide with the few years when I was actually getting my hands dirty (and irradiated) in a molecular lab. In 1987, as part of my dissertation research, I wanted to construct a phylogenetic tree of garter snakes, so I started working in the lab of an evolutionary biologist named Rick Harrison to get the necessary genetic data. For my first year or so in Rick's lab, I was screwing around with something called restriction fragment analysis, a standard method at the time. The technique required a long procedure for isolating the DNA from snake mitochondria—my main memory of this step is being afraid I would destroy an extremely fast and expensive ultracentrifuge by improperly balancing the samples in it—and then using enzymes that would cut the DNA where particular sequences of bases (say, GAATTC or AAGCTT) appeared. What you ended up with, after several more steps, wasn't anything like the full DNA sequence, but instead just bands on a gel that very roughly showed the length, in base pairs, of the cut-up pieces of DNA (called restriction fragments). The process was time-consuming and complex, and, worst of all, it often wasn't clear how to interpret the results. I guess it was fun in a way; there was some satisfaction in working through an elaborate procedure and seeing those bands of actual DNA appear. It was a little like following a difficult recipe to make a nice soufflé. Like a soufflé, though, it was mostly empty; the technique just wasn't going to provide enough information about the differences between species to construct a reasonable garter-snake tree.

Luckily for me, in 1985 the description of PCR had been published. Once Mullis and other scientists at Cetus had worked out all the major kinks, it was obvious that the method would be enormously valuable. Among other things, it would be used to diagnose infections and genetic diseases and to analyze tiny DNA samples from crime scenes, from bits of frozen mammoth and Neanderthal tissues, and from feathers of extinct birds lying in museum drawers. Eventually, it would make possible the sequencing of the entire human genome. I don't remember thinking about any of that at the time. What I knew was that a group of scientists—in the first PCR paper, Mullis's name was buried in the middle of a list of seven authors—had invented a method that would let someone like me, who had no desire to become immersed in molecular biology, obtain actual DNA sequences.

And it did. A bright fellow grad student named Ben Normark was the first one in the lab to learn PCR, and he taught me all I needed to know over a few days. Even the first step showed the advantage of the technique. Since PCR would selectively find the target sequence in a mitochondrial gene and multiply it, there was no need to separate the mitochondrial from the nuclear DNA.* No more balancing samples in that damned ultracentrifuge. Now the first step was just isolating all the DNA, the mitochondrial and nuclear components mixed together, from each garter-snake tissue sample, which was a fairly trivial task. After that I'd load the DNA samples into little test tubes along with the *Taq* polymerase, a bunch of free-floating bases, and some other ingredients, put the tubes into the Perkin-Elmer Cetus thermal cycler, and a few hours later I'd have, relatively speaking, a gargantuan amount of the target stretch of DNA, enough to get the sequences. Even though I understood what was happening in the test tubes, it felt like magic.

Unlike grad students today, who just send their PCR samples to an automated sequencing facility, I had to do the DNA sequencing myself, and that was a pain. I especially remember some unhappy moments with the thin sequencing gels, which tended to fold up on themselves when you were transferring them onto cardboard-like filters; I tossed at least one recalcitrant gel into the trash in disgust, several days' work down the drain. However, at the end of this messy process, if the gel was cooperative, I had a picture with bands in four rows representing the four bases, from which I could read the actual sequences of As, Gs, Cs, and Ts. This result was far more useful to me than the old restriction fragments. The differences between garter-snake species were completely unambiguous—one species would have an A at a particular site while another species would have a T—and, compared to the restriction fragment data, there were lots of differences, which meant more information that I could use to sort out the evolutionary relationships.

With each new set of sequences, I would run a "tree-finding" program on the lab computer (I didn't have my own computer—this was really the dark ages), and out would come a network of dots and lines, the program's best guess for the garter-snake tree. For instance, the program told me that the two slender ribbon snake species (a subset of the garter snakes) were each other's closest relatives (an expected result), and that the highly aquatic Mexican black-bellied garter snake was part of a group of

---

* A complication is that many organisms, including humans, have nonfunctional sequences derived from mitochondrial DNA that have been incorporated into the nuclear genome. When that is the case, using PCR on DNA samples that include both the mitochondrial and nuclear genome will often generate sequences from both the targeted mitochondrial DNA and the untargeted nuclear copies corresponding to the same segment. If, as is often true, the nuclear copies are quite different from the original mitochondrial DNA, this phenomenon can strongly distort the results.

terrestrial species (somewhat surprising). I have to admit that, examining that tree now, it doesn't look too good. Compared to the tree from a far more extensive sequencing study that some colleagues and I did more recently, also using PCR, my initial tree was quite flawed: it was wrong in a few places, and just not all that informative overall. But it was a beginning.

I ran my first PCR in 1989, four years after the method was first described in print. In hundreds of labs, the same thing was happening right about then—that's how fast the technique took off. Suddenly, everyone seemed to be using PCR to "amplify" tiny amounts of DNA into quantities that could be sequenced. In the Harrison lab alone, there were people sequencing crickets, swallowtail butterflies, bighorn sheep, crabs, beetles, and various fishes, all because of PCR. Elsewhere, the work extended to countless other animals along with plants, fungi, protists, and bacteria.

This wasn't the first big pulse of DNA sequencing work: molecular biologists had started obtaining such sequences routinely in the 1970s, when new sequencing methods were invented. However, those studies usually focused on species that were medically important or were for some reason especially suitable for investigating basic issues in molecular biology, such as how genes work. The study species tended to be things like *Drosophila melanogaster* (the common laboratory fruit fly), yeast, *Escherichia coli*, the house mouse, and *Homo sapiens*. Mullis's invention obviously helped such research. For our purposes, though, the key point is that PCR made it possible for evolutionary biologists—people who were interested in the diversity of living things, but, typically, had limited inclinations to learn difficult molecular techniques—to get DNA sequences from their favorite groups, whether they happened to be crickets, garter snakes, or maple trees. Thus, what PCR produced was an explosion of this most fundamental and unambiguous kind of genetic data, taken from many branches throughout the tree of life.

At the time I wrote my dissertation, including a chapter on the garter-snake tree, molecular dating analyses definitely were not in vogue. This was almost thirty years after Zuckerkandl and Pauling had first described the molecular clock, and by then it was clear that the pace of genetic change could be quite different in different lineages; the clock didn't tick at a constant rate. For instance, mammals had a fast clock, sharks an exceptionally slow one. Within mammals, the rodent clock ran especially fast. Because of this inconstancy, the clock idea was considered something of a dinosaur, an idea that had to be jettisoned once the data were in. It was a nice thought, but it didn't pan out. For my dissertation, I never even considered using the garter-snake DNA sequences to place ages on the branching points in the tree. This attitude was about to change, though. The clock was about to make a comeback.

A Big Pile of Bottles

In the Nevada ghost town of Rhyolite, not far from a cartoonish sculpture of a miner and his penguin and a more detailed one of the robes but not the bodies of Jesus and the apostles in their Last Supper poses, sits a house made almost entirely of old bottles embedded in mortar. The house is no artistic masterpiece, but it's striking for the sheer number of bottles involved, reportedly about 30,000 of them. It turns out that in the early 1900s, when Rhyolite was in a gold-mining boom, lumber was scarce in the town, but the place had more than fifty saloons, and thus no shortage of beer and whiskey vessels. I picture some of the townsfolk sitting around at that point, saying, "We've got no wood, but look at all these damned bottles. We could make a house out of these things." I envision these would-be architects a bit less than sober, drinking beers and (somewhat) carefully piling up the empties. In any case, however it came about, the idea of a bottle house seemed reasonable enough that the folk of Rhyolite ended up building three of them, although only the one survives.

I imagine a similar beginning to the explosion of molecular dating studies. Not boomtown rats in the desert this time, but a group of evolutionary biologists sitting around in a brewpub, pint glasses in hand, saying, "Hey, look at all these damned DNA sequences. They've gotta be good for something. Maybe we should use them to revive that old molecular clock idea." This isn't really how it happened. As far as I know, there was no particular get-together that rekindled interest in the molecular clock. However, it does seem that what led to the explosion was this overwhelming resource, all of these DNA sequences from crickets and crawdads, guinea pigs and alligators, mushrooms and magnolias. Typically, the point of this sequencing had been to figure out how groups are related to each other: Are animals more closely related to plants, or to fungi? (*Fungi, unequivocally.*) Where do whales fit into the tree of mammals? (*Right next to hippos.*) But now that these sequences were available, sitting in databases like GenBank that anyone could access with the click of a mouse, the old question about time started insinuating itself into people's minds. It was a pile of bottles too big to ignore.

For someone interested in molecular dating, the pile of sequence data, in addition to being big, was enticing in ways that other kinds of molecular data were not. For one thing, as mentioned above, sequence differences between species are very clear and countable. You can know that the cytochrome *b* gene in a black-necked garter snake differs at exactly 92 nucleotide positions from the same gene in a checkered garter snake. In contrast, some other measures of genetic distances between species are far muddier, relying on indirect and sometimes very imprecise estimates of DNA differences. For instance, consider DNA-DNA hybridization, which I remember as a cutting-edge technique in the early 1980s. This method involved mixing the genetic material from two species, say a finch and a crow, so that double-stranded DNA formed in which one strand was from one species and the matching strand was from the other species. You then heated up this "hybrid" DNA and recorded the temperature at which, say, 50 percent of it had broken apart into the original single strands from finch and crow. That temperature was a measure of genetic similarity between the two species, because the closer the base-pair match between the strands of DNA, the stronger they stuck to each other and, therefore, the higher the temperature that was needed to jostle them apart. It was a measure, but it wasn't a very precise measure. The breaking-apart temperature depended, for instance, on properties of the strands other than just the percentage of matching bases. Moreover, even with samples from the exact same specimens, the results could vary from lab to lab, or even from one trial to the next in the same lab, which was troubling. Some other measures of genetic distance have been similarly muddy. For evolutionary biologists thinking about how they might refine studies using the molecular clock, the perfect precision of DNA sequence differences was a refreshing change. It was a bit like going from a historical record based on hieroglyphics to one that uses a true written language with a large lexicon.

The discreteness of DNA sequence differences—an A in one species, a T in another—coupled with other properties of DNA, also made this kind of data a dream for biologists who wanted to use mathematical models to describe evolution, and this suitability for modeling would turn out to be critical for molecular dating. The modelers liked the discreteness of DNA because it let them treat the evolutionary process as straightforward probabilities—for instance, the probability that an A would switch to a T or a G or a C over some period of time. They were also drawn to DNA because the nucleotide changes could be fairly neatly categorized in ways that reflected differences in how evolution works. For instance, some mutations occur more frequently than others (A to G happens more than A to C or T, for example); some changes in the sequence switch the corresponding amino acids in the resulting protein and others do not; some parts of a gene code for especially critical sections of a protein for which almost any change in the amino acids would be harmful, while other parts code for less critical sections where some amino-acid changes are tolerated. The fact that you can divide up DNA changes into these, and many other, logical categories meant that the modelers had a lot to play with. In essence, DNA sequences gave them something to do that was tractable (a favorite word of modelers), but also satisfyingly complex.

This modeling of DNA changes had actually started before the invention of PCR. What PCR did was provide an enormous number of sequences to which the models, growing more and more complex, could be applied. For molecular dating, the importance of the models was that they provided improved estimates of the actual amount of genetic change separating two species, or, more generally, the amount of change along the branches of the evolutionary tree. This may sound a little counterintuitive; you might think that the amount of evolutionary change separating two species is just the observed number of nucleotide differences between them, like those 92 differences in the cytochrome *b* gene between black-necked and checkered garter snakes. However, this is not necessarily the case. In particular, if nucleotide substitutions (the preferred term among biologists) are happening frequently enough, there will be some sites in the sequence that have switched more than once along the evolutionary path connecting two lineages. Changes will have piled on top of other changes, so to speak. For any one site, you can never know for sure just how many substitutions have occurred—if one species has an A and the other a T, that could mean that their ancestor had an A that changed to a T in one lineage (one substitution), or that the ancestor had an A that went to G then back to A in one lineage and to C and then to T in the other (four substitutions), or, theoretically, an infinite number of other possibilities. You would know there was at least one change, but you couldn't rule out the possibility of so-called *multiple hits* at the site. It's like running into a friend you haven't seen for many years, who previously had brown hair but now has green hair, and wondering whether he went straight from brown to green or took some more circuitous hair-coloring route, such as brown to blonde to orange to pink to green.

A critical thing that the models do is make an educated guess—actually a calculation based on probabilities—at how many extra "hits" have occurred, changes that are hidden from direct observation, like your friend's possible multiple hair-color switches. In many cases, especially when dealing with distantly related species, that number can be substantial; in fact, the inferred number of hidden substitutions for a given sequence can be larger than the observed number of differences between species. The models are almost certainly right about the existence of many hidden hits, which, in turn, means that they are giving much better estimates of the actual amount of change along branches in the evolutionary tree than a simple tally of the differences between species. And, almost certainly, the estimates are getting better and better as the models become more realistic, taking into account more of the actual complexity of evolution. All of this is desirable for molecular dating studies—it's giving us a much better handle on half of the equation for relating genetic change to time.

Today, the Rhyolite bottle house sits locked and empty. Tourists, most of them on their way into or out of Death Valley, circle it, snapping photos under the baking desert sun. However, the house wasn't built just as a curiosity; it actually functioned as a dwelling for a few years and, later, as a trinket shop. All those beer and whiskey bottles really were good for something (apart from holding beer and whiskey). The same might be said for the unexpected information that came out of the big pile of DNA sequences, the outcome of Kary Mullis's insight (and Watson and Crick's discovery of the structure of DNA, and Fred Sanger's invention of an efficient DNA sequencing method, and the work of countless others). The pile of sequences and the evolutionary models that turned those sequences into amounts of genetic change generated a flood of molecular dating studies. For biogeography, these studies have been critical, providing evidence of timing, of what happened when. In my view, this evidence has been the key to extricating biogeography from the intellectual cul-de-sac created by the vicariance school.

However, as noted earlier, many scientists still have doubts about the validity of molecular clock analyses, and not all of these people are hard-core vicariance advocates. For instance, a friend of mine, an evolutionary biologist who I think of as both moderate and reasonable, refers to such analyses succinctly as "bullshit." As the foundation for a new view of biogeography, "bullshit" doesn't really work. Before we go on, then, we need to confront the criticisms of molecular dating. We need to establish that, in the results of these analyses, we're dealing with a functional bottle house, not a dangerous pile of broken glass.

On Sunday, December 26, 2004, at about 8:00 in the morning, Rizal Shahputra and several other workers were laying the foundation for a mosque in Banda Aceh on the Indonesian island of Sumatra when they felt and heard a strong tremor. Not long after, a boy came running toward them, warning that big waves were coming and they should run for cover. But it was too late. The tsunami that was on its way was made up of waves up to 100 feet tall and would become one of the worst natural disasters in history, killing more than 200,000 people. The waters rushed in and swept Shahputra and the rest out to sea.

Drifting in the ocean, Shahputra and many others spotted a floating tree, swam over to it, and held on. Days passed. One by one, Shahputra's companions weakened and slipped into the sea, until he was the only one left. Bodies floated past, some of them of people he knew. He drank rainwater and ate coconuts and packets of a powdered chocolate drink that he found on the water. He recited verses from the Koran. Finally, a container ship passed nearby, and an officer spotted him waving frantically. When rescued, he had been floating on his tree-raft for eight days and was more than 100 miles from Banda Aceh.

A woman known by the single name Melawati had a similar experience, drifting on the ocean for five days after the tsunami hit, clinging to a sago palm. She survived by eating the bark and fruit of the plant. As an example of how humans or other primates might have established new populations after improbable ocean journeys, her experience is especially cogent, for she was three months pregnant, and her unborn child survived the ordeal.

Chapter Six

BELIEVE THE FOREST

"Pretty Sloppy Stuff"

My first serious encounter with the molecular clock had to do with parasites—specifically, tapeworms living in the guts of humans and, in another part of their life cycle, in the muscle tissues of some domestic animals. It was a fascinating study, but also troubling. It was troubling because we had to rely on the clock.

The usual story with these tapeworms was that humans had picked them up, historically speaking, through the domestication of cattle and pigs. In this view, the wild progenitors of cattle and pigs were natural hosts of the parasites and, as a result, humans became infected when we started routinely eating the domesticated versions of these animals. However, the lead scientist on our study, a parasitologist named Eric Hoberg, had the idea that the real story was just the reverse, that we humans had a long history with the tapeworms and had infected our domestic animals with them. In this alternate scenario, cows and pigs would have become hosts when they became associated with humans, and thus began eating food contaminated with tapeworm eggs from human excrement.

My part in this project was to perform a molecular clock analysis using tapeworm DNA sequences downloaded from GenBank (a database of molecular sequences that is the most obvious manifestation of the DNA explosion made possible by PCR). The specific goal was to figure out when two species of tapeworms that infect humans had separated from each other, because, under the reasonable assumption that the common ancestor of these species also infected humans, that divergence date would indicate the latest possible time when the tapeworms had become associated with humans. That time could tell us if humans had acquired tapeworms before the domestication of cattle and pigs some 10,000 years ago.

This was in the late 1990s, but, even back then, there were many different ways to do a clock analysis. I sorted through the possibilities, figured out which approaches seemed to make the most sense, and did the analysis several different ways. None of them inspired great confidence, particularly because the DNA sequences were short, and I had to assume that the tempo of the tapeworm clock fell within the range of rates that other biologists had calculated for organisms only distantly related to tapeworms, like shrimp and mice and sharks. Nonetheless, the calculated ages told a striking story: even the youngest possible ages for the tapeworm split were older than the likely age of domestication of cattle and pigs. It looked as if Eric Hoberg was right and the traditional story was wrong—we had given tapeworms to cattle and pigs, not vice versa. (In a way, one could view this conclusion as one more bit of support for a major theme that came from Darwin, namely, that humans are not intrinsically lords or victims of nature, but are simply parts of the living world, like all other species. At least, that's the way I thought of it.)

I came away from that tapeworm study with a new appreciation for what molecular clocks could provide, but with my skepticism about such analyses intact. I had been reasonably conservative in my approach, but I still wondered whether some new evidence would come to light showing that my estimates were all wrong. Maybe, for instance, someone would discover that the tapeworm clock ticks much faster even than the fastest clock I had used for calibration (the snapping shrimp clock). Or, perhaps, different genes would give a very different answer. Our conclusions about the origins of tapeworms in humans and domestic animals hinged on the clock results, and those results were debatable.

I'm certainly not alone in having used molecular clock analyses while remaining somewhat leery of them. For instance, Michael Donoghue confessed to me, "I've been involved in a bunch of dating analyses and I think they're mostly pretty suspect. And I look at the literature and I'm sort of appalled by the [analyses] people do." Although Donoghue believes that new methods for molecular dating are improvements, he imagines that, down the road, people will look back and think, "We just did a decade's worth of work with BEAST [a commonly used dating program], . . . and now we realize that actually a lot of that's pretty sloppy stuff."

And yet, Donoghue also thinks, as do many other biogeographers, that the vicariance people are crazy to ignore the molecular dating evidence. He sees their dismissal of this evidence as "misguided" and a case of "burying their head[s] in the sand." So there seems to be a bit of a disconnect here: How can a person view molecular dating as "pretty suspect" and "pretty sloppy stuff," but also think that this information cannot be ignored? The point of this chapter is to make sense of this apparent contradiction by establishing that the molecular dating evidence *in general* is not likely to be far from the truth, especially in what it's saying about biogeographic history. It's a case of believing what the forest is telling us rather than getting too hung up on particular disease-ridden trees. In building to that point, though, it will be instructive to first deal with specific problems—the reasons for the diseased trees—to understand why there is such distrust of molecular dating. It will be apparent that the problems are real and substantial, but that, in the end, they don't derail the whole enterprise.

Two Problems and How to Deal with Them

The ideal molecular clock would be one that ticked at the same rate for all branches in the tree of life and could be perfectly calibrated so the rate was known precisely. With a clock like that and the appropriate DNA sequences, we could confidently put an age (with error bars) on any given branching point in the tree. In other words, the ideal clock would produce a reliable historical chronicle—a record of what happened when—with respect to the timing of the separation of evolutionary lineages. We would know, with great precision, when humans separated from chimps, or Australian baobab trees from Malagasy ones, or African from South American lungfishes, and we would be able to relate those evolutionary splits to geologic or climatic events, such as the opening of the Atlantic Ocean or the Pleistocene Ice Ages.

Unfortunately, that ideal clock does not exist. In essence, it doesn't exist for two reasons, a problem with fossils and a problem with molecules. These problems are big, fundamental ones, and, to the likes of Gary Nelson and Michael Heads, they're insurmountable. However, other scientists have not been so pessimistic and have tried to solve them.

The first problem has to do with the fact that one usually needs fossils to assign ages to calibration points. These are the particular evolutionary branching points used to calculate the tempo of genetic change, that is, to figure out how fast the molecular clock is ticking. The basic procedure here is to find fossils of known age that can be connected to given branching points. Just figuring out where a fossil should be placed in the tree of life—that is, inferring the branch to which it belongs—can be difficult, especially since fossils are incomplete specimens, usually missing some parts of the skeleton, shell, or other hard parts, and almost always lacking clear indications of soft parts of the anatomy. For instance, certain apes, known only from very incomplete fossils, fall somewhere close to the evolutionary split of chimps and humans, but cannot confidently be placed in the tree beyond that. They might fall on the chimp side, or on the human side, or they might be ancestors of both, and the exact placement of these fossils makes some difference in how they're used in calibrations. In addition, the age of a fossil often cannot be precisely determined. The vast majority of fossils are assigned ages based on the geological strata in which they were found, and this sometimes means that we "know" a fossil's age only within a range of several million years. Even if a fossil's age is obtained from radiometric dating (using the rate of radioactive decay of one isotope into another), the slop in the estimate can be fairly large, depending on the circumstances.

However, the big problem with fossil calibrations is not about where fossils should be placed in the tree of life or how old they are. The real stumbling block comes from the fact that the fossil record is not even close to being a continuous chronicle of everything that ever lived, but is more like a series of snapshots, some of them close together and others widely spaced in time. (Actually, if the fossil record were complete, there would be no need for molecular dating.) The fragmentary, snapshot nature of the record means that, even if all the fossils ever discovered were placed in the correct evolutionary groups, and even if they were assigned perfectly accurate ages, the greatest difficulty with calibrations would still remain. A particular problem is that the time when a group first shows up in the fossil record only gives us the *latest* possible age for the group's origin; the actual age will almost always be older, and it will often be much older. For instance, the first fossil hummingbirds are from the Late Eocene, about 35 million years ago, but it is likely that the group has a history, so far invisible to us, that extends much deeper in time. Thus, if one used that 35-million-year date as the age of the split between hummingbirds and their nearest relatives, the swifts, the calibration probably would be way off. As a result, one would end up thinking the molecular clock was running much faster than it really was. Calibration errors like this one are especially likely for groups like hummingbirds, with delicate parts that do not easily fossilize, but the problem potentially applies to any group. This kind of inaccuracy, caused by the incompleteness of the fossil record, is the fundamental weakness of calibrating a molecular clock with fossils.

An obvious solution is to use only especially good calibrations, based on parts of the fossil record that paleontologists consider reliable. What qualifies as "especially good" and "reliable" is subjective, but there are some cases that do seem convincing. Consider the branching point between the bird lineage and the crocodile lineage, for example, which has often been used to calibrate bird molecular clocks (see Figure 6.1). A key fossil in this instance is from a creature called *Arizonasaurus* (discovered in Arizona, of course, although it has also been found in Texas) dating from the early part of the mid-Triassic, about 240 million years ago. *Arizonasaurus* was an impressive-looking ten-foot-long carnivore with a sailfin on its back made up of long, upward extensions of its vertebrae. The sailfin made *Arizonasaurus* look superficially like some much older sailfin reptiles related to mammals,* but details of its anatomy indicate that it belongs on the crocodile branch. Assuming that's the case, it is the oldest fossil known to come after the bird-croc split, and as such, it sets the youngest possible age for that branching point.** Just a little further back in time, in the period from about 245 to 250 million years ago, there are many fossils of close relatives of the bird/croc group from all over the world, but nothing from within that group itself. In other words, there are indications that the bird/croc group did not yet exist at that time. We might therefore conclude, conservatively, that the bird-croc branching point could not have occurred earlier than 250 million years ago. Splitting the difference between our minimum and maximum ages for the bird-croc split—240 and 250 million years ago, respectively—we might then come up with a calibration age of 245 million years. There are quite a few other branching points, for vertebrates and other groups, for which similarly reasonable ages can be given.

This bird/croc example brings up another aspect of dealing with the calibration problem, namely, incorporating uncertainty in the age of a branching point into the analysis. In the bird/croc case, the calibration could be entered as the whole range of ages between 240 and 250 million years ago, rather than as exactly 245 million years ago. Using a range of

---

* *Dimetrodon* is the most famous of these older sailfin reptiles, well known to children who play with plastic prehistoric animals.

** There are other fossils almost as old from both the crocodile and bird sides, so the argument does not depend entirely on *Arizonasaurus* (Benton et al. 2009).

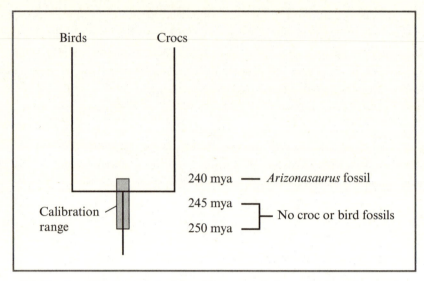

6.1 Fossil calibration of the branching point between birds and crocodylians. *Arizonasaurus*, from 240 million years ago, is placed on the croc branch, indicating that the bird-croc split had already occurred by then. Mya = millions of years ago. Between 245 and 250 million years ago there are many fossils of close relatives of birds and crocs, but none that seem to be within those two lineages, suggesting they did not exist at that time. The bird-croc branching point is therefore assumed to have occurred sometime between 240 and 250 million years ago.

ages instead of just a single value will make the molecular clock calibration less precise but more defensible. By analogy, if you were estimating how long a drive would take and didn't know the driver, it would be prudent to calculate the time based on a range of possible speeds rather than just one. Many recent studies do incorporate such uncertainty for all calibration points.* Also, some investigators have used several to many calibration points, then figured out which ones were out of synch—making the clock run especially fast or slow in comparison to the other points—and removed these outliers in the final analysis. Do these various improvements solve the calibration problem? Not completely. Bad calibrations can creep in, and it's even conceivable that a set of calibration points could

---

* The widely used program BEAST (Drummond and Rambaut 2007) can incorporate age uncertainty by allowing the user to enter the calibration point as a curve that, across a range of ages (e.g., 240 to 250 million years ago, in the above example), specifies the probability that a given age is the actual age of the branching point.

be consistent with each other but consistently inaccurate; by analogy, it is certainly better to have multiple witnesses to a crime than one witness, but it's also possible that all of those witnesses are lying. Nonetheless, in studies that use "especially good" calibrations and remove outliers, calibration points that are far out of line with reality are probably rare.

The second big issue—the problem with molecules—is the well-documented unclock-like nature of the clock, that is, the fact that genetic changes occur at different rates in different lineages.** Basically, the more variable, and therefore unpredictable, the clock is, the more difficult it becomes to estimate the ages of branching points, and it turns out that the clock is pretty variable. I mentioned in Chapter Five that the clock tends to run rapidly in mammals and extremely slowly in sharks, and that, within mammals, it runs especially fast in rodents (about ten times faster in mice than in humans). This is just the tip of the iceberg; in studies that include many groups, it's fairly rare to find that the clock is running at close to the same rate in all lineages.

The reasons for the inconstancy of the clock are complex and only partially understood. Lineages with short generation times often have fast clocks, probably because mutations mostly occur when the DNA is replicating, and short generations mean more replications over a given period of time. That might explain why, for instance, rodent clocks run faster than those of people or rhinos or elephants. Ultraviolet radiation causes mutations, and that could be why at least some plants that live in exposed, sunny places have fast-running clocks. The effect of natural selection differs among lineages and undoubtedly has an effect on overall rates of genetic change in some cases. For instance, selection can be relaxed in parasites, because some of their physiological functions are performed by their hosts, and this means that many mutations that would be weeded out in nonparasites can spread within such parasite populations; that effect can translate to a faster clock. (I worried about that for the tapeworms, and got around it by only using changes in the DNA that should not have been influenced by selection.) Similarly, in small populations, the influence of natural selection, which mostly acts to prevent genetic change, is easily overwhelmed by random substitutions in the DNA sequence (genetic drift), again leading to faster rates of change.

The bottom line, however, is that none of these factors, or any of the many others that might speed up or slow down molecular clocks, have

---

** The clock also runs at vastly different rates in different genes or parts of genes, but those problems are easier to deal with than the variability among lineages.

such clear and consistent effects that they can be relied upon to predict the rate for a particular group. So, when it comes to molecular dating, none of these insights into what might change clock speed have much practical impact; we're still faced with a fickle, inconstant, and unpredictable clock. What this inconstancy means is that, even if you have accurate calibration points, you cannot know for sure how fast the clock is running in *other* parts of the tree, including the parts for which you're trying to calculate ages. To put the problem in extreme terms, it's like trying to predict how long a journey will take without knowing whether you'll be riding in a horse carriage, a car, or an airplane.

The inconstant clock, even more than the problem with fossil calibrations, has been considered the Achilles' heel of molecular dating. To make use of the flood of DNA sequences—that big pile of bottles—people realized that this problem had to be addressed. As a result, over the past ten years or so, there has been a big push to develop so-called *relaxed clock* methods, which do not assume the clock is constant and instead estimate when and how much it shifts. These methods don't use generation times, exposure to UV radiation, and the like, but instead are more like mathematical puzzle-solving devices that search for the set of rates of genetic change that best fits the data at hand, that is, the DNA sequences from the different lineages and the collection of calibration points. In essence, these methods are designed to figure out which lineages are horse carriages and which are airplanes (or anything in between) and calculate ages accordingly.

When these relaxed clock methods have been tested using computer programs that make simulated DNA sequences evolve along branches of a simulated evolutionary tree, they usually give pretty accurate ages. Of course, simulations are not the real thing, and one could always argue that such computer games have left out some important aspect of the actual evolutionary process. Nonetheless, these methods seem to be improvements over the older methods, which assumed a constant clock.* When several different relaxed clock methods are applied to a very large amount of data (meaning many genes and many sampled groups), in conjunction with a large number of calibration points scattered throughout the evolutionary tree in question, the results can be convincing.

Consider, for instance, a 2011 study done by a team of evolutionary biologists, led by a mammalogist named Robert Meredith, to construct

---

* A caveat is that relaxed clock methods may produce less accurate ages than constant clock methods in cases where the constant clock assumption actually holds.

a "timetree" for mammals as a whole. This research team analyzed parts of 26 different genes (a total of about 35,000 base pairs—the bits of information—for each sampled species) and included representatives of all mammalian orders and more than 97 percent of the families. My friend and colleague John Gatesy, who was part of this team, told me that, by using calibrations from different groups—for instance, only whales versus only rodents—one could get vastly different branching-point ages for the whole tree. That kind of problem is expected for a large group like mammals, and it reflects the inconstancy of the clock. However, this group of scientists didn't use just one or a few calibration points; they included 82 calibrations spread throughout the evolutionary tree of mammals, with each one entered as a distribution of possible ages rather than as a single age. Then they analyzed their dataset using several kinds of relaxed clock methods that made different assumptions about the evolutionary process. Basically, they did exactly what you're supposed to do in a molecular dating study: they used many species, many genes, and many fossil calibration points, and they incorporated uncertainty about the ages of those calibrations and about how genetic change happens. No doubt someone will improve on this study, probably in the near future, by using more species, more genes, more calibrations, and yet-to-be-invented methods that take into account more of the complexity of evolution. Nonetheless, it would be surprising if the mammal timetree from this study was severely distorted. It would be surprising if it was very far from the truth.

For our purposes, an important finding from this state-of-the-art mammal timetree is that it validates the biogeographic conclusions of all the mammal cases mentioned in this book, despite the fact that those conclusions were based on earlier studies using less sophisticated methods and far fewer data. Other groups lack such a standard for comparison. However, a massive survey of molecular dating results, called the Timetree of Life project, suggests general agreement among recent studies for most groups. For the taxon I know best, the snakes, the two most convincing studies gave strikingly similar branching ages despite using different sets of genes. This general agreement among molecular dating results is one reason to believe the forest—that is, the overall picture—even if some published studies, individually, remain "pretty suspect."

THE MOLECULAR CLOCK AND THE PITFALLS OF EXTREMISM

The incompleteness of the fossil record has led some biologists to turn to an alternate source for calibrating molecular clocks, either for particular fossil-poor groups or in general. Specifically, these researchers have used fragmentation events, such as the separation of landmasses due to continental drift or sea-level rise, to set the tempo of the clock. The notion here is that if an evolutionary branching point came about because of such a geologic or climatic event—that is, because of vicariance—then the ages of the branching point and the physical fragmentation would have to be the same. This kind of calibration would introduce circular reasoning if the goal of the study was to evaluate the reality of that particular vicariance event, but the method can be useful for other purposes. For instance, a team of Japanese researchers used a fossil-calibrated clock of cichlid fishes to indicate that deep evolutionary separations in this group were caused by Gondwanan breakup. They went on to argue that those deep splits could be used as calibration points for future studies in place of or in addition to fossil calibrations. That all seems perfectly reasonable. However, in other cases, researchers have relied entirely and uncritically on calibrations that assume vicariance, and this approach has produced some very bizarre age estimates for groups. In fact, the ages are so anomalous that they serve as strong warnings against accepting vicariance too readily.

Consider, for example, a 2004 study on amphisbaenians, limbless or two-legged, burrowing reptiles that are often called "worm lizards" because they look like oversized earthworms. To calibrate a molecular dating analysis, the authors of this study chose a branching point separating a group of South American worm lizards from their African relatives, setting the age of the split by the opening of the Atlantic Ocean 80 million years ago. (This age for that geologic event is actually too young, but a more accurate age would only make the example more extreme.) Thus, they were accepting that this South America–Africa branching point in the worm-lizard tree had been caused by Gondwanan breakup. Using that calibration, they estimated the ages of some branching points within a genus called *Bipes* (the single group of two-legged amphisbaenians). These ages within *Bipes* seem surprisingly old for the separation of very similar-looking species—one point was estimated at 69 million years ago, which was before the extinction of the dinosaurs—but there aren't any fossils within *Bipes* that contradict those estimated ages. The analysis also implied, however, that the earliest branching points within amphisbaenians as a whole occurred more than 200 million years ago, and that result is decidedly strange. Amphisbaenians are not an especially early evolutionary branch within the lizard group, yet that age of 200-plus million years is considerably older than *any* lizard fossil. The fossil record of lizards is substantial, so this is a conundrum; finding that worm lizards are older than the oldest known lizards of any kind is sort of like discovering that you're older than your grandmother. The problem here almost certainly lies not with the lizard fossil record, but with that calibration point, based on the opening of the Atlantic. In other words, the problem is with the assumption of ancient vicariance.

An even more egregious example comes from Michael Heads, who has stated explicitly that tectonic and other fragmentation events should be used as calibration points in place of what he sees as horribly unreliable fossil calibrations. In criticizing a molecular dating study of hedgehog-like mammals called tenrecs, Heads recommended calibrating the analysis with the split of Malagasy and African tenrecs, setting the age of that point as the time of the separation of Madagascar from Africa 120 to 165 million years ago. As with the amphisbaenian case, this calibration assumes that vicariance was the cause of the piecemeal distribution, and it leads to a similar kind of paradox: tenrecs are not an early branch within placental mammals, yet this calibration age *within* tenrecs predates the first known placental mammal fossils by at least 55 million years! Another way to look at it is that, if tenrecs are as ancient as Heads suggests, then the placental mammal group as a whole must be about 200 million years old, more than three times older than the earliest known placental fossil. It is probably safe to say that no one with a deep knowledge of the mammalian fossil record believes this is even remotely possible. The problem here is the uncritical acceptance of vicariance and the related notion that tectonic events must be used to calibrate molecular clocks.

What Fossils Say About the Clock

Making comparisons among different molecular studies for the same group isn't the only way to evaluate molecular dating results. Another argument focuses on comparing age estimates from the molecular approach with those based on the fossil record. The key here is to limit comparisons to branching points, such as the bird-crocodile split, for which the fossil record should give accurate ages. These fossil-based ages can then serve as yardsticks to judge how well molecular dating analyses are working.

When such comparisons are made, they show that the molecular age estimates match the fossil-based ones well up to branching ages of about 400 million years. Up to that age, the observed data—the points relating molecular estimates to the corresponding fossil ages—are generally not far from the line that represents a perfect match between the two kinds of estimates (see Figure 6.2). The simplest explanation for this result is that, although there is a fair amount of slop in the data, molecules and fossils are converging on the truth, that is, on the real ages of these branching points. The alternative is that both fossil and molecular estimates are wrong, but, for some unknown reason, they just happen to be off in the same direction and to the same degree. That would be a strange coincidence and, therefore, seems unlikely. Also, and importantly, there is little suggestion from the graph in Figure 6.2 that the molecular estimates are biased to be too young (which would show up as points tending to fall above rather than below the line of perfect match). This lack of bias is critical because it means that, even if molecular estimates are off, which they must certainly be in some cases, there is no tendency for them to *generally* support oceanic dispersal hypotheses by giving spuriously young age estimates. Errors are likely going both ways, sometimes erroneously strengthening the case for recent dispersal but, just as often, mistakenly weakening the case. In short, if it turns out that molecular dating suggests that ocean crossings have been very important in general, that's probably because ocean crossings really have been very important.

For points older than 400 million years ago, a glance at the graph shows that something odd is happening. These points represent events such as the separation of animal phyla or groups of phyla, and here the molecular estimates are consistently older than the fossil ones, in some cases far older. Some scientists argue that this discrepancy reflects an extremely poor fossil record pertinent to very ancient branching points, whereas others claim that it is caused by a one-time, universal shift in

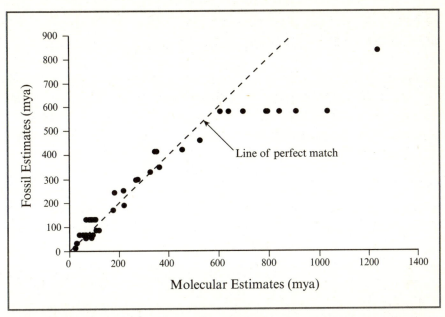

6.2 Molecular estimates compared to fossil estimates for the same evolutionary branching points. The fossil estimates are ones considered to be relatively reliable. Mya = millions of years ago. Redrawn and modified from Hedges and Kumar (2009b).

the process of genetic change that happened early in the history of multicellular animals. However, for our purposes, that part of the graph does not nullify the importance of molecular dating. For one thing, all of the pertinent biogeographic studies in this book involve groups and events much younger than that 400-million-year threshold; at that time, Pangea hadn't even formed, and the beginnings of the breakup of Gondwana were still far off in the future. Also, if the fossils are actually correct, the anomaly in the graph represents a tendency for molecules to produce age estimates that are *too old*; and, as indicated above, a bias in that direction would lead to *rejection* of recent dispersal explanations, not their widespread acceptance. There are a couple of other notable cases, involving early evolutionary radiations of birds and mammals, in which the fossil and molecular age estimates conflict, but the molecular estimates are, if anything, too old in these cases as well. The bottom line is that, if this bias toward overestimating ages really exists, yet it turns out there is still strong evidence for recent dispersal, we should be even more confident in that conclusion.

Messy, Useful Science (or What Would Darwin Do?)

Molecular clocks are easy to find fault with and hard to love. Hence all the pejoratives thrown at these studies—the comments that they're "pretty suspect," or a futile "molecular dating game," or, simply, "bullshit." In some cases, the criticisms seem justified. Even now, one sees cringe-inducing studies based on very small amounts of DNA sequence, or using only one or a few untrustworthy calibration points. However, since the early days of the molecular clock, when Zuckerkandl and Pauling were thinking that the clock ticked at a constant rate, there have also been vast improvements in methods, not to mention an explosion of DNA sequence data to which the new methods can be applied. Moreover, the overall agreement between molecular age estimates and good fossil-based ones indicates that the approach is reasonable in general, even if some particular studies are not. And, finally, if there is a bias in molecular dating, it seems to be toward giving spuriously old ages, which will, if anything, lead to *underestimating* the frequency of oceanic dispersal. It is all too easy to find bad molecular dating studies—the disease-ridden trees—and to extrapolate that the entire enterprise is rotten. But that is a mistake: for biogeography, we need to pay attention to the message from the whole forest.

The idea that we should believe the forest relates to the point I made in Chapter Three about vicariance biogeographers wanting a black-and-white science. This business of molecular dating is messy; it's thoroughly statistical, meaning it gives one probabilities, not absolutes, and it relies on assumptions about calibration points and the way that DNA sequences evolve that can be hard to justify for any particular case. There is a certain kind of mind that revolts against that sort of messiness and finds no solution other than to dismiss the whole practice. If some molecular dating studies are obviously botched and all are based on unverified assumptions, why should we believe anything that comes from this approach? These "anti-dating" scientists never reach the point where they can see that the picture as a whole is highly informative. They never see the forest.

More generally, I would also suggest that it is the business of evolutionary biologists and other historical scientists to draw conclusions from obviously flawed and potentially misleading information. That is simply the nature of historical evidence, especially the kind that bears on ancient events. Evidence degrades through time, and over millions of years, most of it disappears completely. Thus, we are forced to try to unravel deep history by examining incomplete relics, like a fragment of a fossil jawbone, or indirect evidence, as when we use the DNA sequences of living species to build evolutionary trees. Quite often, it is only through the convergence of inferences from different sources that we are persuaded that some evolutionary scenario is correct. For instance, we might become convinced of a vicariance explanation through the dovetailing of support from fossils, DNA, and geology.

Given the nature of historical evidence, it is not surprising that good evolutionary biologists often are notable sponges, sifters, and synthesizers of information. As the archetype, we have Darwin, who absorbed encyclopedic knowledge of taxonomy, animal breeding, anatomy, embryology, geology, the fossil record, and (of course) biogeography to construct his argument for evolution. No one piece of evidence was entirely convincing, and some of it, such as the sudden appearance of many animal phyla in the fossil record, seemed to contradict his views, but he didn't let that blind him to the bigger picture. More recently, evolutionists like Ernst Mayr, George Gaylord Simpson, and Stephen Jay Gould, among others, have followed in that tradition. Those three were all great sponges, sifters, and synthesizers.

These thoughts make one wonder what Darwin would have made of the ages of lineages estimated from DNA sequences. I'm sure he would have recognized that molecular dating evidence—like all evidence that bears on history—is not entirely to be trusted. However, I also believe that, as he looked into the subject further, he would have seen that this dating evidence has the potential to be illuminating, and that to believe it is grossly misleading is unjustified. And then he would have done what he did with so much other information, from the variation in pigeon breeds to the anatomy of eyes: he would have used it.

*Researchers tracked three estuarine crocodiles (*Crocodylus porosus*) by satellite as the animals moved at least 34, 255, and 366 miles in the coastal waters of northeastern Australia, the fastest of the three covering more than 14 miles per day. Crocodiles have very limited endurance, but the movements of these three were consistently aligned with the direction of ocean currents, suggesting that they were simply going with the flow, like giant, scaly message bottles. In one case, when the current slowed sharply, the crocodile in question moved ashore for two days, and only went back into the sea when the current picked up again.*

The researchers noted that estuarine crocodiles might travel vast distances in this way, as they are enormous animals, have low metabolic rates, and can excrete excess salt through glands in their nasal passages. A 22-pound C. porosus *can last up to four months in seawater without eating, which implies that an adult weighing 200 to 400 pounds could survive an ocean voyage of many months. Fossil and molecular genetic studies indicate that, in fact, crocodiles have dispersed across wide ocean barriers several times in their history. Within the past several million years, for example, one species apparently crossed the Atlantic from Africa, giving rise to the four species of New World crocodiles.*

Section Three

The IMPROBABLE, the RARE, the MYSTERIOUS, and the MIRACULOUS

Chapter Seven

THE GREEN WEB

The Lost World, Found

If you get stranded up there, whatever you do, don't try to climb down. That's basic advice for biologists working on the summits of tepuis, the isolated, table-top mountains that dot the savanna of southern Venezuela and nearby parts of Brazil, Colombia, Guyana, and Suriname. Tepuis—"houses of the gods" in the language of the local Pemón Indians—are the monolithic remnants of a sandstone plateau that formed well over a billion years ago, long before the breakup of Gondwana or the formation of Pangea. Much more recently, over the past 70 million years, wind and water have been grinding down the plateau, with only those parts that were capped with more resistant rock remaining as these towering flat-topped mountains, their sandstone cliffs now rising thousands of feet above the surrounding country. On the tops of the tepuis are misty, dripping landscapes of dwarf forests; streams running through pink sand; rock formations eroded into weird convolutions; and fields of plants with naked stalks and tufted heads, like something out of Dr. Seuss. "Otherworldly" is a word that crops up often in descriptions of these places. Adventurous tourists and scientists can get dropped off by helicopter onto tepui summits, but, not uncommonly, clouds or fog will prevent the scheduled pickup. At that point, the thing to do is wait, even if the delay drags on for days, because trying to find a way down the cliffs is likely to get you lost, or stuck halfway down, or inadvertently airborne.

The otherworldliness and inaccessibility of tepuis are exactly what draws biologists to them. Those cliffs that keep people from climbing up or down are also a barrier for other living things and, to many biologists, this has implied that the organisms living on the tops of the tepuis have been evolving in prolonged isolation, perhaps for many millions of years. This idea of tepuis as areas populated by relicts took root in the Western world in the 1860s, before Europeans had even climbed them. In the early 1900s, when Arthur Conan Doyle wrote *The Lost World*, famously imagining a tepui filled with dinosaurs, pterodactyls, ape-men, and other prehistoric creatures, he wasn't inventing a view of tepui biotas so much as taking the accepted notion to a romantic extreme. And it's a notion that persists, albeit without the megafauna. For instance, in the late 1980s, Vicki Funk and Roy McDiarmid, Smithsonian biologists who had explored tepui summits (and gotten stranded on one for ten days), speculated that tepui organisms might be "remnants of an ancient biota that dates back millions of years to a time when South America and Africa were a single land mass." In other words, tepui biotas are like isolated pieces of old Gondwana, reminiscent of what people have imagined for islands such as New Zealand and New Caledonia. Along the same lines, Jesús Rivas, a Venezuelan herpetologist, said of the tepuis, "They are like a place where time stopped."

But, then again, maybe not.

On the boggy plateau of the tepui called Cerro de la Neblina—"Mountain of the Mists," a name that would fit almost any tepui—lives a miniaturized plant that eats animals. The species is *Drosera meristocaulis*, and it's a pygmy sundew, a plant that catches insects in sticky drops exuded from tiny stalks that cover its red, spoon-shaped leaves. On the tops of tepuis, it turns out, carnivorous plants like *D. meristocaulis* are a dime a dozen; tepui soils are infertile, which makes them good places for plant species that can supplement nutrients from the soil by trapping and digesting insects and other small creatures. Some tepui plants may even specialize in eating protozoans, which they trap in corkscrew-shaped structures on root-like underground leaves. For our purposes, however, what is really significant about *D. meristocaulis* is not what it eats, or anything else it does, but where it came from.

Drosera is a big genus, with some two hundred species, concentrated in the Southern Hemisphere, but with at least some species on every continent except Antarctica. Find a nice bog, and there's a good chance some species of sundew will be living there. Despite their wide distribution, though, sundews, because of the properties of their seeds, are not considered great long-distance dispersers. Their seeds don't float and are killed by salt water; they don't have wings or tufts for drifting on the wind, or hooks for latching onto fur or feathers; and they aren't enclosed in fruits designed to be eaten by birds or other animals that might transport the seeds to some faraway place. Basically, sundew seeds just fall to the ground or get blown a short ways, usually ending up close to the parent plant. Over millions of years, members of the genus obviously have moved great distances, as their worldwide distribution indicates, but the assumption is that they accomplished this in small steps, not giant leaps.

In the evolutionary tree of sundews, one can see the signature of their limited powers of dispersal. Specifically, the sundew phylogeny strongly shows *geographic structure*, a fancy term that just means that closely related species tend to occur near each other.* So, for instance, a western Australian sundew's closest relatives are likely to be other western Australian sundews, and a southern African sundew's closest relatives are usually other southern African sundews. However, there are some conspicuous exceptions. For instance, two species found in the eastern United States fall within a group of South American species, indicating that their ancestors dispersed from that continent.

The weirdest exception to the rule of geographic structure in sundews, though, is the Cerro de la Neblina sundew, *D. meristocaulis*. There are many other sundew species in South America, including some that live near the tepui region, but, according to a recent DNA sequence study, none of these are the closest relatives of *D. meristocaulis*. Instead, *D. meristocaulis* is firmly placed by the molecular data within a group of Australian sundews. These Australian species also happen to be miniaturized, like *D. meristocaulis*, and they share unusual pollen and leaf-hair features, among other things, with the tepui species. In other words, both DNA and anatomy indicate a close evolutionary relationship between *D. meristocaulis* and the Australian sundews.

South America and Australia were both once part of Gondwana, so one can imagine a continental drift scenario to explain the connection of a Neotropical sundew to Australian species. However, molecular dating makes that explanation extremely unlikely, if not impossible. According to molecular age estimates, *D. meristocaulis* split from its Australian

---

* When Alfred Russel Wallace argued in his "Sarawak paper" that species with close taxonomic affinities are usually found in geographic proximity, he was basically using geographic structure to argue for evolution—see Chapter One.

7.1 Long-distance voyager: the sundew *Drosera meristocaulis* from the Cerro de la Neblina tepui. Photo by Fernando Rivadavia.

relatives only within the past 12 to 13 million years, long after the breakup of Gondwana, and its ancestors therefore must have reached South America by oceanic dispersal. The tepui sundews are certainly now isolated from any closely related species, but they are by no means ancient relicts. Measured by the span of deep time, they are yesterday's immigrants.

The conclusion that a sundew in some form, whether seed or whole plant, with no obvious means of overwater dispersal, managed to travel from Australia, either east across the Pacific or, more circuitously, west across both the Indian and Atlantic Oceans, and, on top of that, negotiated the towering cliffs of Cerro de la Neblina to end up on the plateau of the tepui, seems, to use a stock phrase from Darwin, "absurd in the highest degree." The authors of the sundew study suggested that Darwin's old mechanism of seeds stuck to the feet of a bird was one possible means of transport, although they also pointed out that there is no known avian migration pathway connecting Australia to northern South America. Fluky events of one sort or another can take birds out of their normal routes, however, leading to what birdwatchers call "accidental" occurrences. The sundew story reminded me, for instance, that I (and several dozen other birders, armed with a thicket of spotting scopes) once saw a Fork-Tailed

Flycatcher, a species that normally gets no farther north than tropical Mexico, sallying for insects over a farmer's field near Rochester, New York. This bird probably had been migrating from southern to northern South America and missed some cue to stop, overshooting its mark by several thousand miles. A mistake like that could lead to a bird transporting seeds outside of any normal migration route, and even, perhaps, to a site on the other side of the world.

Admittedly, this argument has gotten us into the realm of nothing-that-you-would-want-to-hang-your-hat-on. But maybe that's part of the point—that even when no likely mechanism presents itself, nonetheless we often must infer that long-distance dispersal did occur. Rare, mysterious, and miraculous things do happen, even if we may never know exactly how. (I'll take that point up in more detail in Chapter Eight.) The main point here, however, is that if a sundew can voyage thousands of miles, mostly over the ocean, from Australia to the top of a tepui in northern South America, then we should also expect many other plants, some of which have better means of traveling long distances, to have made similar journeys. We should perhaps expect that, if such plant dispersal events were traced on the globe, we would end up with an amazing tangle of lines going every which way, a green web connecting the landmasses of the world.

Fifty-One Bean Plants

I was talking on the phone with a botanist named Matt Lavin, a professor at Montana State University, about the biogeography of plants. Lavin and I have never met, but I know from photos that he's thin, with angular features and slightly unkempt, graying hair. (In the course of our conversation I found out that we know some of the same people and even overlapped at Cornell for a year, while I was a grad student and he was a postdoctoral researcher. We were in different departments, but undoubtedly went to some of the same talks, so we had probably seen each other without knowing it, which might seem surprising, but, in the small world of evolutionary biology, probably shouldn't.) For an academic, Lavin's speech is relaxed and down-to-earth, with the hint of a cowboy drawl. He tends to leave the "g"s out of his "ing"s. When I asked him how he got interested in biology, he said, "I grew up all over the West because my dad was a Forest Service guy, and he moved around every couple years, so we lived all over southern Idaho, western Wyoming, northern Nevada, so kind of livin' in sagebrush country, gettin' out in it, . . . and adjacent forest areas, and learnin' plants." That upbringing explains the rural western shadings in his voice and, maybe, the fact that a big part of his research deals with sagebrush communities. Since I live in Nevada and like getting out in the wilds too, our conversation veered into talk of out-of-the-way places in the West. I asked him, for instance, how he ended up writing a flora of the Sweetwater Mountains, an eastern outlier of the Sierra Nevada, where I've collected alpine plants and arthropods. (Our shared familiarity with this obscure mountain range struck me as more unexpected than our overlapping time at Cornell.) What I really wanted to discuss with him, however, was the other major part of his research, dealing with the worldwide biogeography of legumes, that is, plants in the bean family.

Beans, of course, are enormously important food plants, but, from a biodiversity perspective, they are much more than that. The bean family, Fabaceae, is a cosmopolitan group and the third largest family of flowering plants, with nearly 20,000 described species. (For comparison, there are about 5,400 known species of mammals.) In temperate parts of the world, the family is mostly represented by herbs—lupines, vetches, locoweeds, and the like—but in the dry tropics and subtropics, woody legumes such as acacias, locusts, and mesquites are often the most abundant large plants. If one thinks of the Sonoran Desert, the first plant that comes to mind is the saguaro cactus, but in fact that desert is dominated by bean trees and shrubs of various kinds. The same goes for many parts of the world that have warm, dry climates punctuated by a distinct rainy season.

Matt Lavin had done his PhD dissertation in the 1980s on evolutionary relationships within a genus of woody legumes called *Coursetia*, traveling to Mexico, the Greater Antilles, Argentina, and Venezuela to collect specimens and, in the process, becoming interested in bean trees and shrubs in general. In the mid- to late 1990s, he was contemplating taxa of woody legumes that contain species on both sides of the Atlantic. In particular, he was wondering whether these distributions had been made possible by Cenozoic land bridges that at various times linked northeastern North America with Europe. The idea here was that when climates were relatively warm, tropical and subtropical plants could have spread northward and then moved, via normal, "garden-variety" dispersal, over these bridges from the Old to the New World or vice versa. When the climates cooled, forcing warmth-loving plants south again, the Old and New World lineages would have become widely separated, a classic case of vicariance.* In a paper published in 2000, Lavin and several colleagues argued

---

* The land bridges themselves also eventually disappeared beneath the sea, but

for this scenario for a group called the *dalbergioid legumes*. In the same paper, they also claimed to have found "relatively few instances where over-water [that is, transatlantic] dispersal can be invoked."

I asked Lavin if his whole focus at that time was on vicariance, and he said, "Oh definitely, definitely," which means he was thinking like a lot of other people (though by no means everyone) at the time. Like many scientists with a vicariance mindset, he was interested in searching for common patterns, so he extended his work by "gleaning" the legume family tree as a whole for groups that he thought were especially old, old enough to have had their ranges pushed southward and, thus, fragmented by the cooling climate of the mid-Cenozoic. He turned up some good candidates, ones that, for instance, contained large subgroups in which all the species were entirely restricted to Madagascar, Mexico, or the Caribbean. That kind of geographic confinement seemed to indicate that these plants weren't moving around very much, which, in turn, implied that chance, long-distance dispersal was especially unlikely for them. The large numbers of species in these taxa also suggested to Lavin that they had been around for a long time. Pondering these groups, he thought, "Man, these things have to be old. If anything's old, it has to be these."

Although Lavin was focused on vicariance, he didn't share the notion of some of his colleagues that cladograms, representing the evolutionary branching relationships among groups, were the only definitive kind of evidence in historical biogeography—the only things that could be used to reject or support vicariance. Specifically, he thought that molecular dating could be useful, especially now that relaxed clock methods—ones that did not assume a constant molecular clock—had been developed. So he took DNA data from the legumes and ran them in one of the new relaxed clock programs to get estimates of the ages of separation of Old World and New World lineages. And something odd came out.

The ages were all too young. "We tried to bias 'em and make them old," Lavin said, " . . . by putting the fossil [used to calibrate the clock] on the crown instead of the stem node . . . or taking the oldest possible minimum age that the fossil could be, if that makes any sense." In other words, in trying to make the data support vicariance scenarios, Lavin and his colleagues played with the fossil calibration points to make the ages come out as old as possible, probably making them unrealistically old. But it didn't matter. From the vicariance point of view, nothing worked. Lavin

---

climate would have separated the Old and New World ranges of warm-adapted plants such as the woody legumes before the actual land connections were broken.

continued, "We could hardly make them old, we could not make them older than 20 million years or something," which meant that the evolutionary splits weren't old enough to have been caused by the "land bridge plus cooling climate" hypothesis.

In 2004, Lavin, along with eight other botanists studying legumes, published an extensive molecular dating study with more or less the diametrically opposite conclusion from Lavin's earlier paper on the dalbergioids. From DNA data on fifty-nine groups in the bean family that had distributions broken up by oceans or seas—occurring, for instance, in Africa and Middle America—they found only eight that were old enough to have spread via land bridges (never mind spreading within an unfragmented Gondwana, which was irrelevant for *all* the groups). The other fifty-one likely dispersed from one area to the other by crossing seas or oceans—the Atlantic, the Caribbean, or the Mozambique Channel between Africa and Madagascar. They were fifty-one strands in the green web. In just a few years, Lavin had gone from thinking that for woody legumes there were "relatively few instances where over-water dispersal can be invoked" to concluding that most cases of distributions broken up by oceans *had* to be explained in that way. It was a major shift in thinking for him about a group of plants he'd been studying for years.

More generally, Lavin's whole view of biogeography had changed, essentially because of the molecular clock. Where before he had "definitely" been thinking in the vicariance mode, he now sees things as fundamentally about rates of dispersal, with those rates often being much higher than people had previously imagined. For vicariance biogeography, he predicts "a long slow death," and he clearly sees his new dispersal-oriented view as one in which the pieces have fallen into place. After he had finished the big study on legumes, for instance, he was at the Royal Botanic Garden in Edinburgh and went to an informal talk on the Araucariaceae, a group of conifers that occurs on several Southern Hemisphere landmasses and is considered a prime example of Gondwanan vicariance. The speaker was showing an evolutionary tree for the group, and it struck Lavin that some of the branches were very short, meaning there were hardly any genetic changes in those parts of the tree. The rest of the audience seemed to want to interpret this as a drastic slowdown in the rate of molecular evolution; after all, that was what was required to push those evolutionary branching points deep enough into the past to have been affected by continental drift. Lavin, however, was thinking of a completely different explanation. If you weren't wedded to the notion of Gondwanan vicariance, you didn't need to imagine a precipitous decline in the rate of genetic change. You

7.2 Matt Lavin. Bean plants and molecular clocks changed his view of biogeographic history.

could instead infer that some of the branching points in this evolutionary tree were relatively young. And, in that case, what the data were saying was that Araucariacean conifers, like Lavin's woody legumes, hadn't just stayed put while the Earth and its environments fragmented, but had sometimes found their way across ocean barriers.

Those fifty-one ocean-crossing legumes that Lavin and his colleagues uncovered turned out to be just one wave in a massive flood of evidence—mostly from molecular studies—addressing that big question in biogeography, "What explains distributions broken up by oceans?" Since the early 1990s, there have been studies supporting oceanic dispersal by plants of grasslands, deserts, dry shrublands, temperate forests, and tropical jungles; by grasses, herbs, shrubs, and tall forest trees; by plants in dozens of different families and with seeds of all sizes, dispersed by wind, water, or animals. An especially prolific German botanist named Susanne Renner, on her own and with a variety of colleagues, has carried out molecular dating studies that reveal many dozens of ocean crossings by plants.* In a particularly striking 2009 study, Hanno Schaefer, Christoph Heibl, and Renner (all then at the University of Munich) found evidence for some forty ocean crossings in the single family Cucurbitaceae (cucumbers, squash, and relatives) (see Figure 7.3). Many cucurbits form gourds that will float, an obvious means of ocean transport, and others have barbed fruits that could latch onto bird feathers, so it isn't surprising that they are proficient at crossing ocean barriers. Still, before this study, few botanists would have guessed that members of this family had made forty successful long-distance ocean journeys.

The collection of recent studies still represents only a fraction of the world's plant groups, but a coherent picture is nonetheless emerging from it. To see that picture more clearly, it's appropriate to start with New Zealand, a hub of biogeographic studies, and then move outward to encompass all of Gondwana.

"Explain New Zealand . . . "

We left off the story of New Zealand in 1994, with Mike Pole having bucked the tide of vicariance thinking by claiming that the country's native plants generally are *not* Gondwanan relicts. Pole had argued instead that most, maybe even all, New Zealand plant species are descended from ancestors that arrived from over the ocean, most of them long after Zealandia broke away from Antarctica/Australia. Plants could also, of course, have dispersed *from* New Zealand to other landmasses. Applied to the general issue of ranges broken up by oceans, Pole's view meant that plant taxa found on New Zealand and elsewhere had these piecemeal distributions because they had traveled over water either to or from New Zealand (or both to *and* from New Zealand). As we have seen, his conclusions were based mostly on the turnover of plant groups in the fossil record and on the existence of living "Gondwanan" plants on oceanic islands

---

* Renner, coincidentally, was in the Smithsonian group, mentioned earlier, that was stranded for ten days on the Cerro de la Neblina, the tepui where *D. meristocaulis* grows. Even though she was pregnant at the time, it apparently wasn't a big ordeal for her. "We had plenty of sugar and coffee," she says. (Renner, email to the author, October 4, 2011.)

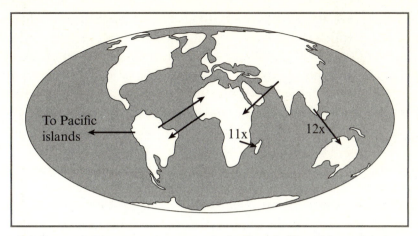

7.3 Some of the many overwater colonizations indicated by a timetree of the cucumber family. The land configuration is for the Mid-Miocene, about 14 million years ago, and all the colonizations shown are estimated to have happened in the Miocene or later; "11x" and "12x" represent 11 and 12 independent dispersal events. Redrawn and modified from Schaefer et al. (2009).

like Norfolk and Lord Howe, places that could only have been colonized by overwater dispersal. As I mentioned in Chapter Four, Pole did refer to molecular dating work that supported his case, but at the time such studies were few and far between. However, since then, as described in Chapters Five and Six, molecular dating has exploded, and a disproportionate number of these studies have dealt with New Zealand, probably in part because the belief persists that if one can explain New Zealand, the biogeography of the world will fall into place around it.

To evaluate the importance of dispersal versus vicariance for the New Zealand flora, we basically need two lists: one for cases in which the split between a New Zealand lineage and its nearest relatives on another landmass is estimated as too recent to be explained by continental drift, and the second for cases in which the split could have occurred at or before the time of fragmentation. One can think of the first list as the set of ocean voyagers—at least one ocean journey to or from New Zealand is required to explain the distribution of each of these groups—and of the second list as the set of Gondwanan relicts. Graham Wallis and Steve Trewick, biogeographers working in New Zealand, recently compiled those exact lists based on published molecular studies. Their numbers show a striking pattern.

On the first list, there are forty-one cases, representing, minimally, forty-one ocean journeys either to or from New Zealand.* (The number is actually considerably larger because some of these cases represent multiple dispersal events.) In most instances, the age estimates strongly favor overwater dispersal, because these estimates are not just a little too young to be explained by Gondwanan vicariance, they are far, far too young; for the majority of them, the ages fall within the past 30 million years, some 50 million years after Zealandia separated from Antarctica/Australia. Additionally, almost all of them represent dispersal to rather than from New Zealand, and the majority involve colonizations by Australian species.

On this list of overwater colonists are many of New Zealand's most abundant and conspicuous plant taxa, ones I remember seeing there even though I was mostly watching birds. They include diverse shrubs and small trees in the genus *Pittosporum*; two lineages of podocarp conifers; the world's largest buttercup species (the one Tara and I saw at Arthur's Pass on the South Island), along with all the other New Zealand buttercups; the many species of *Celmisia* daisies; the ubiquitous, scaly-leaved *Hebe* shrubs; *Sophora* bean trees with showy yellow flowers; and two lineages of southern beeches (*Nothofagus*). Basically, anyplace in New Zealand with lots of native plants is home to many taxa on this list of long-distance colonists.

In addition, more general molecular dating studies for flowering plants imply that the catalog of ocean voyagers must contain many other groups as well. We know this because these dating studies indicate that many *entire* plant families that include New Zealand species are too young to have been around at the birth of Zealandia, and this means that these groups had to cross the Tasman Sea or some other part of the Pacific to get there. Conservatively, there are about three dozen such families, probably encompassing hundreds of cases of oceanic dispersal. One of these young groups, the Asteraceae (sunflowers), is easily the most diverse plant family in New Zealand and by itself represents perhaps several dozen overwater colonizations, only a few of which were included in Wallis and Trewick's list of ocean voyagers. To think of that particular case in another way, consider that there are 24 native genera and some 240 native species of

---

* Crisp et al. (2009) concluded that there have been more than one hundred overwater colonizations of New Zealand by plants. However, since I could not obtain comparable numbers for relict groups from that study, I am using the smaller compilation from Wallis and Trewick (2009). I tallied seed plants from the Wallis and Trewick study but excluded ferns and mosses, as these plants are generally seen as having great dispersal abilities and I did not want to "inflate" the dispersal list.

sunflowers in New Zealand, and that not one of them is a Gondwanan holdover. They are all simply too young. The same goes for all the species in the mustard family, the pink family, the gentian family, the mint family, the mallow family, and many others.

The ocean voyager list even includes some members of two groups that have been viewed as "obviously Gondwanan," the podocarp conifers and the southern beeches. If even these iconic Gondwanan trees are not relicts, one may well wonder what New Zealand plant lineages *are* on the second list, the one for piecemeal distributions caused by continental drift. It turns out that in Wallis and Trewick's compilation, there are exactly four, including one podocarp lineage and that other emblem of Gondwana, the giant kauri tree (the lone New Zealand member of the Araucariaceae), although the ages for the latter vary widely and are only consistent with vicariance when certain debatable fossil calibration points are used. In any case, even if the kauri is counted as a relict species, the studies, to date, suggest there are about ten times as many overwater colonist lineages as Gondwanan holdover lineages. And that ratio is likely to become even more skewed in the future, because the old lineages have likely attracted a disproportionate amount of study up to this point. In short, the evidence from molecular dating clearly supports Mike Pole's contention that the flora of New Zealand overwhelmingly derives from long-distance colonists.

That conclusion is startling, given the earlier belief that New Zealand is a prime example—maybe *the* prime example—of the dominance of vicariance, a so-called "Moa's Ark" on which Gondwanan lineages have been transported by the engine of seafloor spreading. That thought takes me back to our 2006–2007 visit to New Zealand, to an evening boat trip from the town of Oban on Stewart Island to an isolated peninsula, where we would see a kiwi picking sandhoppers (amphipod crustaceans) off a sandy beach. On the boat, about ten of us were crammed together in the dark on wooden benches just behind the captain, and among the other passengers was a couple, both fisheries scientists at a university in Wellington, both obviously sharp people. At some point, they were remarking on the primordial nature of New Zealand forests, the old story of Gondwana. As an outsider, I felt uncomfortable about correcting them and kept my mouth shut. Also, at the time maybe I wasn't quite sure myself. Now I am. That view of the New Zealand biota as ancient and Gondwanan is clearly wrong, at least for plants. Instead, it makes more sense to think of New Zealand as an obvious long-distance destination (and, to a lesser extent, a starting point) in the web of relatively recent ocean crossings.

"... And the World Falls into Place Around It"

Most scientists spend a good deal of time reading published papers (and manuscripts before they are published), and at some point, for many, this becomes a chore as much as a pleasure. There are various symptoms—you read fewer and fewer entire papers, and instead merely skim the summaries; you turn down requests to review manuscripts; if you have an anti-institutional streak, as I do, you cut off your subscriptions to journals (they're almost all online now anyway, but somehow their greater availability makes you read less rather than more). You find, above all, that even the "take-home message" of most papers gets lost somewhere in the dim neuronal passageway between short-term and long-term memory.

Every once in a while, though, you read something that rises above the gray miasma, that shakes you out of slumber, something that seems to make the ground shudder a little beneath your feet. Instead of glancing at the summary and forgetting what it said as soon as you turn the page, you read the entire paper and then find yourself going back to it again and again, picking up nuances you missed before. In the course of researching the subject of disjunct distributions, I've been lucky enough to read several papers or passages like that. The long fourth chapter of Lars Brundin's midge monograph, for instance. A paper by George Gaylord Simpson on the mechanics of dispersal. Mike Pole's article on the origins of the New Zealand flora.

Another was written by a young Spanish biogeographer named Isabel Sanmartín and a Swedish scientist, Fred Ronquist, who are both known for devising quantitative biogeographic methods. Out of all the biogeographic papers written in the past twenty years, this one, published in 2004, may have produced the biggest shudder of all.

The details of Sanmartín and Ronquist's approach are complex, but basically they were trying to see if the order of branching in evolutionary trees for Southern Hemisphere taxa matched the sequence of breakup of Gondwanan landmasses.[*] In that sense, they were doing what Lars Brun-

---

[*] In a nutshell, they tried to come up with the best combination of vicariance (splitting of lineages because of splitting of areas), dispersal, and extinction of lineages to explain, for each group, how the particular evolutionary tree, with species found in the observed geographic areas, could have come about. Each vicariance, dispersal, or extinction event was assigned a cost, and the combination of those events that explained the evolutionary/geographic pattern and had the lowest total cost was considered "best." Their method may have been biased to choose vicariance

din had done in the 1960s with his chironomid midges, and what Gary Nelson and others had advocated in the 1970s and 1980s as *the* way to approach the geographic history of life. By including as many groups as they could, Sanmartín and Ronquist also were following in the footsteps of Léon Croizat, who, for all his looniness, had admirably pushed the idea of looking for generalities rather than focusing on the histories of single taxa. But Sanmartín and Ronquist were (and are), in a significant way, not at all like Nelson or Croizat. In particular, they were open to the idea that long-distance dispersal might be important. They wanted to find out what evolutionary trees of Southern Hemisphere groups really had to say.

For plants, what Sanmartín and Ronquist found is that, by and large, those evolutionary trees do *not* match the order of Gondwanan fragmentation. For instance, the closest relatives of plants in Australia tend to be found in New Zealand, whereas continental breakup predicts that Australian plants should be closest to those in southern South America. Similarly, Madagascan plants tend to have their closest relatives in Africa, yet those two landmasses actually separated very early and, thus, the vicariance explanation predicts only distant relationships between the groups that live on them; Madagascan plants ought to be closer to Indian ones, and African plants closer to those in South America. In short, the results do not support a Gondwanan vicariance explanation as a good generalization for plants; the expected evolutionary links, for the most part, are not there. Sanmartín and Ronquist concluded instead that oceanic dispersal has been frequent, especially over the Tasman Sea between Australia and New Zealand and across the Mozambique Channel between Africa and Madagascar, and even, in some cases, over much longer distances, such as the entire Indian or Pacific Oceans. And, if anything, their results were somewhat biased against finding dispersal, because they left out plant groups found widely in both hemispheres, that is, the kinds of plants most likely to move long distances. In other words, the deck was probably stacked to find support for vicariance, and yet that hypothesis still did not hold up.

Sanmartín and Ronquist's study was significant for its generality, reminiscent of Croizat, and for the fact that it had essentially used the approach of the vicariance scientists to refute vicariance. Like the cladists, Sanmartín and Ronquist had used only the branching *order* in

---

scenarios over dispersal scenarios, because vicariance events were given a cost that was much less than that of dispersal events, with the result that scenarios that relied on vicariance tended to be preferred.

evolutionary trees, not the ages of branches, and that was important, because it meant that even people who weren't sold on molecular dating could believe the results. In short, it was a broad study about the biogeography of the Gondwanan landmasses that almost everyone had to take seriously. If there was one paper that marked the shift in thinking in the field, the watershed moment, this was it. An Australian botanist named Michael Crisp, who has done some of the key biogeographic studies on southern beeches and other Southern Hemisphere plants, described Sanmartín and Ronquist's study as signaling "the last great gasp of the vicariance paradigm."

Crisp and his colleague Lyn Cook provided some complementary evidence, from molecular dating studies, on the origins of Southern Hemisphere plants in a 2013 review of the flora of Australia. Interestingly, for one subset of Australian plants, those that have their closest relatives in South America, they found a strong signal of Gondwanan vicariance. Specifically, in fourteen out of twenty-one such cases, the branching points were old enough to suggest that the plants in question had persisted in South America and Australia since before their connection through Antarctica was severed some 30 million years ago. (At that point in my immersion in the botanical literature, I was actually shocked to see such a clear indication of Gondwanan relicts.) However, the rest of the Australian flora reflected a story like New Zealand's, or like Matt Lavin's bean plants: there were twenty-eight Australian plant taxa with sister groups on other Gondwanan fragments (six in New Caledonia, eight in New Zealand, and fourteen in Africa), and all twenty-eight had divergence ages too young to be explained by the breakup of Gondwana.

To sum up, evolutionary branching patterns, as in the Sanmartín and Ronquist study, and a large and rapidly growing number of molecular dating studies indicate that Gondwanan fragmentation and vicariance in general cannot explain most plant distributions broken up by oceans. Instead, it has become increasingly clear that over tens of millions of years, the collective flora of the Earth has jumped ocean barriers willy nilly, probably often in the form of seeds—blown on the wind, on the feathers or feet or inside the guts of birds, on rafts of vegetation, or simply floating on the water. Modest barriers like the Tasman Sea have apparently been surmounted hundreds, if not thousands, of times, and even enormous ones like the Atlantic and Pacific have been successfully crossed by numerous plant lineages.

In Matt Lavin's eyes, many plants, given some thousands or millions of years, can move about the Earth rather easily, undeterred by apparent

7.4 Isabel Sanmartín. Her study, with Fred Ronquist, on the biogeography of the Southern Hemisphere opened a large crack in the vicariance worldview.

geographic barriers. "I don't see the formation of a mountain range or a river or the separation of continents . . . I don't see those sorts of historical events as imposing any sort of barrier to migration for plants," he says, expressing a view that would have been foreign to him earlier in his career. "Plants seem to get around. It's a question of whether they're going to find opportunity in the new area they land in." Quite often, the opportunity has been there, as it was for many of the woody legumes that, after crossing a sea or ocean, managed to find a hot, dry region with a pronounced rainy season, that is, an environment similar to the one from which they had come. The ultimate result of such colonizations is the world of the green web. If one began with a vicariance perspective, it's a world turned upside-down.

The Plant Dispersal Thread

A conversion from vicariance to dispersal, like the one Matt Lavin experienced, was a common outcome for botanists as the molecular dating results accumulated. However, it would be misleading to suggest that transformations like his were close to universal. There are, for instance, some botanists who remain unconvinced by the new results, and are steadfastly committed to the vicariance view, people like Michael Heads. In addition, a fairly large contingent of botanists didn't need to be convinced of the great importance of long-distance dispersal, because they already believed in it. This latter group formed part of an intellectual tradition that can be traced back to the origins of an evolutionary view of biogeography, back to Darwin himself.

"I must now say a few words on what are called accidental means, but which more properly might be called occasional means of distribution." That's Darwin, in *The Origin of Species*, introducing the section on what we would now call chance, long-distance dispersal. The reader expecting tales of ballooning spiders, rafting iguanas, or swimming elephants, however, will be disappointed. Darwin probably figured that, if he was going to run through just a few examples, it would be best to stick with ones that his skeptical readers wouldn't find far-fetched, ones that he himself found convincing. And so, he continued, "I shall here confine myself to plants." In fact, he pretty much confined himself to seeds. He went on to describe the likelihood of small seeds being transported on the feet of birds, observations of seeds lodged behind stones embedded in the roots of floating trees, the germination of seeds collected from bird excrement, the possible oceanic dispersal of seeds on icebergs, and, of course, his famous seeds-in-seawater experiments.

Darwin's emphasis on plants for his examples of long-distance dispersal is part of an intellectual thread that runs without any complete break to the present. Basically, since Darwin, there have always been people who believed that plants have often made successful ocean crossings and other long-distance journeys. It was a belief that persisted even when intellectual fashions went against it. Perhaps it did so because it was derived in fairly straightforward ways from botanical knowledge.

Part of the belief came from simply thinking about the properties of plants and, especially, of seeds. Those properties ought to make plants proficient long-distance dispersers, much better, for instance, than nonflying vertebrates. Most flightless vertebrates, with their need for fresh water and vulnerability to overexposure and drowning, almost require a substantial raft to cross a wide stretch of ocean; the tortoise that apparently floated from Aldabra to the African coast is a major exception, not the rule. Many seeds, in contrast, have been molded by natural selection to be borne on the wind, by birds, or in water itself, as in the case of gourds and other plants with air-filled fruits or seed capsules. Also, and crucially, even seeds that are not particularly adapted for wide dispersal can lie dormant for long periods, "awakening" to germinate when they encounter the right conditions; seeds are essentially in suspended animation, enabling successful long-distance journeys to occur in a multitude of ways. To an extent, arthropod eggs—many of which are laid in the fall and lie dormant over the winter—have the same advantage, but they are generally more susceptible to being killed by drying out or by exposure to heat or cold than are plant seeds. Arthropod eggs that are well designed for prolonged dormancy—like the ones from "sea-monkeys" (a kind of brine shrimp) that you can leave in an envelope for years and then revive in salt water—are fairly unusual. Among plants, that kind of dormancy is very common, which is why it seems totally unremarkable that you can buy many kinds of seeds in packets at the local nursery and then let them sit around for months or years before planting them. The results of the seed survival experiments carried out by Darwin and later researchers were striking demonstrations of how this property could facilitate dispersal.

Apart from the obvious properties of seeds, there were other observations that convinced Darwin and his intellectual descendants that chance, long-distance dispersal by plants is especially common. For one thing, plants have populated oceanic islands more readily than animals. For instance, the proportion of the world's plant lineages that have colonized the remote Hawaiian Islands is far greater than the proportion for any group of land animals, even insects. (The number of insect lineages that have reached Hawaii is somewhat larger than the number of plant lineages, but worldwide there are far more insect than plant species to draw from as potential colonists.) In addition, the fossil record implies that many flowering plant groups with distributions broken up by oceans are too young to have gone through ancient events such as Gondwanan breakup, thus implicating recent ocean crossings. This observation of groups being too young, now based on molecular data, is obviously critical today, but fossil evidence was used to make the same argument long before the molecular dating explosion.

The perceived importance of long-distance dispersal by plants waned (but did not disappear completely) with the rise of land-bridge explanations, and waxed with the general spread of the dispersalist thinking of the New York School. Then came the revolution—the validation of continental drift and the rise of vicariance biogeography—whereupon many botanists were *still* arguing for the great significance of ocean crossings and other dispersal events. For instance, in the early 1990s, with both plate tectonics and vicariance biogeography well established (but before the flood of molecular dating studies), plant experts contributing to a book called *Biological Relationships Between Africa and South America* frequently invoked ocean crossings between those continents. None of these botanists denied the fact of plate movements, but they apparently hadn't absorbed the message that continental drift plus cladograms had to change their whole worldview.

In short, since Darwin's time, knowledge of the properties of plants, their distributions, and the fossil record has been sufficient to convince many botanists that long-distance dispersal by plants is frequent and important. In the past fifteen to twenty years, these were the people who did not require conversion; they didn't need it, because they were already there. Susanne Renner, who has probably done more than anyone else in recent years to convince people of the ubiquity of plant dispersal, was one of them. Her graduate school adviser, Klaus Kubitzki, had a strong belief in the importance of long-distance dispersal, and from him she picked up that conviction and never let it go. During a stint as a postdoctoral researcher at the Smithsonian Institution in the 1980s, she was in contact with the other side, the vicariance biogeographers, but she says she always found their extreme views "dogmatic and a bit silly." For Renner and others, molecular dating simply provided the final corroboration of what they already strongly suspected was true. When it came to plant dispersal, Darwin's long shadow had never come close to being erased, even if it had at times grown faint.

On to Animals

When Isabel Sanmartín and Fred Ronquist examined plant studies, they found that the branching order of evolutionary trees for Southern Hemisphere groups generally did not match the sequence of Gondwanan fragmentation. However, they also examined animal groups—mostly vertebrates and insects—and for these, the story was very different. Evolutionary trees of animals *did* tend to match Gondwanan breakup; for instance, when a group was found on New Zealand, Australia, and southern South America, the New Zealand branch tended to be outside of a lineage that included both the Australian and South American branches,

7.5 Susanne Renner. She didn't have to be converted to a belief in the importance of dispersal; she was already there. Photo by Pierre Taberlet.

presumably reflecting the more recent land connection between those two areas. (That particular pattern was the one found by Lars Brundin for the chironomid midges, way back in the 1960s [see Figure 2.5]). In fact, Brundin's study was included in Sanmartín and Ronquist's compilation.)

There are problems with jumping from these results to the conclusion that Gondwanan vicariance is a good general explanation for piecemeal distributions of Southern Hemisphere animals. Recall, for starters, that Sanmartín and Ronquist's study excluded groups found widely in both hemispheres, thus perhaps skewing the sample away from animals that are good dispersers, that is, exactly the kinds of animals that might cross oceans. There are other methodological issues as well, such as that the particular algorithm they used may be biased to support vicariance over dispersal, and that, since the ages of branching points were not used, some cases that seem consistent with vicariance might turn out not to be. Also, there are conspicuous exceptions to the pattern. For instance, animal lineages on New Caledonia and New Zealand typically are not closely linked, despite the fact that they were recently connected as parts of Zealandia.

Nonetheless, the results do suggest that animal distributions reflect Gondwanan vicariance more so than do plant distributions. Obviously, both plants and animals must have been carried as passengers on the moving fragments of Gondwana. The difference seems to be that the signature of continental drift for plants has been obscured by subsequent ocean crossings (and by extinction), whereas the drift signal has remained more nearly intact for animals. In short, the green web has overwritten the ancient pattern of continental breakup. Subsequent work indicates that this distinction holds for the Northern Hemisphere as well, with movements between Eurasia and North America being much more common for plants than for animals (although some of these intercontinental movements may represent normal dispersal over land bridges).

The conclusion that animals don't get around nearly as easily as plants do is a fundamental, if unsurprising, inference.* However, this doesn't mean that animals have an insignificant role in the story of chance dispersal. While plants show the importance of long-distance journeys through sheer weight of numbers, recent studies of animals also have something to teach. As I will show in the next two chapters, what they reveal is that, in the long course of evolutionary time, very strange things happen, stranger than fifty-one ocean-crossing legumes, stranger even than an Australian sundew colonizing a South American tepui. Ultimately, what animals show in a particularly clear way is that the history of life is extremely serendipitous and unpredictable. A frog, for instance, might make a journey that no frog seems to have any business making.

---

* This dichotomy between plants and animals is obviously an oversimplification that obscures a lot of variation in dispersal ability within those groups. For instance, some animals, such as certain kinds of dragonflies, spiders, and birds, are very adept at long-distance dispersal, and are thus "plant-like" in that sense.

Fossils of elephants have been found on many islands, including Malta, Crete, Cyprus, Sulawesi, Flores, Timor, Okinawa, and the Northern Channel Islands off California. Through the 1970s, many researchers surmised that the existence of elephant fossils meant that these islands must have been joined to continents by land bridges at some point; otherwise, how could a large animal like an elephant have reached them?** Such land connections were sometimes assumed even when nothing in the geological record seemed to indicate their existence.

However, in 1980, a biologist named Donald Johnson, as part of his doctoral dissertation, compiled a large number of reports indicating that both African and Indian elephants are very good swimmers, thus calling into question the need for land bridges to explain the island species. Elephants apparently swim with a porpoise-like, lunging motion, using their trunks as snorkels, and can do so for hours at a time. One observer recounted how elephants in a Cambodian lake, being hunted for use as work animals, swam "wildly in every direction, their glistening black heads and bodies emerging like those of sperm whales," and noted that they could swim for several hours or more. These elephants were harpooned through the ears and chained to the tops of trees emerging from the flooded landscape, producing a final spectacle of "elephants swimming in circles round every tree over a large area."

Johnson found many firsthand accounts of elephants swimming to islands very close to the east African and Sri Lankan coasts. Another report had seventy-nine elephants traversing parts of the very wide Ganges River, in one case swimming for six hours, resting on a sandbank, and then continuing for three hours more. Using swimming speeds taken from other accounts, Johnson estimated that these seventy-nine elephants covered more than five miles during the six-hour segment of their Ganges journey. The record distance for a swimming/floating elephant was considerably farther; an individual lost overboard from a ship off the South Carolina coast managed to reach land thirty miles away, in a heavy gale no less.

** Most of these island elephants were pygmy species that probably evolved to be smaller in size after colonizing the islands. They represented instances of the "island rule," whereby small mammals, such as mice, tend to get larger on islands, and large mammals, such as elephants, tend to get smaller.

Chapter Eight

A FROG'S TALE

The "Yellow Snake" of São Tomé

John Measey was looking for the cobra bobo. His six-foot-two frame was folded onto a tiny 125 cc motorbike, knees up to the handlebars, and he was riding around in the rainforests and cocoa plantations of São Tomé, an island off the west coast of Africa. The place seemed like paradise, with stunning forested mountains, beaches with turquoise water, and friendly locals who would engage Measey in philosophical discussions late into the night, despite his fractured Portuguese (Brazilian Portuguese at that, quite different from the Portuguese creole spoken on the island). And, best of all, there was the possibility, the *certainty*, according to Bob Drewes, of finding not just one but many cobras bobos.

Measey, a young British biologist who was working at the Institut de Recherche pour le Développement (IRD) in Paris, had talked to Drewes at a scientific meeting in Kenya in April 2002. Drewes has been a curator of herpetology at the California Academy of Sciences in San Francisco for more than thirty years, but, despite his job location, is one of the world's leading authorities on African reptiles and amphibians. He had recently begun a massive project on islands in the Gulf of Guinea, especially São Tomé and Príncipe (the two islands making up the country of the same name), bringing together experts on amphibians, reptiles, fishes, mammals, insects, spiders, sea slugs, corals, barnacles, echinoderms, fungi, diatoms, and plants to catalog the biota of the islands and, down the line, study its evolution. São Tomé and Príncipe rival the Galápagos in their

8.1 John Measey. A legless, burrowing amphibian called the cobra bobo lured him to São Tomé and Príncipe and, eventually, to a study of chance, overwater dispersal.

assemblage of unique species, from the world's smallest ibis to the world's largest begonia. They do not rival the Galápagos in how well they've been studied, however; before Drewes and his colleagues started their work, very little research on the islands' biota had been done since the nineteenth century. Correcting this oversight had become Drewes's passion, a "mission from God," as he says.

At the meeting in Kenya, Drewes lounged by the hotel swimming pool, beer in hand, and talked for hours about São Tomé and how easy it was to find the cobra bobo there, applying some of his missionary zeal to convince Measey to make a trip to the island. *Cobra bobo* means "yellow snake" in Forro, the island's Portuguese creole, but the animal in question is actually a caecilian, a kind of legless, burrowing amphibian. That was the weak spot that Drewes was counting on: a snake would not have drawn Measey to São Tomé, but a caecilian, now that was something else entirely. Measey has a passion for caecilians and is one of a handful of biologists in the world who specialize in studying these obscure amphibians.

After the meeting, Drewes continued his assault on Measey with emails about the cobra bobo. Eventually, maybe inevitably, Measey gave in and booked a flight to São Tomé. So there he was, six months after the Kenya meeting, puttering around on his undersized motorbike and making long treks on foot through the forest. He had a few troubles—he came

8.2 Bob Drewes, director of a massive study of the unique biota of São Tomé and Príncipe, with a cobra bobo in hand. Photo by Dong Lin. Courtesy of California Academy of Sciences.

down with an especially serious form of malaria, and his field assistant was briefly arrested for disobeying a police order—but he was not disappointed. Almost everywhere he went, he found the cobra bobo, *Schistometopum thomense*, which, being bright yellow, stood out against the dark soil. To most of us, these animals would look like giant earthworms, admittedly strange, but probably not worth a special trip to see. Measey, however, sounding like a lovestruck teenager, says he "had eyes for nothing else." There's a common frog on São Tomé, *Ptychadena newtoni*, that he would eventually have great reason to remember, but on this trip, *P. newtoni* and the other amphibians barely registered.

Measey set up bags filled with earth and added some cobras bobos and soil invertebrates to see what effect the caecilians would have on populations of their prey, but all but one of the caecilians escaped, ruining the experiment. Measey didn't seem to care that much. Just seeing so many cobras bobos was enough.

When Measey got back to Paris, he couldn't get something out of his mind, and it wasn't the image of dozens of bright yellow worm-like amphibians slithering on the dark earth, or at least it wasn't just that. It was the fact that São Tomé and Príncipe even have amphibians at all, including the cobra bobo. How in the world did they get there?

To understand why this question intrigued John Measey, we first have to consider the deep history and geography of the Gulf of Guinea islands. That history begins with a series of volcanoes, called the Cameroon Line, running roughly southwest to northeast and crossing both the Equator and the west coast of Africa. Over the past 80 million years, these volcanoes have poured out lava at various regions along the line, now in one place, now in another. On the continent, that lava formed a discontinuous arc of highlands (some still containing active volcanoes), while, in the Gulf of Guinea, the lava piled up to form a set of four islands. Bioko is the largest of the four, the farthest north, and the closest to the African coast. Several kinds of amphibians live there, but their existence on the island is no great mystery. Bioko is separated from the mainland by a channel less than two hundred feet deep, so, during various ice ages, sea level dropped far enough to connect the island to the mainland; amphibians could have walked, hopped, or slithered over to the island at those times. The island on the southwestern end of the line, Annobón, presents no problem for amphibian dispersal either, because it doesn't have any amphibians.

The middle islands in the chain, Príncipe and São Tomé, are the enigmas for Measey and other amphibian biologists. These two islands are more than 130 miles from the African coast and are separated from the continent by seas more than a mile deep. They are true oceanic islands, meaning that, since their emergence from the ocean—some 31 million years ago in the case of Príncipe and about 13 million years ago for São Tomé—they have never been connected to the mainland. Amphibians had to cross that saltwater barrier to get to these islands, and therein lies the problem: amphibians are not supposed to be able to disperse that far across salt water. At least that was what Measey thought until he ran into the amphibians of the Gulf of Guinea islands. That was what Charles Darwin, among many others, had thought also.

"Batrachians (frogs, toads, newts)," Darwin said in *The Origin of Species*, "have never been found on any of the many islands with which the great oceans are studded." Darwin made this statement at a time when many British intellectuals had rejected the story of Noah's Ark (and its implication that all species had dispersed from Mount Ararat) and instead believed that God had placed each species in its native range, an area where that species was especially fit to live. Buffalo were fit for roaming the plains of North America, tigers for stalking Asian jungles, and amphibians, well, each amphibian species was fit to live wherever it occurred naturally. Under this view, Darwin argued, the absence of amphibians from oceanic islands was a paradox. Why, for instance, should continental rainforests be teeming with frog species, whereas apparently similar island forests had no native frogs at all? One couldn't argue that the island environments were unsuited for frogs, because many frog species flourished when introduced on islands. Instead, to rescue the "special creation" view, one had to imagine a quixotic God conjuring up continental amphibians left and right but deciding that these creatures just didn't belong on remote islands. However, the absence of amphibians from oceanic islands made perfect sense if such islands were originally devoid of life and had to be colonized naturally from continents. Amphibians aren't out there for a simple reason: they can't get there. This was and remains a compelling argument for a naturalistic explanation of the distributions of living things. It's also one step in a powerful argument for evolution, because, if island forms (all those various nonamphibians) originally came from continents, and yet are obviously distinct from any continental species, then descent with modification is implied, either of the island forms, the mainland forms, or both.

But I'm digressing—although almost unavoidably, since it would be difficult to quote Darwin's observation about "batrachians" without explaining the context. What we're really interested in here is not Darwin's argument for evolution, but one of the assumptions of his argument, namely, that amphibians are hopeless ocean voyagers, so dismal at making ocean journeys that all of the "many islands with which the great oceans are studded" are beyond their reach. In *The Origin*, all Darwin said by way of explaining this claim was that "these animals and their spawn are known to be immediately killed by sea-water," a statement that isn't strictly true but was good enough for his purposes. There is probably no amphibian that could survive a journey from, say, the West African coast to São Tomé or Príncipe if it were simply floating in the ocean. Still, there is a bit more to say about amphibians as poor oceanic dispersers.

The problem, in a nutshell, is that the skin of most amphibians is very permeable to water. My adviser in graduate school, Harvey Pough, a herpetologist and physiologist, was fond of saying that a frog placed on a table would dry out as fast as an uncovered bowl of water. This means that for most amphibians, a long voyage on a dry raft is out of the question. Many kids who keep frogs and salamanders as pets (as my brothers and I did) have experienced the vulnerability of these creatures firsthand; if their pet gets out of its moist terrarium, it will often be found, perhaps only a couple of days later, as a dried-up husk of an amphibian. (The husk I particularly remember was a California newt.) Drifting in the ocean itself or on a raft with ocean waves breaking over it is no good, either, because, when immersed in salt water, most amphibians rapidly lose water by osmosis across the skin. In other words, like air, salt water quickly dehydrates them.

There are amphibians on islands, to be sure, but for the most part, they seem to be the exceptions that prove the rule that these creatures are poor ocean travelers. If the amphibians are native, the islands are ones like Madagascar, Borneo, or the Seychelles, which were once connected to a continent. If the islands were never connected to a continent, the amphibians turn out to be introduced species. For instance, Hawaii has a poison-dart frog, a giant toad, and a handful of other amphibian species, but they were all introduced by humans, some to control insects and others accidentally. In short, when you find amphibians on an island, it's usually easy to explain how they got there without having to invoke an implausible natural ocean crossing.

As John Measey had learned, however, São Tomé and Príncipe seem to be real exceptions—they are islands that have never been connected to a continent and yet have amphibian species that are apparently native. The existence of these species contradicts Darwin's confident assertion—there *are* native "batrachians" on some oceanic islands. More to the immediate point, they're an affront to vicariance biogeographers, who have used amphibians and other hopeless oceanic dispersers as indicators that vicariance *must* have occurred. Specifically, the argument goes, if you find related, native amphibians on landmasses separated by salt water, you can assume that there was once a land connection between those regions. The history of connection and disconnection might have something to do with the movement of tectonic plates (that's why there are frogs and caecilians on the Seychelles with relatives in India), or rising sea levels (that's why there are frogs and caecilians on Borneo with relatives in mainland Southeast Asia), but, one way or another, a formerly continuous distribution was broken up by the formation or spread of oceans or seas. In other words, the explanation for these cases is vicariance, not oceanic dispersal. However, if frogs, caecilians, or salamanders sometimes do cross saltwater barriers, then the assumption of vicariance, even in such apparently cut-and-dried cases, is not an entirely safe one. It would be yet another blow to the vicariance worldview.

Without planning it, Measey had found himself thinking about a key case in the debate over dispersal versus vicariance.

The case for the natural dispersal of amphibians to São Tomé and Príncipe wasn't entirely convincing though, not yet. In a way, the extremeness of the case was both its strength and its weakness, what made it both hard to deny and hard to completely believe. It was extreme because of the number of amphibian species involved—six frog species plus the cobra bobo, all of them unique to the islands and, in total, requiring five independent colonizations. This was a strength because it was difficult to imagine that all of these species existed but remained undiscovered on the mainland; therefore, it was unlikely that all of them had been introduced by humans. And yet, if it was hard to believe that even one amphibian species could have reached the islands naturally, a moderate handful of such dispersals sounded borderline absurd. From the start, Measey was thinking that natural overwater dispersal was the answer, but he also figured that he needed more evidence to strengthen the case.

Not long after returning from São Tomé, Measey was standing in the lunch line of the canteen at the IRD in Paris and got to talking about his trip with the director of the IRD's Bondy office, Alain Morlière. Measey (now speaking fractured French instead of fractured Portuguese) started telling Morlière about amphibians dispersing to the islands and, in particular, about the cobra bobo and the curious fact that no close relative of this species was known from the nearby West African coast. The nearest place where another specimen of the genus *Schistometopum* had been found was the eastern part of the Congo River drainage, closer to the Indian Ocean than the Atlantic. Measey was explaining that it was hard to imagine how a burrowing amphibian could get from the eastern Congo Basin to the Gulf of Guinea islands. Morlière is an oceanographer and, coincidentally, had done research on Rolas, a small island off the south coast of São Tomé. It's simple, he said, the plume of water coming out of the Congo River delivers lots of debris right to the Gulf of Guinea. I picture Measey's pupils suddenly dilating at that point. Fantastic! he thought. Here was the explanation, or at least the beginning of an explanation, to the mystery of how the cobra bobo and other species crossed to São Tomé and Príncipe. Natural rafts, perhaps clots of earth and vegetation dislodged from the banks of the Congo, could have been carried downriver to the ocean, then drifted north with the prevailing current to eventually make landfall on the Gulf of Guinea islands.

One can imagine Measey running to his office and immediately getting to work on a paper about the dispersal of amphibians to the islands. That's not what happened, though. Instead, he sent an email to some oceanographers who Morlière said would know more about the Congo plume and the currents off the west coast of Africa. Then he waited for the responses. Nothing happened. Nobody replied, probably because oceanographers, by and large, couldn't care less about how the cobra bobo and a bunch of frogs ended up on São Tomé and Príncipe.

Measey, stymied for the time being, went back to thinking about the effect of caecilians on their soil-dwelling prey.

Seeing the Unexpected

On the other side of Africa, a half-Spanish, half-German biologist named Miguel Vences was studying frogs on Madagascar, the Seychelles, and other islands in the Indian Ocean. Vences is one of those scientists who's involved in so many projects that one wonders when he finds time to sleep. Among many other things, he and his colleagues have discovered something like three hundred new species of frogs on Madagascar in the past twenty years. (Apparently he does have his limits, though, because so far *only* about a hundred of the new frogs have been formally described in scientific papers.)

For his doctoral dissertation at the Rheinische Friedrich-Wilhelms-Universität in Bonn, Vences studied the geographic origins of Indian Ocean frogs. When he began the project, he was thinking what almost any biogeographer would have thought about frogs on the Gondwanan islands of Madagascar and the Seychelles: frogs can't cross salt water, so these island species must be Gondwanan relicts. His aim was to work out the details of the vicariant origins of the frogs. Overwater dispersal wasn't even in his thoughts. Then a funny thing happened, the same thing that has happened to many biogeographers over the past twenty years or so: he started looking at the frogs and sequencing their DNA—that is, he started gathering *data*—and the frogs told him a different story. Actually it was several stories, but all with the same basic conclusion.

Vences's most straightforward study didn't focus on frogs on former pieces of Gondwana, but on the island of Mayotte, known for its white sand beaches, turquoise seas, and French ambience (Mayotte is a territory of France). The island lies some 190 miles west of the northern tip of Madagascar and, geologically, is part of the Comoros Archipelago. The Comoros are volcanic, with successive islands probably forming as the Somali tectonic plate slid eastward over a "hotspot," a plume of magma rising up from the Earth's mantle (the same kind of process that formed the

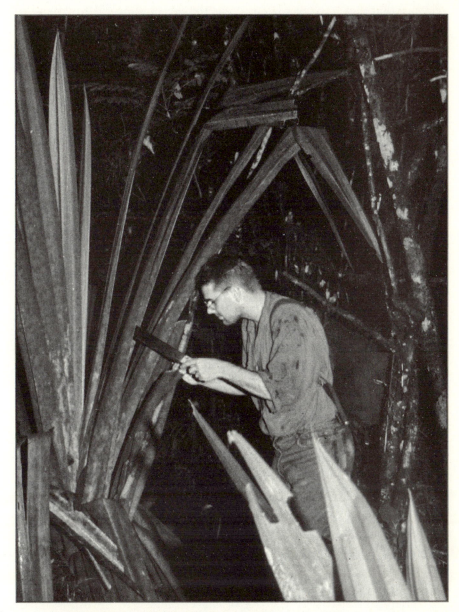

8.3 Miguel Vences looking for frogs in 1991 on the first of many trips to Madagascar. At the time, Vences's thoughts on biogeography revolved around vicariance, but that would change. Photo by Katharina Wollenberg-Valero.

Hawaiian Islands.) All of the Comoros are thought to have been formed in the past 9 million years, and, as with São Tomé and Príncipe, there is no evidence that they have ever been connected to a continental landmass. In this case, the point to remember is that they have never been connected to the continental island of Madagascar.

Mayotte has two species of frogs, and, since amphibians are not supposed to be able to get to oceanic islands on their own, it was assumed that they had been introduced. In fact, biologists describing the Mayotte frogs had concluded that they were *Boophis tephraeomystax*, a pale gray-green treefrog, and *Mantidactylus granulatus*,* a nondescript little brown frog, both of them common Madagascan species. Maybe they had arrived on Mayotte as accidental stowaways on a boat from Madagascar. Or maybe some kids had brought them over as pets and later released them.

When Miguel Vences and his colleagues examined the Mayotte frogs, though, they immediately found a problem with this story: the frogs didn't actually look like the species they were supposed to be. The *Boophis* on Mayotte was larger, had eyes of a different color, and had bumpier skin than its Madagascan counterpart. Mayotte's *Mantidactylus* was even more distinct. Compared to the "same" species on Madagascar, it was smaller and had a single white vocal sac rather than two blackish ones. Its mating call was especially different, sounding more like the rapid "scritch-scritch-scritch-scritch" of a Bank Swallow than the measured croak of the *Mantidactylus granulatus* on Madagascar. The differences in vocal-sac anatomy and mating calls were especially telling: female frogs are not attracted to males with the wrong call, so frog populations with dissimilar calls are almost always different species. The two Mayotte frogs were also unlike any of the hundreds of other frog species on Madagascar. So, even by looking at fairly obvious physical traits and behavior, Vences was thinking the Mayotte frogs were species unique to that island.**

Vences and his colleagues also sequenced several genes—one from the nucleus and three from the mitochondrion—of the Mayotte frogs and forty-five related Madagascan species. When they constructed an

---

* This species is now known as *Gephyromantis granulatus*.

** If new Malagasy frog species are being discovered at an incredible rate, how could Vences be confident that the Mayotte frogs were not members of some as yet unknown species on Madagascar? He was sure in large part because the frogs from Mayotte are lowland species found in second-growth forests, while nearly all of the new species on Madagascar are from undisturbed mountain forests. Lowland, second-growth environments on Madagascar have been extensively surveyed, and frogs that could be the same as the two Mayotte species have not been found.

evolutionary tree from these gene sequences, they found that the Madagascan species that were supposed to be the same as the Mayotte frogs are apparently not even the closest relatives of the latter. More significantly, they also found that the Mayotte species are genetically distinct from all the other species they sampled. Both anatomy and genes thus pointed to the same conclusion: the two frog species on Mayotte are indeed new species only found there. The evolutionary tree clearly shows that their ancestors came from Madagascar (and not, for instance, from mainland Africa), but the ocean voyages those ancestors made to Mayotte must have happened long ago, long enough to allow the current differences in anatomy and gene sequences to evolve. The upshot of all this is that these voyages could not have been made with the help of people, because they happened well before anybody knew how to make a boat. In fact, these journeys almost certainly took place before there were any people at all, if, by people, we mean members of *Homo sapiens*. In short, batrachians, Darwin's hopeless ocean voyagers, had made at least two natural passages (one for *Boophis* and one for *Mantidactylus*) between Madagascar and Mayotte.

The story of the Mayotte frogs reminds me, oddly enough, of a quail that Tara and I raised some years ago. Motivated, maybe, by some stirring of parental instincts, we bought a dozen or so mail-order California Quail eggs and put them in an incubator meant for reptiles. Predictably, most of them didn't develop, but we successfully raised one chick, whom we named Fiver. One day our friend Janet Bair, an accomplished birdwatcher, was at our house, and we brought out Fiver the California Quail, who by then was grown up and tame and would entertain people by hopping on their shoulders and pecking food off the table. As I've mentioned, I'm a birder, and I think of myself as being, not an expert, but at least competent, to the point of having led birdwalks for the Audubon Society and other organizations from time to time. Anyway, Janet took one look at the bird and said, "That's not a California Quail." I looked at Fiver, as if for the first time, and realized she was right—our "California Quail" had a white throat and a big rusty stripe on the side of his face, coloring that no bird of that species would have. Checking the field guides, we decided that Fiver was probably a cross between a Gambel's Quail and a Northern Bobwhite, which wasn't too unusual for a bird hatched out of an egg from a quail farm.

The point of this story is that, having been told that we had California Quail eggs, I simply accepted that Fiver *was* a California Quail, despite the obvious evidence to the contrary and my years of birdwatching experience. In the same way, I think the biologists who originally described the frogs of Mayotte were blinded by the notion that amphibians could not have reached that volcanic island on their own, and that, therefore, people must have brought them there. I suspect that these biologists saw them as species from Madagascar because that was what they expected to see, despite the obvious physical traits declaring that the Mayotte frogs were distinct.

Being blinded by preconceptions is such a common thing, not just in science but in all human affairs, that you can hardly blame someone for falling victim to it. (Maybe that's why I'm not too embarrassed to tell the story of Fiver.) In dealing with the history of poor dispersers, though, we have stumbled into a particularly egregious area, a kind of country of the blind. Rule out even the possibility of oceanic dispersal from the start, and one must then force the evidence to fit one of the remaining alternatives, either human introduction or vicariance (involving some kind of former land connection), or, in a bygone time, separate creations of species. In a sense, when Darwin did his seed experiments, he was thinking about the need to overcome this very preconception, the notion that living land organisms cannot cross wide stretches of ocean. That battle, in slightly different guise, continues today.

Along with the case of the Mayotte frogs, Miguel Vences and his colleagues discovered two other instances of amphibians naturally crossing seas to colonize Indian Ocean islands, both involving treefrogs in a family called the Hyperoliidae. In these cases, the island species were known to be quite distinct, so nobody thought they had been introduced. Instead, the accepted explanation for how they arrived on islands was continental drift; that is, the ancestors of these frogs were supposedly on the landmasses in question before those areas became islands.

Vences addressed these cases through molecular dating using relaxed clock methods, that is, ones that do not assume that the clock ticks at a constant rate. In one case, he found that a lineage of treefrogs on Madagascar had split from its closest relatives in Africa sometime between 19 and 30 million years ago. In the other, a treefrog species on the Seychelles had split from its closest relatives on Madagascar between 11 and 21 million years ago. The ages of these branching points in the frogs' evolutionary trees are not even close to being old enough to be explained by continental drift: the landmasses of Africa and Madagascar have been separated for about 130 million years, and Madagascar and the Seychelles have been separated for more than 80 million years. In other words, by the time the two frog species in question reached Madagascar and the Seychelles, those landmasses had long been islands.

Vences, who had begun his research in the Indian Ocean thinking only about vicariance, had come up with four separate instances of frogs crossing to and from islands. As with Matt Lavin and his bean plants, the evidence—in the frog studies, meaning anatomy, behavior, and DNA sequences—had forced Vences to change his mind.

Of Frogs and Floating Islands

Vences and his colleagues reported their results in a 2003 paper called "Multiple Overseas Dispersal in Amphibians" published in the *Proceedings of the Royal Society of London*. I remember that paper well—the discovery that frogs had dispersed to and from Indian Ocean islands not just once but four times was one of the key findings that made me think that something unusual was happening in the science of biogeography. At the time, we were living in Ely, a copper-mining town in eastern Nevada and a place that, for one reason or another, often had me thinking about long-distance journeys. For starters, Ely is one of the most isolated communities in the lower forty-eight states, which meant that, when Tara and I needed to take a plane flight, we usually had to drive at least three and a half hours just to get to an airport, the nearest large one being in Salt Lake City. Also, much of the human history of the area involved almost inconceivably strenuous journeys. For instance, forty miles north of Ely is Schelbourne, the site of an old Pony Express station, and seventy-five miles farther north lies the Hastings Cutoff, the wagon train "shortcut" that delayed the Donner Party long enough to eventually get them trapped in the snows of the Sierra Nevada. And, finally, the Basin and Range landscape, stark and expansive, with its seemingly endless rhythm of valley after mountain range after valley after mountain range, made me think about the long-distance dispersal of plants and animals.

I spent a lot of time hiking and, occasionally, awkwardly cross-country skiing in two nearby ranges, the Snake Range and the Schell Creek Range, high mountains built of sand and shells laid down in shallow seas 500 million years ago. Having grown up taking many trips to the classically scenic Sierra Nevada, I at first found these Great Basin mountains, with their scruffy vegetation, few lakes, and no rivers, pale in comparison, but I soon came to love their subtle beauty. And as I wandered in these ranges, separated from each other and from the mountains east and west by desert valleys, I thought about mountains as ecological islands, as countless biologists had before me (although not many of them did their thinking in eastern Nevada).

I especially thought about that when I hiked above 11,000 feet or so, where gnarled old bristlecone and limber pines and mountain sagebrush give way to the ground-hugging plants of the alpine tundra. These tundra environments, widely separated from each other by much hotter, drier environments, are the most island-like of all the altitudinal zones in the mountains. How do alpine plants and arthropods—especially plants that have no special means for dispersing their seeds, and arthropods that cannot fly—get from one of these tundra islands to another? The old story is that alpine species moved by normal dispersal among the mountain ranges during the ice ages, when the valleys that they had to cross were much cooler and wetter, and therefore more alpine-like, than they are today. Vences's study, however, along with my own research with Robin Lawson on the Baja California garter snakes and some other works I had encountered, were making me wonder: maybe biologists had underestimated the ability of alpine species to travel long distances across inhospitable terrain, just as they had underestimated the ability of frogs to cross salt water.

In any case, living in such isolation, I was very glad that Al Gore had invented the Internet (although our "high-speed" connection would break up if rain or snow was falling between our house and the mountaintop transmitter), and that I could go to the website of the *Proceedings of the Royal Society of London* and download any article I wanted from the journal. I read about the frogs of Mayotte and wondered how a dwarf alpine paintbrush could get from the top of the Snake Range across the desert to the top of the Schell Creeks.

In Paris, a place that, unlike Ely, could reasonably be considered the heart of the civilized world, John Measey also read Vences's article. Measey was convinced by Vences's evidence, but he was also struck by a sentence in the paper's introduction that was meant to emphasize the uniqueness of the results. That sentence read: "No endemic amphibian species are known from truely [*sic*] oceanic islands." ("Endemic" means found only in the area in question.) Vences and his colleagues had mentioned oceanic islands, like Hawaii, that have only nonnative amphibians, and also islands, like those of the Sunda Shelf (Borneo, Sumatra, and others), that have endemic amphibians but were once connected to continents. They never mentioned the amphibians of São Tomé and Príncipe. Bob Drewes was right: the Gulf of Guinea islands really were overlooked biological treasures; even Miguel Vences, an expert on island amphibians, had never heard of the cobra bobo and the other endemic amphibians of São Tomé and Príncipe.

Measey and Drewes briefly considered writing a paper that would have pointed out that Vences and his colleagues were wrong about the lack of known endemic amphibians on oceanic islands. The paper would have been short, basically just saying that several endemic amphibian species from the oceanic islands of São Tomé and Príncipe had been described beginning in the late 1800s, and that the ancestors of those species probably arrived by oceanic dispersal. But Drewes knew Vences and liked him, and he was a little uncomfortable with the idea of writing a paper that took Vences to task, however mildly. Instead, Measey and Drewes ended up recruiting Vences into a more involved project on the Gulf of Guinea amphibians. I'm guessing they didn't have to twist his arm. Vences was by now the "king of amphibian oceanic dispersal," so the new study was right up his alley. Plus, he was the guy who had discovered three hundred new species of frogs while being involved in dozens of other projects at the same time; if "overextended" was in his vocabulary, he probably thought it was a state of grace.

The new study focused not on the cobra bobo, but on a fairly generic-looking green and brown frog called *Ptychadena newtoni*, one of the species that Measey had barely noticed in passing while he was looking for the bright yellow caecilians on São Tomé. The goal was to figure out if *P. newtoni* was indeed a species unique to São Tomé by comparing its DNA to that of other *Ptychadena*. Bob Drewes will tell you that *P. newtoni* is exciting because it's the largest species in the genus *Ptychadena* and is, therefore, an example of the tendency for island organisms to become either giants or dwarfs (although, topping out at three inches long, *P. newtoni* is only a giant among smallish frogs). However, the main reason they chose to study *P. newtoni* was more practical and prosaic: between them, Vences and Drewes already had mitochondrial DNA sequences or tissue samples from most species of *Ptychadena*, including the ones that might conceivably be the same as *P. newtoni*.

8.4 *Ptychadena newtoni*, one of the endemic amphibians of São Tomé and Príncipe, and the subject of Measey et al.'s key 2007 biogeographic study. Photo by Andrew Stanbridge.

After getting more DNA sequences from the tissue samples, the group used the usual computer programs to crank out an evolutionary tree of the genus *Ptychadena*. Sure enough, the specimens of *P. newtoni*, from São Tomé, were all together on their own branch, well separated genetically from any other species. So, just as with the frogs on Mayotte, *P. newtoni* is apparently a unique island species and therefore likely arrived on São Tomé by a natural overwater crossing long before there were any boats to carry frogs there. This conclusion wasn't too surprising, because Measey and the others already knew that *P. newtoni* was anatomically different from all other species of *Ptychadena*.

The really odd thing about the *Ptychadena* results was where the *P. newtoni* branch arose in the evolutionary tree. You would think that the nearest relatives of *P. newtoni* would be from the Gulf of Guinea coast, because that location would require the shortest journey to São Tomé. As a parallel, the closest relatives of the Baja California garter snakes that Robin Lawson and I studied turned out to be almost directly across the Sea of Cortés from the peninsular populations. That is not the case, though, for *P. newtoni*. Instead, Measey and colleagues found that its closest relative is an undescribed species from Kenya, Tanzania, Uganda, and Egypt, that is, most of the way across the continent in East Africa. The next closest related frogs in the evolutionary tree are from West Africa,

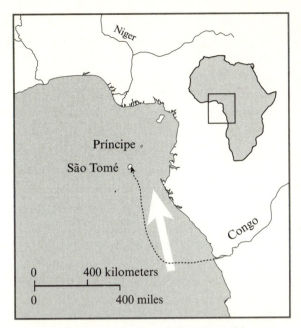

8.5 A possible route (dotted line) for rafting amphibians from the mouth of the Congo to São Tomé. The white arrow indicates the direction of the Congo Current. Redrawn and modified from Measey et al. (2007).

so it's plausible that the ancestor of both *P. newtoni* and the undescribed species was found in West Africa, and that those two lineages dispersed independently to their current ranges. Still, that undescribed East African relative raises the possibility that the ancestors of *P. newtoni* instead made an epic journey from the eastern part of the Congo Basin many hundreds of miles downriver to the Atlantic, and from there were carried by ocean currents to São Tomé (see Figure 8.5).

Forced to choose between those two possibilities, I would normally choose the first one, the one that does not require an astounding voyage down the Congo. But that would be forgetting the cobra bobo. As Measey had mentioned to the oceanographer, Alain Morlière, the closest relatives of the cobra bobo are in Central and East Africa, not West Africa. This eastern connection could just be a coincidence, and a reflection of the fact that there are many amphibian species still to be discovered in Africa; maybe some as-yet-unknown West African frog and caecilian will turn out to be the actual closest relatives of *P. newtoni* and the cobra bobo.

However, two of the other frogs of São Tomé and Príncipe also have their closest known relatives in East Africa. That means that, to the best of our knowledge, four of the five amphibian lineages on São Tomé and Príncipe (representing the five presumed colonizations from the mainland) are most closely related to species that live in the eastern half of the continent, all either in or close to the drainage of the Congo River. (The fifth lineage has not been studied, so it is unknown whether its affinities also lie in the eastern Congo Basin.) One or two species with this eastern connection could just be a fluke arising from our incomplete knowledge of African amphibians, but four must be more than a coincidence. It means that we should look for an explanation of how amphibians dispersed from East Africa to São Tomé and Príncipe.

For skeptical biogeographers, the example of the Gulf of Guinea amphibians may bring to mind Gary Nelson's contemptuous phrase that dispersalism is "a science of the improbable, the rare, the mysterious, and the miraculous." Nelson's words were wrapped up in his search for generality and his adherence to Karl Popper's notion of falsification, but, in another sense, they were more transparent and commonsensical: he was pointing out the absurdity of many dispersal hypotheses, such as, say, amphibians traveling on a giant raft down the Congo and up the west coast of Africa to colonize São Tomé and Príncipe. I have to admit that these stories sometimes do sound ridiculous. However, as I have learned more and more about biogeography, a word for strange dispersal explanations—other than *improbable*, *rare*, *mysterious*, and *miraculous*—often comes to mind. The connection between the Gulf of Guinea amphibians and East African species makes me think of that word. The word is *necessary*.

With the *Ptychadena* project in full swing, John Measey finally called up Bernard Bourles, one of the oceanographers who had intimate knowledge of the plume of water emanating from the Congo. What Bourles knew about the Congo plume was even better than Measey had hoped. It was the last big piece of the puzzle.

A small river can create an area of brackish water around its mouth, but a really big river can do much more than that. Fresh water tends to float on top of the denser salt water of the ocean, and a river the size of the Congo can alter the salinity at the surface of the ocean hundreds of miles from the river's mouth. In the rainy season, the combined output of the Congo, south of the Gulf of Guinea, and the Niger, which empties into the gulf itself, substantially lowers the salinity at the surface of the ocean from the Congo's mouth through the entire gulf. That was the data that Bourles had, showing arcs of lowered salinity radiating out from the mouths of those two big rivers. What's more, at various times, both historically and prehistorically, rainfall in tropical Africa has been much greater than it is today, which means more water would have poured out of the Congo, and the "freshening" of surface seawater would have been even stronger. In short, there have been times when the surface waters from the Congo's mouth all the way to São Tomé and Príncipe probably have been heavily diluted with fresh water.

Amphibians crossing ocean barriers need all the help they can get, and it may be that being carried on a natural raft floating on brackish water, rather than seawater, was the last piece that made the voyage from the Congo to São Tomé and Príncipe doable. Waves might be crashing over the raft, and, in any case, water would be seeping constantly into the matrix of earth and plant roots, even on a calm sea. In those circumstances, it would be far better for a frog or a cobra bobo if that water was much less salty than seawater. This is what Measey and his collaborators were arguing.

Let us now imagine the whole story, with some help from images that Bob Drewes has collected. It's some time within the past few million years, and it's the rainy season in the upper eastern reaches of the Congo River Basin. The banks of the Congo are soaked with rain, weakened to the point that big chunks, complete with tall forest trees and undergrowth, are breaking off and sliding into the river. Most of the chunks break up without getting very far, but some of them, maybe the biggest ones, stay intact, or at least partially intact, and float far down the river. It's not a stretch to think that most of these natural rafts have some amphibians on them, lodged in a burrow or among the roots of a tree—that would be the case for almost any sizable chunk of African rainforest.

In places, the Congo forms rapids, and it passes over several waterfalls. Most of the rafts don't make it past those gauntlets, but a few bump along through the whitewater stretches and crash over the falls, losing some large pieces and a few unhappy frogs in the process, but remaining mostly intact. After that, it's a long drift down to the ocean.

To make the point that very large natural river rafts are a reality, Bob Drewes points to aerial photos of the Amazon's mouth, where giant rafts called *camalotes* have run aground in the delta. In one case, the camalotes have piled up over the years to form an island roughly the size of Belgium. The Amazon delta actually is not a great place for a large raft to begin an ocean journey, because the shallows are vast and most rafts get stuck, but Amazon camalotes occasionally do make it out to open water. A raft reaching the mouth of the Congo is much more likely than one in the Amazon to reach the open ocean, because the shallows there are not so extensive.

Once a raft makes it offshore, the prevailing Congo Current will push it northward toward the Gulf of Guinea. At this point, for a frog or caecilian on a raft, the odds of making it to São Tomé or Príncipe are still not looking very good. The raft is traveling in the general direction of the Gulf of Guinea islands, but it still has about six hundred miles to go, miles in which it could easily break up or become so inundated that even the brackish water produced by the Congo plume will dehydrate any amphibian beyond recovery. Even most rafts that stay intact will not bring live amphibians to the islands. Bourles, the oceanographer, estimated that under perfect conditions, a raft might drift from the mouth of the Congo to São Tomé or Príncipe in two weeks, but it might take considerably longer. Some rafts will take too long to reach the islands, and the animals will starve or desiccate. Most of the rafts, like ships without pilots, will simply miss making landfall on the islands altogether. The ocean clearly acts as a filter that some terrestrial organisms have a hard time passing through. Almost all amphibians that find themselves on the ocean get caught in that filter and wind up dead.

Measey, Drewes, and the rest are not claiming, though, that a lot of rafts carrying amphibians have reached São Tomé and Príncipe. They know that, for any particular raft one might choose as it slides off into the water from the banks of the Congo, the odds of delivering live amphibians to the Gulf of Guinea islands are almost nil. But in a high flood year (which during some periods might be every year), dozens of rafts with this remote chance of succeeding might form. Multiply that by the number of such years since the Congo and the northward-flowing ocean current have existed, and you're talking about many thousands of rafts with amphibian passengers. It would take only a few successful ones to account for the six frog species and the cobra bobo now found on São Tomé and Príncipe.

Bob Drewes became sufficiently enthralled with this scenario that he got an artist friend to make a painting of a natural raft drifting north in the Gulf of Guinea (see Figure 8.6). In Drewes's mind and in the painting,

8.6 The colonization of São Tomé and Príncipe intrigued Bob Drewes so much that he commissioned an artist friend to paint a floating island, a likely means of dispersal for amphibians and other groups. Painting by Richard E. Cook.

the raft is more like a floating island, complete with a small forest and a meadow atop tan-colored cliffs. There's even a stream flowing off the back end of the island. The way the artist has painted the flow of ocean water makes it look as if the raft-island has a hidden motor, but other than that, the depiction is realistic. On its current course, the raft-island seems to be headed straight for a real island on the horizon. Presumably that island is São Tomé, and the floating island is carrying a few bright yellow caecilians and generic-looking frogs, the ancestors of the modern cobra bobo and *Ptychadena newtoni*.

The Importance of the Possible

"We cannot prove this happened," Bob Drewes wrote of the hypothesis that amphibians rafted from East Africa down the Congo and then up to São Tomé and Príncipe. I don't think he meant this in the philosophical sense of nothing ever being known with absolute certainty in science. I suspect he meant it in a more practical sense along the lines of "Since we can't build a time machine, the hypothesis will never become an accepted fact." We may accept that some sort of oceanic dispersal is required to explain why there are amphibians on São Tomé and Príncipe, but it is a big step from there to buying the entire convoluted scenario described by Measey and his colleagues. There are other ways to explain the distributions of the Gulf of Guinea amphibians and their relatives, and we will probably never be able to rule them all out. For some scientists, this might sound like a reason not to pursue the subject. Why spend a lot of time on a question you can never answer definitively? In fact, the whole enterprise of fleshing out this dispersal story doesn't even quite seem like "Science" with a capital "S," at least not in the way we were taught in our high-school or college classes. Where's the list of alternative hypotheses and the critical tests to try to falsify each one?

To see where this and many other dispersal scenarios fit into a scientific argument, it's worth consulting a philosopher. Not, as it turns out, a well-known philosopher of science like Karl Popper or Thomas Kuhn (and that pretty much exhausts the list of well-known philosophers of science), but a fairly obscure philosopher of history named William Dray. In particular, we'll consult a 1957 book Dray wrote with the straightforward, academic-sounding title *Laws and Explanation in History*. In that book, Dray pointed out that it's very common for an explanation of an event to consist of simply showing how the event *could* have occurred, without ever claiming that is how it actually *did* occur. That kind of explanation is useful when the event otherwise seems unbelievable; the explanation converts the listener from thinking that something could not possibly have happened to thinking that it might have happened. Dray called this kind of account a "how-possibly explanation," as opposed to the usual "how-actually explanation."

Whether one knows the term or not, how-possibly explanations are familiar to all of us in one form or another. For instance, consider the following: *Witnesses say that the suspect was at a party when the murder took place. But it was a costume party and everyone was wearing a mask until the end of the evening, when identities were finally revealed, so no one can say for certain exactly when the suspect was there and when he was not. An accomplice could have been behind the mask, while the suspect was off killing the victim.* That's a standard sort of argument used by prosecutors and mystery writers to show that a suspect could have done something that at first seems to have been ruled out. It is a how-possibly argument. A bit closer to our subject, the famous "origin of life" experiments by the chemist Stanley Miller in the early 1950s, in which he mixed together molecules thought to have been present in the Earth's early atmosphere, ran an electric current through the mixture, and ended up with organic compounds, including amino acids, is part of a how-possibly argument. Miller didn't claim that he had recreated an actual step in the origin of life, but his experiments helped overcome disbelief about one part of that process, the natural generation of organic compounds from inorganic ones. Once we know how to recognize a how-possibly argument, we see them all over the place.

In the case of the Gulf of Guinea amphibians, the event we want to explain is how caecilians and frogs reached those islands all the way from East Africa, a journey that on the face of it seems absurd. The explanation, the one I have just described, is essentially a series of arguments aimed at overcoming the incredulity of the audience. *It sounds like the amphibians would have needed an extremely large natural raft.* Well, extremely large rafts form all the time. Look at the aerial photographs of the Amazon delta. *Wouldn't a raft reaching the mouth of the Congo just get stuck in the shallows or drift off into the heart of the Atlantic?* The shallows in the Congo delta are not all that extensive, and the Congo Current would push a raft north toward the Gulf of Guinea, not immediately west toward South America. *How could amphibians, with their vulnerability to salt water, survive a long sea voyage?* The oceanic part of the voyage might only take two weeks, and the fresh water from the Congo and the Niger during the rainy season could make the surface waters of the ocean much less salty than normal seawater.

How-possibly scenarios like this have a long history in evolutionary biology. In fact, as a biologist/philosopher named Robert O'Hara pointed out in 1988, *The Origin of Species* can be read as one long how-possibly argument aimed at erasing the reader's disbelief. To drive this point home, O'Hara makes a list of how-possibly questions and Darwin's answers to them. For instance: "How possibly could evolution have occurred, since there is no force to drive change? Darwin removes the objection with the introduction of natural selection. How possibly could evolution have happened in so short a time? Darwin tells us that the earth is older than we thought. How possibly could evolution have taken place if we don't see all the intermediate stages? Darwin tells us about extinction and the imperfection of the fossil record."*

---

* I would argue, uncontroversially, that *The Origin of Species* is not just a set of how-possibly arguments. For instance, a critical aspect of the book is that it shows that many biological phenomena make sense under evolution but not under creationism. In other words, Darwin shows not only that evolution is possible, but that the alternative explanation is untenable.

The last entry in O'Hara's list brings us back to our specific subject, ocean crossings: "How possibly could species isolated on islands be descended from other species? Darwin tells us about the powers of dispersal." This is where Darwin brings in his seeds-in-seawater experiments, the snails on the duck's feet, the icebergs. In so doing, he becomes the most conspicuous contributor to a long line of how-possibly arguments in the study of dispersal. For instance: Alfred Russel Wallace wrote of accounts of large natural rafts covered with vegetation drifting among the Moluccas and the Philippines; in the 1930s, a South African biologist named John Muir speculated that seeds might be carried on pieces of floating pumice; a primatologist named Anne Yoder recently suggested that lemur ancestors might have gone into a kind of hibernation to survive the trip from Africa to Madagascar; Robin Lawson and I—only dimly aware that we were following in the footsteps of many others—pointed out that the garter snakes that crossed the Sea of Cortés are especially resistant to the desiccating effects of salt water. This tradition, I think, is no fluke; it's not a coincidence that scientists studying dispersal have come up with more than their share of how-possibly scenarios. Rather, it seems to me that advocates of dispersal, especially when invoking *oceanic* dispersal, are constantly faced with a very basic and widespread disbelief: How in the world do beeches, snails, lemurs, frogs, caecilians, freshwater snakes—you name your unlikely dispersing group—make ocean voyages without the aid of humans? It sounds improbable, mysterious, even miraculous. How possibly? We'll show you how.

Today's how-possibly dispersal stories are usually just the last step in an argument that begins with evidence from fossils, the anatomy of living species, or DNA showing that long-distance dispersal is required to explain the distribution of some group. Many of these how-possibly stories probably couldn't stand on their own. For instance, if there was no genetic or anatomical evidence to suggest that amphibians had dispersed from East Africa to the Gulf of Guinea islands, nobody would care about Measey and his colleagues' arguments about floods in the Congo Basin, and camalotes, and the "freshening" of the surface waters of the ocean. (For that matter, Measey et al. would never have bothered to make those arguments in the first place.) A bizarre dispersal scenario like that only becomes really interesting when there's at least a reasonable chance that it could be true.

However, this doesn't mean that the how-possibly story is just the speculative fluff tossed in after the real work has been done, like a meringue after a meal of steak and potatoes. By overcoming incredulity that a thing could even *possibly* happen, a compelling how-possibly story can strengthen the case that the event actually *did* happen. That is obviously the intent of a prosecutor using a how-possibly argument: it's a step toward a conviction, not just an intriguing story. In the case of the Gulf of Guinea amphibians, the how-possibly argument helps convince us that the small holes in the evidence for dispersal—for instance, the possibility that the island species are actually present but undiscovered on the mainland—are just that, small holes. If it were impossible to imagine how a frog or a caecilian could get from East Africa to São Tomé or Príncipe, those small holes would start to look bigger.

The Ladder of Improbability

The study of the Gulf of Guinea amphibians isn't finished yet, at least not in John Measey's eyes. Measey has been contemplating a trip to the Democratic Republic of the Congo to look for caecilians, to get a better idea of exactly where the ancestors of the cobra bobo came from, but the realities of the location have dampened his enthusiasm. Based on secondhand reports of bright yellow caecilians, the place to look is Ituri Province, in the northeastern corner of the country. Biologically, Ituri Province might have a connection to São Tomé and Príncipe, but in human terms these areas are like night and day. Bob Drewes says he loves São Tomé and Príncipe partly because, unlike most of the rest of Africa, there is no "we vs they" there. Ituri Province, in contrast, is a region of violent tribal conflict, a place few foreigners visit unless they are missionaries or United Nations peacekeepers.

Whether or not Measey ever makes it to Ituri Province and finds the relatives of the cobra bobo, he, Miguel Vences, Bob Drewes, and their colleagues have already made a big contribution to the study of oceanic dispersal. They've shown that Darwin underestimated the ability of "batrachians" to cross saltwater barriers. At the same time, they've put a dent in the armor of vicariance biogeography, although just how big of a dent might depend on who's looking. To my mind, the amphibian cases, taken together, are pretty significant. Amphibians are supposed to be the worst oceanic voyagers among all the major groups of land vertebrates, and yet we now have strong evidence from the work of Measey, Vences, Drewes, and their collaborators that these animals have reached Madagascar, the Comoros, the Seychelles, and São Tomé and Príncipe without the aid of people. Other phylogenetic studies extend that list of oceanic dispersal by amphibians—toads probably crossed from North America to South America when those two continents were still separated; fanged frogs reached Sulawesi, the Philippines, the Malukus (Moluccas), and the Lesser Sundas; slender salamanders colonized the California Channel Islands; and several frog lineages dispersed to and among the islands of the Caribbean. These studies show that one can no longer assume that related amphibian species separated by salt water must have been introduced by humans, or, as vicariance scientists would usually have us believe, reflect a former land connection. Oceanic dispersal is in play for amphibians, where before it was not.

Still, vicariance scientists can point to the fact that none of these cases involves a really long ocean voyage. If the Congo raft story is right, the oceanic part of the journey to Príncipe would have been about 630 miles, and that would be the longest known oceanic dispersal event for any amphibian. One might even say that the larger pattern fits the vicariance view pretty well. Frogs haven't crossed the Atlantic. Salamanders haven't rafted from North America to Australia. Caecilians, for their part, are found in East Africa, yet they haven't even made the three-hundred-mile journey across the Mozambique Channel to Madagascar. Amphibians are certainly better at dispersing across saltwater barriers than people thought, but they don't seem to be great at it. Furthermore, at least some of the deep branching points in the amphibian evolutionary tree occurred at times that indicate distributions related to continental movements. For instance, the caecilians and some of the frogs on the Seychelles are most closely related to groups in India, the expected connection for a continental drift explanation, and these Seychelles and Indian lineages separated before those areas split from each other, which is also consistent with vicariance.

So where does this leave us? It leaves us with some other cases of oceanic dispersal to contemplate, ones that are more extreme than the amphibian ones. Think of the Gulf of Guinea amphibian examples as one rung on a ladder of increasing improbability, with each rung representing a kind of dispersal event. The rungs near the bottom of the ladder are relatively easy events, such as birds reaching Japan from the Asian mainland, or gourds drifting to many islands. The middle rungs are less likely events, maybe anole lizards rafting among Caribbean islands, or garter snakes crossing the Sea of Cortés. With the amphibian cases, we've reached a high rung on the ladder, but not the top. Call that amphibian rung the "improbable and rare." The next rung up might be the "mysterious," and the final rung, the "miraculous."

We're about to keep climbing up the ladder.

*"That insects can be transported into the upper regions of the atmosphere by ascending air-currents was long ago remarked by Humboldt, and the subject has been discussed with his usual acumen by Whymper (*Travels amongst the Great Andes of the Equator*). Carried along in the higher air-currents these insects might finally be deposited at places far distant from their home. One reads occasionally extraordinary accounts of a rain of insects. A very circumstantial account was given to me when I was on Keeling Atoll of a shower of dragon-flies that fell on the islands, their remains being found in quantities in the lagoon. Dragon-flies, it is known, are often found at sea far from land, and one species has been observed nearly all over the world, including the Pacific islands."*

—H. B. Guppy, Observations of a
Naturalist in the Pacific, vol. 2

Chapter Nine

THE MONKEY'S VOYAGE

The Long Route to Fernando de Noronha

Off of Brazil's eastward-jutting shoulder, a few degrees south of the Equator, lies the small island of Fernando de Noronha along with about twenty even smaller islands and islets, the emergent heights of a mostly submarine range of volcanic mountains. Like many islands, Fernando de Noronha once served as a penal colony; people have recognized for a long time that water can be an effective barrier to dispersal, at least for other people. In fact, especially unruly inmates at the prison on Fernando de Noronha were sometimes banished without guards or assistance to a small island called Rapta, northeast of the main island, to scrabble for survival on their own. No one was worried that they would escape; there were no trees on Ilha Rapta to fashion a raft from, and the prisoners weren't going to swim a couple of hundred miles to the mainland.

Today Fernando de Noronha is a World Heritage Site and something of an ecotourist destination, a haven for seabirds, marine mammals, and fishes, all taking advantage of the nutrient-rich waters. It's also a place that demonstrates the rule that oceanic islands tend to have few species of native land vertebrates; Fernando de Noronha has had only three that anyone knows about—a skink, a legless amphisbaenian (or worm lizard), and a rat that's now extinct. The presence of just two lizards and a rodent jibes with the view that land vertebrates are lousy oceanic voyagers; Fernando de Noronha is a mere 220 miles from the Brazilian coast, yet only these three species have managed to colonize the island.

For the worm lizards and the rats, there isn't much more to tell. Both species are (or were) closely related to South American taxa and apparently came from that nearby continent. However, starting more than sixty years ago, biologists have recognized something odd about Fernando de Noronha's skinks. (Something other than the fact that they're extremely common and extremely tame: One English visitor in the nineteenth century reported that, while he was cautiously climbing a steep bluff, using his hands to keep from falling, one of these lizards ended up crawling around in his trousers for nearly an hour. He described the lizard as having become "offensively familiar.") Specifically, these skinks (*Mabuya atlantica*, sometimes called *Trachylepis atlantica*) lay eggs and have keels, or ridges, on their scales, which are unremarkable features except for the fact that South American species in the same genus give birth to live young and have smooth scales. However, some African species of *Mabuya* do lay eggs, have keeled scales, and have several other traits suggesting they are close cousins to *M. atlantica*. In other words, it looks like *M. atlantica* is part of an African branch of the genus, not the South American branch.

It's possible that the obvious similarities of *M. atlantica* to African species are misleading, and that the Fernando de Noronha skinks don't really belong with the African ones. For instance, egg-laying is probably ancestral for the whole genus (and therefore doesn't indicate an especially close relationship *within* the genus), and perhaps keeled scales evolved independently in *M. atlantica* and African *Mabuya*, like wings in birds and bats. Maybe *M. atlantica*, despite appearances, is actually most

9.1 *Mabuya atlantica*, the skink from Fernando de Noronha. Photo by Jim Skea.

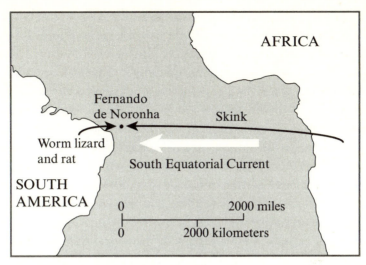

9.2 Fernando de Noronha is only 220 miles from the Brazilian coast, but the skink came from Africa.

closely related to South American *Mabuya*. It wouldn't be the first time that obvious traits in common have led biologists astray. However, *M. atlantica*'s African roots have been verified by recent work: several studies using DNA sequence data have strongly supported the close relationship of the species to African rather than South American members of the genus. Thus, the ancestors of *M. atlantica* must have reached Fernando de Noronha by crossing almost the whole width of the Atlantic, presumably by rafting on the South Equatorial Current, which heads more or less due west from the central African coast toward the island. It's a case that shows that identifying the distance required for colonization based on the nearest continent can be misleading—a relatively short journey for the rat and the worm lizard was a much longer one for the skink, coming from the opposite direction (see Figure 9.2).

The ancestors of *M. atlantica* apparently made their transatlantic voyage within the past 3.3 million years (the maximum estimate for the age of Fernando de Noronha), crossing the ocean in roughly its modern configuration. They must have traveled at least 1,800 miles over water, and in fact, based on likely routes, using westward-flowing currents, the distance traveled was probably closer to 3,000 miles. A final irony is that, even if *M. atlantica* had been derived from South American species of *Mabuya*, it would have had a transoceanic journey in its recent evolutionary past: the same DNA studies that show that *M. atlantica* came from Africa also indicate that the ancestor of all other New World *Mabuya* arrived from the Old World within the past 9 million years or so, probably also dispersing from Africa across the Atlantic.

Molecular dating analyses indicate quite a few other ocean crossings by lizards and snakes, including two skink lineages and two gecko lineages traversing most of the breadth of the Indian Ocean, four different gecko lineages dispersing from Africa to the West Indies or South America, worm lizards making the journey from Africa to South America, and both blindsnakes and threadsnakes crossing from Africa to some part of the New World. The cases of the worm lizards, blindsnakes, and threadsnakes are especially surprising because these animals spend almost their entire lives underground, and such creatures are thought to be especially unlikely to cross oceans.* These burrowing reptiles can't fly or float on the air or stick to the feet of a bird, and, unlike geckos (and to a lesser extent, skinks), they don't often cling to trees or other vegetation that might get blown out to sea. About the only way a burrowing reptile can get across an ocean without human assistance is on a raft containing chunks of earth, and the chunks might have to be large enough to remain unsaturated during the long voyage. For these reasons, such subterranean vertebrates have often been placed in the category of animals that simply do not cross oceans. A burrowing lizard or snake dispersing over the Atlantic is, in this view, as imaginary as a unicorn. But that view is apparently wrong.

These examples of lizards and snakes crossing entire oceans can stand for a high rung on our ladder of improbability, a step up from the amphibian cases. (However, the number of crossings by geckos—at least six and counting—suggests that this particular group may represent a lower rung.) Such events seem sufficiently implausible that many biologists have been very reluctant to invoke them, sometimes banishing them to the realm of the impossible. These ocean crossings seemed implausible or impossible, that is, until the evidence made them unavoidable.

Nonetheless, in hindsight one could argue that it isn't *so* surprising that lizards and snakes have crossed the Atlantic and other oceans. After all, they're small enough that a modest-sized floating island might contain

---

* Incidentally, the ocean-crossing worm lizards eventually gave rise to the species that made the much shorter but still substantial voyage from South America to Fernando de Noronha.

9.3 *Amphisbaena alba*, a member of a large group of South American worm lizards descended from another transatlantic voyager. Photo by Diogo B. Provete.

many of their prey, provisions for the long journey. Also, they have relatively low metabolic rates, meaning that, even when they run out of food, they can last a long time before they starve or die of thirst. That's especially true if they're simply lying around, with their body temperatures equilibrated to the cool maritime air, their metabolic motors running at a very low idle. In such a cold state, many reptiles can go for months without food or water, certainly long enough to raft across the Atlantic or Indian Oceans.

Our next example, however, has none of these things going for it. At last, we are about to ascend to the top of the ladder, a case of oceanic dispersal apparently so absurd that it has, at times, provoked ridicule not only from biologists but also from creation scientists. People who have no vested interest in the topic have laughed at me when I've brought up this case.

We have reached the centerpiece example of the book.

The Mysterious, Miraculous Voyage

It's a hot summer day and my wife Tara, our one-year-old daughter Hana, and I are having a picnic at Pyramid Lake, in the dry, sagebrush-and-greasewood country north of Reno. The white, pebbly beach is deserted, one reason we sometimes come here instead of crowded Tahoe. Gulls cruise along the shoreline. The occasional flock of white pelicans wheels in the bright sky. On the dun-colored, treeless mountain slopes surrounding the lake, one can make out faint bands, the remnants of higher, ice-age shorelines, looking like poorly spaced contours on a topographic map.

The water is cool, calm, and inviting. Some large black-and-white Western Grebes are drifting about a hundred yards offshore, and Tara and I take turns swimming out toward them while Hana, squatting on the shore, happily flips pebbles into the lake. The wind changes direction, now blowing out from the shore, and Tara comes back from her turn in the lake, saying it was hard getting back in. I glance up at the sky, feel the wind on my face, and figure I'll take one more quick swim, trying to get as close as I can to the grebes.

Heading out from the shore is easy, and I end up so close to a grebe that I can see its devilish red eyes fixed on me, as if it's wondering what to make of my disembodied head and shoulders appearing above the water. Swimming back in is a different story. The wind has picked up and the water is getting choppy and, although I put my head down and take what I'm sure are powerful strokes, when I look up the shore seems just as far away as before. It crosses my mind that many people have drowned in Pyramid Lake, often because they were out on the water when the wind turned. It also crosses my mind that I'm a dumbass for not thinking about that earlier. I'm not in a full panic, but I'm pushing hard now, trying to swim the way you're supposed to swim, with deep, full strokes and regular breathing. Almost imperceptibly, the shore starts edging closer. Finally, after what feels like a long time (but is probably about five minutes), my toes touch sand. I wade up onto the beach, not quite exhausted, but definitely winded, and relieved to feel the earth under my feet.

I think about my vulnerability, just a hundred yards out from the shore of Pyramid Lake. If I had been stupid enough to swim a few hundred yards farther, I might have drowned right there, with the devil-eyed grebes drifting idly by. Take that vulnerability and extrapolate, and you see how unlikely it must be for a person to cross a large stretch of ocean without some kind of watercraft. How many times has someone made a long ocean journey—from California to Hawaii, say, or between Africa and the West Indies—without a boat? As far as anybody knows, never. Natural selection hasn't molded us to survive for a long time in water, or even on a natural raft. It wasn't just paranoia that made me think, as the waves pushed against me, "I better get back *now*."

We have the sense, correct I think, that our primate relatives are about as bad as we are at surviving an ocean journey. The history of primate dispersal generally bears that out. For instance, no primate is known to have naturally colonized islands extremely far from other landmasses; the distance record was apparently set by lemur ancestors some 50 million years ago, when they reached Madagascar from Africa, a distance of perhaps three or four hundred miles, or possibly by monkeys that dispersed to the Greater Antilles from South America 20 million years ago or so. Macaques seem to have a special propensity for overwater dispersal, having colonized several Southeast Asian islands, such as Sulawesi and the Nicobar Islands, that are separated from other land by deep water. However, none of these macaque journeys were demonstrably more than about a hundred miles. Similarly, *Homo floresiensis*, the so-called hobbit whose fossil remains were discovered a decade ago on the Indonesian island of Flores, probably dispersed without boats between islands, but again the distances were short, on the order of tens of miles, not many hundreds or more.

All of this—that is, our own experiences and common sense, and the long, generally non-ocean-voyaging history of primates—contrives to make my next statement sound ridiculous: monkeys crossed the Atlantic. Think about it: I'm in danger of drowning after swimming a couple hundred yards in a slightly choppy lake, but monkeys somehow crossed the entire Atlantic Ocean. I have to admit that this case gives me pause. When I think about it, I hear Gary Nelson mocking the dispersalists for invoking the mysterious and the miraculous. I'm not one of those scientists who thinks that everything he has ever written or said is right. I'm by nature a doubter, not only of other people, but of myself, and I wonder if somehow this monkey example will show that the new dispersalism I've been pushing is a house of cards. I pause, then I go on to the evidence.

We can start, as we often do, with an evolutionary tree. This time, it's the tree of primates, so most of the groups will probably sound familiar. One part of the tree, a part that won't concern us much, contains lesser known primates, things like lemurs, lorises, and galagos. The part we're especially interested in is essentially the monkey branch, which includes not only monkeys in the usual, vernacular sense, but also apes. A lot of scientists have spent a lot of time working on this primate tree because, of course, it is *our* tree. As a result, it is now a very reliable one, with much supporting evidence from anatomy and, especially, from many different gene sequences; the upshot is that the ideas about primate relationships aren't likely to change much as new evidence accumulates.

For our purposes, a key is that one branching point in the tree leads to New World monkeys on one side and Old World monkeys (including apes) on the other. The New World monkeys include, for example, spindly spider monkeys; big, grunting howlers; and acrobatic little marmosets. It's usually easy to recognize a New World monkey, even if you see one in an Old World zoo, because they have flat noses with nostrils that point to the sides (the scientific name for the group is Platyrrhini, which means "flat nose"), and many of the larger ones use their tails as prehensile "fifth limbs" as they make their way through the trees. The Old World monkeys and apes have a narrower snout with nostrils that point down (the name for this group is Catarrhini, which means "downward-pointing nose"). Some are short-tailed or tailless, and even those that have a long tail can't grab hold of things with it. While all platyrrhines are highly arboreal, some catarrhines, such as baboons, chimps, gorillas, and, of course, humans, spend much of their time on the ground.

The sister-group relationship of platyrrhines and catarrhines is so well supported by a combination of anatomical and DNA sequence data that no serious student of primates doubts it. And that branching point in the tree implies a biogeographic conundrum. The platyrrhine branch leads to species only found in the New World tropics and subtropics, while the catarrhine branch leads to species only found in the Old World (with the single exception of humans). It's another example of a disjunct distribution on a global scale, and it looks like a pattern we can explain by Gondwanan vicariance. Simply imagine that monkeys once lived in the western part of Gondwana, and that the New World and Old World lineages became isolated and went their separate ways, evolutionarily speaking, with the opening of the Atlantic.

Gondwanan vicariance—the process, if not the phrase—is what laypeople have come up with as an explanation when I've mentioned this monkey example. However, unlike many of the other cases I've described, almost no biogeographers seem to believe this scenario for monkeys. The problem with the vicariance explanation, as we have seen so often before, is that the timing is all wrong. If the opening of the South Atlantic caused the separation between platyrrhines and catarrhines, then that split in the evolutionary tree should have occurred on the order of 100 million years ago. To put this in some perspective, such an old date would imply that the New World and Old World monkey lineages, which we know are not early branches in the primate tree, are actually about 50 million years

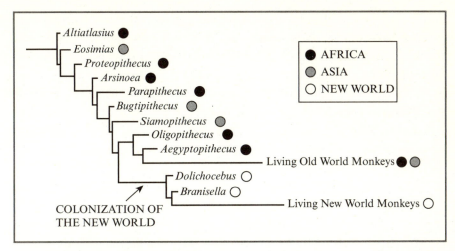

9.4 Part of the primate evolutionary tree. All of the taxa are extinct except those marked as "Living." Note that all of the close relatives of New World monkeys (Platyrrhini), living and extinct, are from the Old World, indicating dispersal from the Old World to the New World along the branch identified with an arrow. *Dolichocebus* and *Branisella* are fossil New World monkeys. The evolutionary tree is redrawn and modified from Ni et al. (2013). Some groups have been left out to simplify the diagram, but this exclusion does not change the interpretation.

older than the earliest known primate fossils of any kind. In fact, these monkey lineages would have to be some 35 million years older than the first known fossils of *any* placental mammal. I know of only one biologist who is seriously arguing that the platyrrhine and catarrhine lineages are that old and were therefore subject to the opening of the Atlantic, and that one biologist is Michael Heads, the panbiogeographer who doesn't believe in long-distance dispersal. Tangentially, it's a bit odd that other diehard vicariance biogeographers, notorious for their disdain for almost any evidence about the ages of lineages, aren't arguing for the Gondwanan breakup explanation in this case. Maybe this is because the timing of primate origin and diversification has been so thoroughly scrutinized that anyone arguing for the Gondwanan vicariance explanation would be immediately ridiculed.

All of the evidence about timing (described below) indicates that the platyrrhine-catarrhine split occurred when South America was either an island continent or was connected only to Antarctica/Australia (a connection that isn't relevant to the origin of New World monkeys). Furthermore, although there are quite a few fossils in the Old World—in Africa and Asia—that might be close to the ancestor of all monkeys, there are no such fossils in South America. In the New World, monkeys appear in the fossil record as if out of thin air about 26 million years ago, which is after the platyrrhine-catarrhine split by anyone's estimate. What this suggests is that monkeys originated in the Old World and dispersed to South America (see Figure 9.4). They had to cross water to get there, but when did they do it, and what water did they have to cross? As is so often the case, the answer to "when" is crucial to the inference of "what."

The dating of branching points in the primate evolutionary tree—the construction of a primate timetree—has received a lot of attention among evolutionary biologists. In fact, the age of divergence between New World monkeys and Old World monkeys has been estimated from molecular data in at least twenty separate studies. As I sifted through this literature, I sometimes felt like I was wading through deep mud and that there would be only confusion at the end of it. The studies estimated the age of that split at anywhere from 31 million to 70 million years, which is a very wide range of ages. In fact, it's the kind of disappointingly wide range that might make one think that Gary Nelson was right to mock the "molecular dating game."

Among these investigations of divergence dates, the study that especially stood out was the one that gave the age of the platyrrhine-catarrhine split as 70 million years old. That study was by a Swedish evolutionary biologist named Ulfur Arnason and several of his colleagues and used a fairly large set of data, complete sequences of eleven mitochondrial genes. It also used what, at first glance, seem like reasonable fossil calibrations for the timetree. For our purposes, the important thing is that 70 million years ago, the Atlantic was very narrow and possibly contained a large island spanning most of its breadth. So, if Arnason et al.'s age estimate is correct, and if that island really existed, monkeys could have crossed the ocean by making just two relatively short overwater journeys—one from Africa to the island, and one from the island to South America. This island-as-stepping-stone scenario would strain almost no one's beliefs about the overwater dispersal abilities of monkeys.

The 70-million-year age is probably way off, though. In particular, Arnason et al. did two things that, in combination, probably made their age estimate far too old. First, they used only mitochondrial genes, which are notorious for evolving in a non-clock-like manner, shifting their rate of evolution strongly and frequently. They could have gotten around this problem, at least to some extent, by using several calibrations *within* primates; in other words, the analysis could have been done so that divergence dates were estimated using primate rates of genetic change. But Arnason et al. had a deep distrust of any primate fossil calibrations and didn't use them, perhaps mistakenly thinking that, because the primate fossil record is generally not very good, all parts of that record must be unreliable. Instead, they calibrated their primate tree using three nonprimate branching points—whales and even-toed ungulates; horses and rhinos; and toothed and baleen whales. Those calibrations indicated slower rates of genetic change than are realistic for primates, and thus pushed branching points within the primate tree too far into the past. It was as if you estimated how long a trip would take on Amtrak when you were actually traveling on a Japanese bullet train.

If we eliminate this Arnason study along with others he did that had similar problems, the range of estimates for the platyrrhine-catarrhine split becomes 31 to 58 million years ago. Furthermore, the best studies—the ones that use large amounts of genetic data (including nuclear genes), many groups of primates, and reasonable methods to account for non-clock-like evolution—all yield estimates of 51 million years or younger. Although 31 to 51 million years ago is still a 20-million-year range, it's a lot better than 31 to 70 million years.* The single estimate I consider the best currently available, because of the amount of data used and the sophistication of the analyses, gives an age right in the middle of the range, 41 million years ago (with a 95 percent confidence interval of 33 to 50 million years).

Armed with these reasonable dates from molecular analyses, we can now infer the window of time in which monkeys must have dispersed to the New World (see Figure 9.5). We know that there were monkeys in the Americas by about 26 millon years ago, because this is the age of the oldest known fossil of a monkey in the New World (from the Late Oligocene of Bolivia); this sets the younger boundary of the time window. The older boundary is set by the oldest reasonable estimate for the platyrrhine-catarrhine split, 51 million years ago. Since fossils indicate that this branching event occurred in the Old World (in Africa, or possibly Asia), the dispersal to South America must have happened at some later time.

---

* This range of 31 to 51 million years is from the lowest confidence limit to the highest confidence limit from these studies.

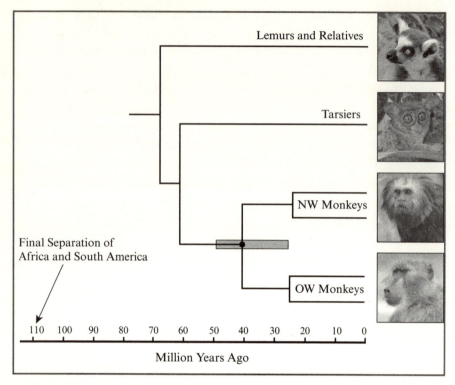

9.5 A primate timetree indicating the probable period for the colonization of the New World by monkeys (shaded bar). The oldest possible age of colonization is set by the oldest reasonable estimate for the separation between the New World and Old World monkey lineages. The youngest possible age is set by the oldest fossils of monkeys in the New World. Timetree redrawn and modified from Springer et al. (2012). Photos, top to bottom: Clément Bardot (ring-tailed lemur), Sakurai Midori (undescribed tarsier species), Bjorn Christian Torrissen (golden-headed lion tamarin), Graham Racher (chacma baboon).

The upper limit of this time window, 51 million years ago, clearly rules out Gondwanan vicariance as an explanation, a surprise to almost no one. It also rules out Arnason's scenario of short oceanic journeys using a giant Late Cretaceous island as a stepping stone. However, on its own, the time window does not force us to accept the scenario of monkeys crossing a wide Atlantic. There is one other hypothesis to deal with, one that's been kicking around for decades, namely, that monkeys reached the New World tropics by crossing from Asia over the Bering Land Bridge and then moving down through North America. This "via North America" hypothesis requires a journey across the Caribbean to South America, but the distance of the required ocean journey is unclear because of uncertainties about the geologic history of that region.

That uncertainty, however, is a moot point. The bigger problem with this hypothesis is that there are absolutely no fossils to support it. For a good part of that window of time from 26 to 51 million years ago, North America was a warmer and more heavily forested place than it is today, and primates are fairly common in the continent's fossil record for much of that period, especially in the Eocene. However, they're not the right primates. There are tarsier-like and lemur-like species of various sorts, clearly outside of the monkey clade, but there are no monkeys of any kind. So, to accept this hypothesis, we have to imagine that monkeys passed through North America but, unlike many other primates of that time, inexplicably left no trace. We know that the fossil record is incomplete, but that it should be incomplete in such a peculiar way—as if monkey fossils were selectively plucked out by some divine hand—seems extremely unlikely. This situation stands in great contrast with the fossil records of Africa and southern Asia, which contain many early monkeys or near-monkeys, some of them potentially quite close to the ancestor of the New World monkeys. Basically, the fossils strongly indicate a dramatic leap from the Old World to South America rather than the long, slow route through North America.

This rather involved consideration of the molecular and fossil evidence leaves us with a seemingly mysterious, miraculous explanation: monkeys crossed the Atlantic Ocean to colonize the New World. Here, as in many other cases in this book, we've come to a conclusion by refuting other reasonable explanations. In particular, molecular dating rules out vicariance via the opening of the Atlantic (or any other Mesozoic event), and the fossil record argues strongly against colonization through North America, pointing us to the path across the Atlantic.* In reaching that point, we are following the philosopher Karl Popper, who identified refutation (or falsification) as the essence of science, or maybe Sherlock Holmes, who famously said, "Once you eliminate the impossible, whatever remains, no matter how improbable, must be the truth."

---

* The fossil record by itself is somewhat equivocal about whether monkeys reached South America from Africa by crossing the Atlantic or from Asia by crossing the Pacific. However, the Atlantic route seems far more likely, both because it is much shorter and because of the moderate handful of cases of Atlantic crossings by other land vertebrates. The only plausible case for a land vertebrate crossing the Pacific is that of the ancestor of *Brachylophus* iguanas reaching Fiji and Tonga from the New World, but this example remains controversial (Noonan and Sites 2010).

Narrowing the Gap

Old World and New World monkeys share several obvious features that indicate they are "sister groups," for instance, the fusion of the two halves of the lower jaw where they meet in front, and the formation of a bony cup surrounding the eye. (In cladistic parlance, these are shared derived traits, or synapomorphies.) Nonetheless, for much of the twentieth century, many primatologists believed that these two groups were not closely related, but instead had converged on the same anatomical features from different starting points. That belief was motivated by biogeography: if Old World and New World monkeys were in fact closest relatives, their distributions seemed to require colonization across the Atlantic, and that simply could not be. Better to interpret their several shared anatomical traits as an odd coincidence than to have to invoke such an improbable dispersal event.

The problem those scientists had with the Atlantic-crossing hypothesis is one that still seems troublesome, even with all the DNA and fossil evidence in hand. The problem, in a nutshell, is *distance*. It's fairly easy to believe that macaques rafted from Borneo to Sulawesi, or that "hobbits" crossed the narrow channels separating some of the Lesser Sunda Islands. However, the Atlantic Ocean is almost 1,800 miles wide at its narrowest point, and it's hard to imagine how monkeys could survive such a voyage. Even on a large "floating island," wouldn't they die of thirst or starvation long before reaching South America? Despite the evidence for this journey, shouldn't we be trying to come up with some other, better route to the Americas?

One thing to remember is that the Atlantic, which formed as magma pushed up and spread east and west from the Mid-Atlantic Ridge, is continually getting wider. The rate of spreading is on the order of the growth rate of a fingernail, which may not sound very fast, but obviously has added up through deep time. Rewind 40 million years, to the time of the presumed simian voyage, and the ocean would be considerably narrower than it is today. In fact, at that time, the Atlantic probably was only about 900 miles wide, still a seemingly mind-boggling journey for a rafting monkey, but not as improbable as 1,800 miles.

In 1999, a Canadian primatologist named Alain Houle built up a how-possibly argument for this dispersal event by estimating how long it might take a natural raft to cross that narrower, Eocene Atlantic. Assuming wind and current speeds typical of the modern Atlantic, and with the wind adding greatly to the hypothetical raft's speed by pushing against the crowns of trees growing on it, he came up with a surprisingly brief trip length of seven to eleven days. There are plenty of things to argue about in Houle's calculations, particularly the very strong effect of the wind, as if the raft were a sailboat, and the assumption that a crossing would take the shortest possible route across the ocean. Still, even if his calculations underestimate the journey's duration by a factor of two or three, they suggest that the crossing of a fairly wide Atlantic Ocean by monkeys is not out of the realm of possibility.

Another thing to consider is that the Atlantic is getting deeper as it gets older, because ocean floor subsides as it cools down from its beginnings as hot magma. This raises the possibility that certain areas that are now submerged might have been islands at some point. A Brazilian biology graduate student named Felipe Bandoni de Oliveira and two of his colleagues recently used a model that takes subsidence into account, along with plate tectonic movements and sea-level fluctuations, to reconstruct the "landscape" of the Atlantic over the past 50 million years. Their model indicates that between 40 and 50 million years ago, there were many more islands in the Atlantic than there are today. As they pointed out, those islands could have served as stepping stones for monkeys and other organisms; instead of making one extremely long voyage, monkeys could have island-hopped shorter distances, building up populations on each

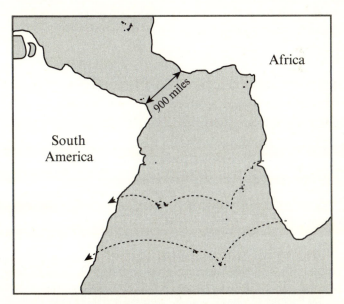

9.6 Possible island-hopping routes across the Atlantic based on a reconstruction of land configurations for 40 million years ago. Redrawn and modified from Bandoni de Oliveira et al. (2009).

successive island before another chance rafting event transported some of them to the next stepping stone (see Figure 9.6). The process would be analogous to the way Polynesians gradually colonized the islands of the South Pacific, although the monkeys' voyages would have been far more sporadic and difficult (since they didn't have boats), and probably would have taken millions rather than thousands of years.

The location of these ancient Atlantic islands actually would have required a much longer overall voyage from Africa to South America than would a nonstop crossing at the ocean's narrowest point. Thus, it is far from clear that the stepping-stone journey would have been easier than the nonstop voyage.* The importance of this exercise, then, is not so much to suggest what probably happened, but rather to add plausible routes, thus expanding the possibilities for what was undoubtedly an unlikely crossing no matter how it might have taken place. One can think of the stepping-stone idea as enhancing the how-possibly story through the accretion of small probabilities, just as buying more tickets to the lottery will increase one's chances of winning (without actually making winning likely).

The general point is that the problem of distance is not as daunting as it might at first have seemed. Forty million years ago, the Atlantic was apparently dotted with islands and roughly half as wide as it is today. It was obviously a huge barrier to the dispersal of land organisms, but not as huge as it would later become.

Patterns Within the Improbable

My argument to this point for monkeys crossing a wide Atlantic can be summarized as follows: Timetrees and fossils rule out alternative hypotheses, indicating that the Atlantic crossing did happen, and how-possibly

---

* An intriguing factor in the stepping-stone explanation is that it suggests the possibility of a "smoking gun," an unusual piece of evidence that could strikingly corroborate the reality of an ocean crossing (see Cleland 2002 for discussion of the "smoking gun" concept). Bandoni de Oliveira et al. (2009) noted that that deep-sea drilling in the South Atlantic has recovered fossils of shallow-water organisms as well as rocks that must have been formed either in shallow water or above the water in places that are now up to a mile below the ocean's surface. These findings could indicate the location of ancient islands reconstructed in Bandoni de Oliveira et al.'s model. Imagine that a sample from one of those submerged islands contained the fossilized bits of a monkey, and that this fossil turned out to be the closest relative of New World monkeys. That would be the geographic equivalent of a missing link, the intermediate piece that corroborated the larger sequence, in this case the stepping-stone journey across the ocean.

scenarios suggest how that crossing could have been accomplished. We can add here a third class of arguments, namely, observations of ocean journeys by other groups that suggest that the monkey case actually fits into some larger patterns, and is therefore not as inexplicably singular as it might at first appear.

The first observation is simply that land vertebrates have crossed the Atlantic far more often than is generally recognized. There is now strong evidence, mostly from molecular dating studies, for at least eleven such cases, including the monkeys. Admittedly, most of these ocean crossings have been by lizards or snakes, but one other crossing was by a mammal, the ancestor of the caviomorph rodents (guinea pigs, capybaras, and their relatives), an event that rivals the monkey's journey in its apparent improbability. Furthermore, in all eleven cases, the crossing has gone from the Old World to the New World rather than the other way around. The reason may lie in the nature of the ocean currents flowing in the two directions; in particular, a raft from Africa can pick up a current heading west right from the coast, whereas the only major current running eastward (the North Equatorial Counter Current) originates well off the coast of South America. It is thought that the configuration of those currents has been in place since the opening of the Atlantic, and therefore they would have influenced ancient crossings such as that of monkeys. In short, the Atlantic crossing by monkeys is not a completely singular event, but is one of at least eleven such crossings, all explicable by the paths of ocean currents, and these eleven crossings include another seemingly implausible voyage by mammals, that of the caviomorph ancestors.

The monkey crossing also fits a taxonomic pattern. I argued above that primates without boats are not great ocean voyagers. That is certainly true if one compares them to really proficient long-distance dispersers, such as many plants, strongly flying insects, ballooning spiders, tiny snails, birds, bats, geckos, tortoises, and crocodiles, among others. However, it is not true in relation to most other mammals. In fact, among orders of mammals (other than bats and marine taxa), primates are probably second only to rodents in their ability to cross significant expanses of ocean. Thus, the Atlantic crossings by monkeys and rodents, although obviously improbable events, actually fit an overall pattern; if you had to pick two kinds of land mammals to make a transoceanic voyage, those would be the two you would pick.

Ironically, Darwin, in the course of arguing against special creation in *The Origin of Species*, also categorically ruled out ocean-crossing monkeys. "Why, it may be asked," he wrote, "has the supposed creative force produced bats and no other mammals on remote islands? On my view this question can easily be answered; *for no terrestrial mammal can be transported across a wide space of sea*, but bats can fly across" (italics added). One wishes that Darwin could have seen the patterns just described and the evidence from the primate fossil record and timetrees, not to mention the evidence for improbable ocean journeys in general. In 150 years, much new evidence has come to light, and many things unknown to Darwin have been uncovered. As the Darwinian tree of life has been built up, using fossils and the anatomy and DNA of living species, strange and miraculous events have been revealed. Among many other things, the evidence shows that mammals have in fact been transported naturally across wide spaces of sea.

The Monkey Was a Swan

An obvious question arises from considering the monkey's transatlantic voyage and other improbable ocean crossings: Do these rare occurrences really matter in the history of life, or are they just inconsequential curiosities, bits of biogeographic trivia? In his 2007 book *The Black Swan*, the writer, stock market analyst, and self-described flaneur* Nassim Nicholas Taleb made the case that, at least in human affairs—whether social, political, economic, or personal—rare events are often extremely influential. In fact, Taleb argued that *most* important human events are unusual and cannot be predicted. Forecasting the fates of nations, the vicissitudes of the world economy, or your own personal destiny is next to impossible, because you cannot depend upon your experience of how things usually are or of what usually happens to anticipate what *will* happen. Things may carry on predictably for a while, but then something big and completely unforeseen takes place—Archduke Ferdinand is assassinated and the world tumbles into the Great War, the stock market rolls along happily on Monday and inexplicably crashes on Tuesday, two hijacked passenger jets explode into the World Trade Center. Taleb called such events "black swans" because Europeans thought such a creature was an impossibility until they ventured to Australia and found the native and mostly black *Cygnus atratus*. He defined "black swans" by three characteristics: they're rare, they have great impact, and they're not predictable except in retrospect.

---

* A flaneur is a person who strolls about aimlessly.

DINOSAURS TOO?

Seventy million years ago, near the end of the Cretaceous, a dinosaur died for some unknown reason and ended up buried near the shores of an ocean, on ground that would eventually become northeastern Italy, near the city of Trieste. This creature was closely related to the well-known duck-billed dinosaurs and, like them, it walked on two legs, had a long snout, and ate plants that it ground up between thick batteries of tightly packed teeth. Compared to the duck-bills and other near relatives, it was conspicuously dwarfed, being a "mere" twelve feet long, and it was probably a fast runner, judging by the proportions of its hindlimb bones. Its hands were like narrow, bony pillars, useless for grasping but perhaps used for balance when moving over rough terrain, as a person might use a pair of walking sticks. After it died, this particular dinosaur was quickly buried by sediment, and, through the vagaries of geological history, it ended up as a beautiful, nearly complete fossilized skeleton. Today the specimen is prominently displayed in a geological museum in Bologna.

This species of dinosaur has been dubbed *Tethyshadros insularis*. As the name implies, it lived on an island that, in broad context, was part of a large European archipelago in the Tethys Seaway, which separated Eurasia from Africa and the island continent of India (still many millions of years away from its collision with Asia). More specifically, the island was a piece of the Adriatic-Dinaric Carbonate Platform (ADCP for short), at the time an arc of islands similar to the modern Bahamas. The ADCP had perhaps been connected to an Afro-Arabian continent early in the Cretaceous, which could explain the existence of *T. insularis*, just as Gondwanan dinosaurs persisted on the islands of India, Madagascar, and Zealandia long after those areas had separated from the rest of the supercontinent. However, that sort of vicariance scenario almost certainly does not hold for *T. insularis*, for a simple reason: the record of sediments indicates that, toward the middle of the Cretaceous, high sea levels submerged the ADCP, wiping out its entire land biota. From a biogeographic point of view, this "drowning" transformed the ADCP into oceanic islands, as was the case with some other continental fragments. Under this scenario, any Late Cretaceous land organisms on the ADCP must have colonized the area by crossing a significant expanse of ocean.

As gigantic land animals with little known propensity to swim in the ocean, dinosaurs are usually viewed as thoroughly landlocked, and consequently their distributions have almost always been explained through overland movements and vicariance. However, if the "drowning" scenario for the ADCP is correct, *Tethyshadros* must be an exception to the rule that dinosaurs only dispersed by land. The progenitors of *T. insularis* are thought to have arrived by island-hopping from Asia, with each "hop" presumably involving a voyage on a natural raft, or, somewhat less plausibly, dinosaurs swimming or floating in the ocean.

Like monkeys crossing the Atlantic, these ocean voyages by *Tethyshadros* may sound absurd. In addition to the geological evidence for the drowning of the ADCP, however, the collection of other fossils found with *T. insularis* hints that the fauna was that of an oceanic island. In particular, the only other vertebrates identified to date are pterosaurs, which could have flown to the island, and crocodylians, a group known to have crossed saltwater barriers many times. Furthermore, in Hungary, in Late Cretaceous strata representing another part of the European archipelago, paleontologists have unearthed a species of ceratopsian (horned) dinosaur called *Ajkaceratops kozmai* whose occurrence also seems best explained by an island-hopping journey from Asia. The examples of *Tethyshadros* and *Ajkaceratops* and a few other, more tentative cases indicate the need to revise our views on the dispersal abilities of dinosaurs, just as work on living species has forced a reappraisal of other groups, such as frogs, burrowing reptiles, and mammals.

Whether or not one buys Taleb's argument for the overarching importance of black swans in human affairs, there is no denying their occurrence, if not their fundamental significance, in evolutionary history. The most notorious such event is the extraterrestrial impact that probably finished off the dinosaurs and many other taxa, thus providing opportunities for other groups, such as mammals. That event was in some sense a chance occurrence and wiped out more than half the species on Earth. We are likely a result of that black swan from space. Stephen Jay Gould and others have argued that evolutionary history has been punctuated by a series of such unpredictable events, determining the kinds of organisms that have come into being, which groups have survived and which have not, all the way back to the origin of life.

In this chapter and in Chapter Eight, we have dealt with high rungs on the ladder of increasingly improbable dispersal events, ending with the rarest of the rare, monkeys crossing an ocean. Can we think of these cases as black swans? These dispersal events certainly qualify as highly unexpected; nobody with a knowledge of the characteristics of worm lizards and monkeys, for instance, would have guessed that those animals would cross the Atlantic Ocean. But rare does not necessarily mean impactful. Are we dealing with the equivalent of Leif Ericsson's voyage to North America, which seems to have had little influence, or Columbus's, which deeply altered the course of world history?

Many rare dispersal events, such as skinks reaching Fernando de Noronha, probably fall into the Leif Ericsson category. However, others have had a much larger impact. In the case of worm lizards, for instance, one colonization event apparently generated a radiation of nearly a hundred modern species spread from southern South America to the West Indies. There's a good chance that some of them will be burrowing through the soil of the New World long after we humans have disappeared from the Earth.

For monkeys, the single colonization of the Americas produced some 130 modern species in a startling array of forms. Monkeys are such a common sight in the forests of the New World tropics, and are such a key part of ecological communities—as foragers, prey, and seed dispersers, for instance—that it seems incongruous to think that their presence is a fluke, the result of a single epic journey in the face of almost unfathomable odds. A raft from Africa reaches the South American shore, some exhausted monkeys amble off, and, over millions of years, give rise to squirrel monkeys, howlers, and capuchins, owl-eyed night monkeys, and bald-headed uakaris. Or the raft founders, the monkeys drown, and we end up with, well, who knows exactly what, but something very different; maybe some of the monkey niches are eventually filled by rodents or marsupials, or maybe not. If you were around in the early Eocene and had to make a bet, you'd definitely take the second scenario—the monkeyless Americas—over the first. It just goes to show that in evolutionary history, as in human lives, odd and unpredictable things happen, and sometimes they change the world.

Cerithideopsis *is a genus of small snails that live in intertidal mangrove and mudflat environments.* The North American sister species Cerithideopsis californica *and* C. pliculosa *from the Pacific and Caribbean coasts, respectively, are quite distinct genetically, having been isolated since the emergence of the Isthmus of Panama 3 million years ago. However, in certain areas on both coasts, one can find snails with mitochondrial DNA sequences that are geographically out of place, indicating that some of their recent ancestors were from the "wrong" ocean. This misplaced DNA suggests that* Cerithideopsis *snails jumped over the land barrier twice within the past million years, and that they did it both ways, once from the Pacific to the Caribbean and once the other way around.*

The researchers who did this genetic work speculated that the snails could have been transported alive in the guts of birds. They pointed in particular to a study published in 1993 showing that common shorebirds called Willets often eat Cerithideopsis *snails. Of course, Willets don't swallow small shelled creatures for the purpose of moving them from one ocean to another; the usual outcome is that the snails are ground to pieces in the birds' gizzards, with the broken bits of their shells and other undigested remains being regurgitated as pellets. However, death by grinding is not the fate of every ingested snail. In that 1993 paper, the author reported placing four Willet pellets in dishes of seawater, whereupon a total of twenty-nine snails came out of their shells and began to crawl about the dishes.*

Chapter Ten

THE LONG, STRANGE HISTORY OF THE GONDWANAN ISLANDS

The Lost Worlds of the Southern Seas

Madagascar. New Zealand. The Falklands. The Seychelles. New Caledonia. The Chatham Islands. From my vantage point in landlocked Nevada—at this particular moment, sitting in a Starbucks in Reno with a coffee, an apple bran muffin, and a view of the somewhat shabby Circus Circus casino—they all sound like wildly exotic destinations, these islands of the southern seas. Images pass through my mind, unfortunately most of them daydreams rather than memories: a giant, mottled gecko camouflaged against a tree trunk in New Caledonia, ring-tailed lemurs bounding through the forests of Madagascar, a kiwi picking out sandhoppers from the seaweed on a beach in New Zealand, flightless steamer-ducks bobbing in the waves off the rocky coast of the Falklands. More than just backdrops for a parade of exotic creatures, however, these islands have come to represent a pivotal point in the debate over vicariance and dispersal, about why living things are found where they are.

All of these islands were once part of Gondwana and all have been disconnected from other major landmasses for many millions of years, thus making them natural test cases for comparing the importance of ancient vicariance and long-distance dispersal. Nearly all have been pointed to, at one time or another, as supporting the vicariance worldview, in the sense of harboring relict Gondwanan biotas. However, as we have seen, that view clearly does not hold up for the flora of New Zealand, which is overwhelmingly made up of the descendants of overwater colonists. Explain New Zealand, Gary Nelson had said, and the world would fall into place around it, and, for plants at least, the world did, a world in which the effects of continental drift have been obscured by a web of ocean crossings.

Nonetheless, my focus to this point on taxa—specifically, plants and various kinds of land vertebrates—leaves open the question of whether the biotas of certain areas might still be best viewed as products of Gondwanan vicariance. Specifically, are the floras and faunas of some or most of the Gondwanan islands dominated by groups that have persisted in place since the breakup of the supercontinent? Are these islands, as some would have us believe, "lost worlds," where the evolutionary histories of ancient lineages parallel the ancient geologic connections? We have already seen that Gondwanan breakup is not *the* great general explanation for Southern Hemisphere distributions. But might it at least be paramount for the ancient continental islands?

An Ark to the End of the Earth

The Falkland Islands, or Islas Malvinas, lie some 350 miles east of southern South America, at roughly the same latitude as the Straits of Magellan, which is to say, pretty close to the end of the Earth. By all accounts it is a harsh place: bleak, treeless, and more or less unrelentingly cold, damp, and windy, the muted landscape dissected by great rivers of quartzite boulders called *stone runs*. Some people find the Falklands beautiful, but it is, at best, a severe beauty.

Biologically, the archipelago is a kind of poor man's Galápagos, with many unique species, including that flightless steamer-duck and, until the 1870s, a native canid called the Falkland Islands wolf (or warrah), but nothing as bizarre as, say, a seagoing iguana or a tool-using finch to capture the imagination. The Falklands are remembered, of course, for the three-month-long war in 1982 between Great Britain and Argentina, but these days, they're probably best known as a place where cruise ships stop for the dolphins, seals, and gigantic colonies of Black-Browed Albatrosses and Rockhopper Penguins. In the history of biogeography, these islands are just a couple of footnotes, one for the excessively tame and hence extinct wolf, and the other for the fact that Darwin visited the archipelago during the *Beagle* voyage and left with a bad impression, finding it a "wretched place," where the country was all "more or less an elastic peat bog."

Bob McDowall visited the Falklands in 1999 for the first and, as it would turn out, only time. As with Darwin, who had wondered if the wolf might have reached the islands on an iceberg, the Falklands made McDowall think about long-distance dispersal. That wasn't unusual; wherever McDowall happened to be, that subject was often on his mind.

This was the same Bob McDowall who, as a young ichthyologist and biogeographer in the 1970s, had baldly challenged the vicariance scientists, arguing, among other things, that we know that long-distance dispersal occurs because people have actually seen it happen. It would be a significant understatement to say he had drawn some ire for his views: the American Museum of Natural History ichthyologist Donn Rosen, after reading one of McDowall's manuscripts, had vowed to "blast him out of the water," and an anonymous reviewer of the same paper had threatened to "destroy this person [McDowall] as a scientifically minded naturalist." (I'm not exactly sure what that threat means, but it doesn't sound good.) Croizat was even more blunt: he said he wanted to "execute" McDowall.

It wasn't easy being a dispersalist in the mid-1970s, and even in the late 1990s it wasn't exactly fashionable. But McDowall had never wavered from his old-fashioned, dispersalist viewpoint. After getting his PhD at Harvard, he had returned to his native New Zealand and had fashioned an impressive career out of research on fishes called *galaxioids*, combining "pure" science with studies that were important for the country's fisheries. Galaxioids spend the early part of their lives at sea—in fact, they're harvested as immature "whitebait"* as they return to the rivers where they will live as adults—and this makes them good candidates for dispersing across oceans. It only takes one life stage to disperse, and that stage doesn't have to be, and often isn't, the adult. Indeed, McDowall had argued throughout his career that these fish, in the whitebait stage, had crossed oceans not just once or a few times, but over and over again. Freshwater fishes are not supposed to be able to cross ocean barriers, since they quickly die in salt water, but galaxioids, McDowall had repeatedly pointed out, are not exactly freshwater fishes. It was this kind of thinking—damned benighted dispersalism filled with assumptions about what kinds of creatures can and cannot cross oceans—that had gotten him into hot water with the vicariance biogeographers back in the 1970s. Never mind that he made a lot of sense.

---

* In New Zealand, whitebait are usually mixed with beaten eggs, milk, and flour, and served as patties. The patties are said to be better if there's more whitebait than batter.

10.1 Bob McDowall seining for fish on the Chatham Islands, east of New Zealand. McDowall thought the vicariance viewpoint ignored biological realities. Photo by Brian Stephenson.

McDowall had been thinking about the Falklands for a long time. Given his pointed criticisms of vicariance biogeography over the years, one might have thought that he saw the archipelago as a testing ground for his dispersalist views. At first, however, that wasn't on his mind at all. McDowall was a "fish guy," and he was dreaming about the Falklands because he had heard that there were galaxioids in the islands' streams (Darwin had even collected some), and that nobody had ever done a survey to figure out exactly what species were there. He wanted to find out. Sometimes the motivations of biologists are pretty simple. Or, at least, they're simple before they get complicated.

McDowall got some funding from the National Geographic Society and the Falkland Islands government and took off for the archipelago. Unfortunately, his luggage didn't reach the islands with him, but that didn't prove too much of an obstacle. He bought some new clothes, and the locals helped him out at every turn, even replacing a lost electric-fishing electrode—used to stun fish so they could be collected—with a jury-rigged device partly made from an old golf club.

For three weeks, at first by himself and then with two assistants, McDowall raced around collecting fish on the two main islands, East Falkland and West Falkland, initially using the golf-club device ("We were terrified of getting electrocuted," he said) and, later, once the lost luggage caught up with them, with the bona fide electric-fishing unit. They collected electroshocked fish at a breakneck pace, hitting 151 sites, mostly in pebbly or gravelly streams with water stained dark from the peat. Not surprisingly, they found just a few species, including one kind of galaxioid. The three biologists ended up writing a book, which they straightforwardly called *Falkland Islands Freshwater Fishes: A Natural History*. With only a few native and introduced species to write about, the book was very short. ("My wee book," McDowall called it, betraying his Scottish ancestry.) Nothing they found out would set the scientific world on fire, not even the very small world of galaxioid fish biologists.

However, while on the islands McDowall talked with Tom Eggeling, an environmental officer working for the Falklands government, who said something about the geology of the archipelago that got McDowall thinking about the deep biogeographic history of the place. The Falklands, Eggeling had heard, were once part of Africa. Somehow they had broken away from that continent and wandered, over many millions of years, to their current position off the east coast of southern South America. More specifically, as McDowall later found out, the small tectonic plate that contained the Falklands had been attached to southeastern Africa back in the Jurassic, back when Gondwana was still Gondwana (see Figure 10.2). The main evidence for this ancient connection is that various geological strata of the Falklands—Devonian sandstones, Permian shales, and Jurassic basaltic volcanics, among others—match those of southeastern Africa so closely that a former connection seems inescapable. (It's the same kind of evidence that Alfred Wegener had emphasized in his original formulation of the theory of continental drift, and it still makes perfect sense.) As the pieces of the supercontinent drifted apart, the Falklands microplate broke off on its own, made a right turn around the tip of Africa, and headed out into the newly forming Atlantic Ocean, rotating 180 degrees in the process and eventually ending up entrenched in the continental shelf of South America. In essence, the bedrock of the Falklands had started out as a piece of Africa and wound up riding on South America's submerged coattails.

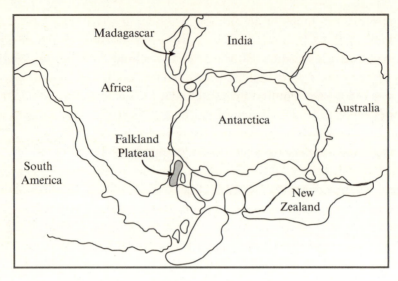

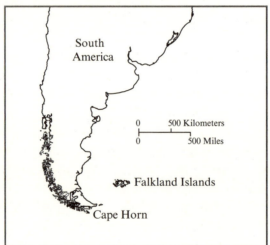

10.2 Movement of the Falklands. Upper: position of the Falkland Plateau before the Middle Jurassic. Lower: present position of the Falkland Islands. Jurassic reconstruction redrawn and modified from Thomson (1998).

Wheels started turning in McDowall's head. On the one hand, if the Falklands had been connected to Gondwanan Africa, then they must at one time have had plants and animals from that area. On the other hand, the archipelago had been sitting off the coast of South America for a long time now, long enough to have received many immigrant species from that continent. Did the modern biota still show a strong signature of its ancient Gondwanan connection, as a vicariance view might suggest, or was it dominated by more recent colonists from South America? Under the Gondwanan vicariance hypothesis, you would expect Falklands' species to be most closely related to groups that were found in Africa, or spread widely among the fragments of Gondwana. (The related groups didn't have to be confined to Africa, because, when the Falklands were part of the supercontinent, Africa was still connected to other Gondwanan landmasses, potentially allowing easy movement of land organisms between Africa and those other areas.) Also, these evolutionary relationships would be distant ones, because the Falklands had been separated from the rest of Gondwana for something like 150 million years; for instance, you wouldn't expect to find many of the same species or even the same genera in the Falklands and elsewhere. Under the overseas immigrant hypothesis, you would expect most of the island species to have their nearest relatives in South America, and many of these relationships would be very close, reflecting recent dispersal from the continent. McDowall knew that the native galaxioid, *Galaxias maculatus*, favored the overseas immigrant hypothesis—this same species is found in South America, but not in Africa. But what about the rest of the biota? He had a good idea what the answer was, but he needed to hit the library and make the case concrete.

The case, as it turned out, was very clear. Consider, for instance, the islands' only native mammal, that extinct wolf, *Dusicyon australis*. The consensus among taxonomists was that it was closely related to some South American canids and not to any African (or more widespread Gondwanan) group of carnivores (see Figure 10.3). (When McDowall wrote his paper, the best guess was that its nearest relative was a living species called the culpeo. Recent DNA analyses using museum specimens place *D. australis* as the sister to *D. avus*, another extinct South American canid.) So count one case for the overseas immigrant hypothesis.*

---

* A complication here is that, according to the latest molecular dating analyses, the ancestors of *D. australis* reached the Falklands during the most recent glacial period, when sea level was considerably lower than it is today and the islands were separated from the mainland by a strait that may have been as little as twelve miles wide (Austin et al. 2013). That strait could have been filled at times with sea ice or glacial ice from the nearby continent, which means that these animals might have made it to the islands under their own power. It's unclear whether a dispersal event of that sort would qualify as normal or long-distance dispersal. An alternative is that *D. australis* might have colonized the islands as passengers on drifting ice, the same idea that Darwin entertained when he encountered the living wolves on his *Beagle* voyage. (The wolves became extinct in 1876, some forty years after Darwin's visit.)

10.3 The extinct Falkland Islands wolf, a recent immigrant from South America, not a Gondwanan relict. Drawing by George R. Waterhouse.

Or take a group on the other end of the spectrum in terms of diversity, the insects, with a few hundred species on the Falklands. The great majority of them have their closest relatives—either populations of the same species or species in the same genus—in Patagonia or on various sub-Antarctic islands. The Falklands insects have no obvious connection to Africa or to Gondwanan landmasses in general. Jumping over to another large group, flowering plants, McDowall found that there are about 140 known Falklands species and, again, their relatives are overwhelmingly in South America and/or on sub-Antarctic islands, with no clear ties to Africa or Gondwana as a whole.

In his paper, McDowall ran through all the evidence he could find—for birds, spiders, crustaceans, snails, annelid worms, mosses, and lichens, among other groups. Taken together, the relationships left no doubt that the main evolutionary connections of the Falklands are to Patagonia and other cool-temperate lands of the Southern Hemisphere. The relationships are close, too, with many Falklands species and almost all the genera shared with Patagonia or other areas, indicating relatively recent arrival on the islands. Potential African or Gondwanan connections are few and far between, and even most of these are debatable. In McDowall's words, "There are only, at best, fragmentary residues in the Falklands biota that may point to an African connection." In short, during the Falklands' long journey to the edge of South America, the original African/Gondwanan biota had disappeared and been replaced, for the most part, by immigrants from nearby Patagonia. McDowall imagined the history of the archipelago as two converging trajectories, the islands themselves drifting westward to eventually intercept the eastward dispersal of species coming from South America. One can picture the second trajectory as a combination flotilla and cloud of organisms variously swimming, rafting, flying, and drifting on the wind, with the ancestors of the wolf and the flightless duck as the flagships.

On one level, McDowall's paper was not surprising at all. Almost everything that had ever been written about the flora and fauna of the Falklands had indicated links to South America and not to Africa. For instance, Darwin, who had hardly made a thorough study of the subject, wrote to his friend Joseph Hooker, "The Falkland Isld. flora seems to combine the Patagonian with the Fuegian," the latter referring to Tierra del Fuego, at the southern tip of South America. Similarly, it wouldn't take a deep investigation to figure out that Falklands birds were mostly connected to South America rather than Africa; a casual comparison of field guides to the birds of South America and Africa would tell you that. Even hard-core vicariance scientists—including Léon Croizat himself, the originator of panbiogeography—had linked the Falklands biota to Patagonia.*

The significance of McDowall's study then wasn't really in the concrete results, which would have been anticipated by anyone with even a passing knowledge of the natural history of the Falklands, but in how he placed the results within the deep history of the islands. Anyone who understood vicariance biogeography and knew that the Falklands had once been part of Africa might have expected to find a significant African/Gondwanan element in the islands' biota. From that perspective, it was a bit of a shock to learn that there was nothing or next to nothing left from those Gondwanan beginnings. By emphasizing the African roots of the Falklands as a landmass, McDowall was setting up the contrast between the ancient geologic history of the islands and their current, obviously South American biota, not to mention taking another jab at his old nemeses, the vicariance scientists.

---

* In a 2005 paper, Juan Morrone and Paula Posadas, two vicariance biogeographers (although not extreme ones), suggested that the Falklands have been connected by land to South America for most of the past 130 million years. However, the plants and animals of the Falklands argue against such an extended connection. In particular, the biota has a very island-like character, with relatively few insects, no amphibians, and only the one native mammal, the extinct wolf. This depauperate biota indicates colonization mostly or entirely by overwater dispersal.

Many Paths to Oblivion

In the biblical story, Noah brought onto the ark the cattle and beasts, every bird and creeping thing, and, after five months on the ocean, these animals, apparently none the worse for wear, disembarked on Mount Ararat. By analogy, New Zealand has been dubbed "Moa's Ark," and the Gondwanan fragments in general have been called "life-rafts." However, the history of the Falklands calls into question the suitability of that analogy. In particular, Noah and his animals were on the ocean for 150 days, whereas the Falklands separated from southeastern Africa 150 million years ago, a rather striking difference. For our purposes, the important point is that a lot can happen in 150 million years and, from the perspective of the original passengers on the "ark," most of it is not good.

Think again about the world as seen through the lens of vicariance. Begin with one landmass, filled with living species. If we start, say, in the Jurassic, those species might include mosses, ferns, and cycads; cockroaches and dragonflies; frogs and shrew-like mammals; pterosaurs and dinosaurs. Imagine that this landmass rifts apart and the two pieces begin to wander off in different directions. Now jump forward 150 million years or so to see what we have. In the vicariance view, we would expect to find related organisms on the two, now well-separated landmasses. All that is required is a kind of biological version of inertia—the lineages in question have to persist in both areas.

This argument is persuasive. It appeals to our sense that a straightforward explanation is better than a convoluted one. It's the reason there has been such a strong expectation among biogeographers that the Gondwanan continents and islands should still retain a clear biological signature of their common origins. The problem is that it ignores a hugely significant aspect of evolution, namely, extinction.

Of course, every biologist knows that species and entire large groups go extinct, and no one thinks that all or even most of the lineages that existed when Gondwana began to break up are still around. Obviously, there are no living pterosaurs or dinosaurs (except for birds). Still, the implications of extinction haven't always penetrated as deeply as they should. In particular, biogeographers haven't given enough thought to the disappearance of lineages when contemplating the fate of old continental islands like the Falklands.

The great metaphor of evolution is the tree of life. The branches are lineages, and the places where one branch grows out of another are where one species splits into two. Drawn as a living tree with a robust trunk and healthy green foliage, the metaphor suggests the exuberant proliferation of life. That image seems to agree with our present knowledge of the living world—the fact that a tropical rainforest might have hundreds of species of trees with hundreds of kinds of insects unique to each one, that new life forms are constantly being discovered in seemingly inhospitable places like boiling springs and sulphurous ocean vents, that nematodes are so numerous that their bodies alone could give a detailed outline of most of the landscape of the Earth. Life *is* exuberant. And yet, nearly all species that have ever existed have left no descendants that are alive today. Most of the exuberance has gone down a one-way path to oblivion.

For most small islands, attrition by extinction is unavoidable. Specifically, if an island does not provide a species with an opportunity to speciate, that is, to form additional new species, then the set of lineages descended from the original inhabitants can only get smaller; species will inevitably go extinct and, without having given rise to any other species, that will mean the end of the lineage. The life span of a species is typically only a few million years, so the impact of extinction on the original lineages of such islands would be huge over the long life of Gondwanan islands such as the Falklands.

To make matters worse, that impact will be exacerbated on islands by the simple fact that a species' chance of going extinct is connected to the amount of area it occupies; the smaller the area, the greater the chance of extinction. Biologists have come up with several explanations for this relationship. One is that, if a species is found at only a few sites, there's a greater chance that some event or events will make all the sites uninhabitable. Think of it as not having your eggs spread in enough baskets. For instance, consider the Devil's Hole pupfish, a Mojave Desert species whose entire native range is a warm-spring pool with a surface area about the size of two bowling lanes. An earthquake or flood that disrupted the pupfish's pool could wipe out the entire species. Another risk is that species with small ranges will generally consist of few individuals, and this makes them vulnerable to what ecologists call *stochastic extinction*. The idea here is that, even without any unusual environmental disruption, all populations experience random fluctuations in size (number of individuals), and the smaller the average size of the population, the more likely that one of its downward swings will hit zero. These random fluctuations are sometimes referred to as a *drunkard's walk*; for small populations, you can picture the drunkard (representing population size) lurching back and forth close to a precipice, the precipice being extinction.

Ecologists and conservation biologists continue to argue about which of these factors, or others I haven't mentioned, are driving the connection between small area and high extinction risk, but no one doubts that the connection exists. This is why almost every species that has a very small geographic range is considered "at risk"—from the Kirtland's Warbler, which breeds only in stands of young jack pines in a small part of northern Michigan, to the Devil's Hole pupfish in its single spring pool, to many of the endemic species on Hawaii or the Galápagos.

For large islands, the effect of small area on extinction will be less severe than it is for small islands, but the effect is probably still present. Even a very large island cannot harbor extremely widespread endemic species; the area of the entire island of Madagascar, for example, is far smaller than the range of continent-spanning species like the coyote or the American Robin. However, for large islands, or archipelagoes of smaller islands, speciation becomes important and can counter attrition by extinction; if the original species form new ones, then the extinction of a species does not necessarily translate to the loss of that particular lineage. Speciation, on islands and in general, can greatly prolong the life of taxa; in fact, almost all living species probably owe their existence to the evolutionary branching of their ancestors. One can imagine that, through this process of species proliferation, many of the original lineages on Gondwanan islands have managed to persist to the present. Maybe such lineages even dominate the biotas of some of these islands.

However, paleontologists tell us unequivocally that there is more to extinction than just slow attrition. Sometimes, there are cataclysms, and there are reasons to believe that these events may be especially devastating for the biotas of islands.

One stereotype of a paleontologist is a rough character in jeans, cowboy hat, and boots, somewhere out in the sagebrush-covered backcountry of Wyoming, trying to unearth a *Tyrannosaurus* with a pickax. The paleontologists I'm thinking of are not that guy; they tend instead to be cerebral, East-Coast types, more comfortable with programming languages than digging implements. They use the fossil record built up by the *Tyrannosaurus* hunters and other field paleontologists to see grand patterns in the history of life. Some of them have essentially spent their entire careers entering reams of data concerning fossils into computers, subjecting these data to arcane and ever more complex analyses, and seeing what the results tell us about the origin and demise of species and larger groups.

The reams of data and arcane analyses have corroborated with hard numbers what was known before in a qualitative way: rates of extinction are highly variable through time. Graphs plotting the number of extinctions (of genera or families usually) through the geologic stages show a chaotic series of peaks and valleys of wildly different magnitude, with the especially prominent peaks representing mass extinctions, when many groups disappeared within relatively short spans of time. The great extinction that finished off the dinosaurs and many other taxa at the end of the Cretaceous is the best-known peak, one of the so-called Big Five, but it's not the biggest. That honor goes to the end-Permian extinction—what Stephen Jay Gould called "The Great Dying"—which the number-crunching paleontologists tell us knocked off more than 90 percent of all species on the planet.

Paleontologists argue about exactly what constitutes a mass extinction, but there is at least general agreement that the worst of them represent real catastrophes, times when something more than just "extinction as usual" was going on.* These great pulses of extinction suggest some kind of rare event or set of circumstances, the impact of the comet or asteroid at the end of the Cretaceous being a prime example. We may be witnessing a mass extinction right now—one that's not at the level of the Big Five yet, but unfortunately has no end in sight—and it is certainly the result of an unusual set of events, namely, the origin of *Homo sapiens* and our series of technological "advances."

---

* Some scientists have claimed that even the greatest extinctions are not really outliers, because they are predicted by a simple power law describing the frequency of extinction events of different magnitudes. For at least some compilations, extinctions twice as large as some baseline are about four times less frequent, extinctions four times as large are roughly sixteen times less frequent, and so on up to the most extreme events on record. The fit of extinction data to a simple power law has been used to argue that even the largest mass extinctions have the same underlying causes as background extinction (Buchanan 2001). The notion here is that small perturbations can have consequences ranging from very small (few extinctions in a time period) to very large (mass extinctions), the latter occurring when small perturbations cascade unpredictably to produce disproportionately large effects. A common analogy is that dropping a single grain of sand on a sandpile can cause just a few other grains to move or can set off a massive avalanche involving millions of grains. However, making this connection between the power law and underlying cause is something of a leap of faith. In particular, the power-law distribution of extinction frequencies could be caused by a parallel distribution of underlying causes; that is, it could be that the perturbations are not all small, but instead have a great range of magnitudes—from single grains of sand to giant boulders, so to speak—that fit the power law. Along with most paleontologists, I am following this latter notion in assuming that mass extinctions have unusual causes.

In discussing the slow attrition of lineages and the area effect on islands, we've essentially been dealing with extinction during periods when nothing extraordinary is happening. Now let us consider the cataclysmic. What effects might mass extinctions have had on the Gondwanan islands?

Returning to the Falklands, we know that, after their separation from Africa, they experienced one of the Big Five, the end-Cretaceous mass extinction. Since that event was global (perhaps caused by the sun-blocking dust cloud raised by the asteroid or comet impact), there is no reason to think the biota of the Falklands could have avoided it. In fact, it's very possible that the proportion of species that went extinct was even greater on islands than on continents, simply because of the stochastic extinction numbers game; whatever caused the mass extinction in general might have reduced the already small populations of island species to the point where they were dangerously close to suffering extinction from random fluctuations in population size. In other words, during the end-Cretaceous event, a disproportionate number of island species might have taken the drunkard's walk off the cliff of extinction.

The Falklands also passed through several other minor global mass extinctions, including one at the end of the Eocene, about 34 million years ago, and another in the middle of the Miocene, some 14.5 million years ago. However, it seems very likely that the islands' biota also suffered from what might be called "local mass extinctions," events that wiped out many species, but only in a small region. In his paper, McDowall mentioned events of this kind, connected to the Pleistocene Ice Age cycles, to explain the disappearance of the original African biota of the Falklands. He suggested that these oscillations in climate, repeated several times, could have inflicted "a severe sorting of the Falklands biota." What he meant was that these climatic shifts might have generated pulses of extinction, episodes during which many species got "sorted" either into the category of survivors or into the trashbin of history. On continents, the ice-age cycles apparently did not cause mass extinctions, because species that couldn't handle the climate change could move. For instance, during the most recent ice age in North America, the ranges of many species, from voles to pine trees, shifted to the south. However, on islands like the Falklands, this would not have been an option. If the climate of the Falklands became too cold, or the growing season too short, for a particular species to survive, the members of that species would have nowhere to go. Terrestrial species can't shift their ranges into the ocean.

In a way, the Falklands, viewed over their long history, look like a perfect storm of extinction. To begin with, they're small enough, at least in their current configuration, to have experienced a strong effect of area on extinction rate. Also, in their recent history, at least, the islands haven't provided much opportunity for the formation of new species, the other variable in the attrition formula; nearly all of the endemic lineages consist of just one or a few species. Perhaps most importantly, they've had such a long history of isolation that they have almost undoubtedly been battered by multiple mass extinctions, including global ones, like the end-Cretaceous event, and local ones, such as those connected to the Pleistocene Ice Age cycles. Maybe it's no wonder that there's not a single clear example of a living Falklands species whose ancestors were on board the microplate as it separated from Africa. The perfect storm has obliterated almost any trace of the islands' Gondwanan origins.

From the Black Robin of the Chathams to the Lemurs of Madagascar

Compared to the other Gondwanan islands, the Falklands may be a somewhat unusual case. However, of the remaining islands, we can say one thing right off the bat: even with their perfect storm of extinction, the Falklands cannot possibly be more extreme than the Chatham Islands.

The Chathams, which lie some 420 miles east of New Zealand, are a small archipelago with an area less than that of New York City. They're cold and windy, but perhaps not quite as severe as the Falklands; the Chathams do have patches of native forest, for instance, and not so many peat bogs. As with many islands, the human history of the Chathams has been defined by remoteness. Maori from New Zealand colonized the archipelago at least five hundred years ago, but the descendants of those colonists, the Moriori, afterward remained isolated until the British reached the islands in 1791. By then the Moriori had developed a distinctive, peaceful culture—they had literally outlawed warfare—which unfortunately made them easy victims for the bellicose Maori, their distant brethren, who followed in the wake of the British. The Maori ended up killing or enslaving nearly all of the Moriori. Today, the population of the Chathams—a mix of European, Maori, and Moriori—is dwindling through other effects of remoteness, namely, the high cost of living and the desire to be connected to the larger human world.

All of that, of course, happened within a thin slice of geologic history. In deep time, the Chathams haven't always been isolated from other landmasses. Geologically speaking, they are part of the Chatham Rise, most of which is currently underwater and shows up on ocean maps as a pale blue swath of shallow water running back to New Zealand's South Island. (Parts of the Rise are more than 3,000 feet below the ocean's surface, but, compared to surrounding ocean regions, this is shallow.) However, at one point, more than 70 million years ago, most of the Chatham Rise was above water and formed part of Zealandia, and, a little further back in time, Zealandia made up a large chunk of the eastern edge of Gondwana.

From this deep history, vicariance scientists have supposed that at least some species on the Chathams are Gondwanan relicts. That there *was* a Gondwanan biota on what is now the Chathams is undeniable: there are Late Cretaceous strata on the islands that contain the fossil remains of a typical Gondwanan flora, including southern beech trees, araucarian conifers, and podocarps. These beds even contain some unimpressive but unmistakable bones of theropod dinosaurs, relatives of *Tyrannosaurus*. But the big question, as with the Falklands, is, "Did any of those Gondwanan lineages persist to the present?"

The rocks of the Chathams tell us the answer is, "Not likely." Strata about 6 million years old are made up of sediments from lava flows and volcanic ash that indicate the emergence of a volcanic island. However, for most of the Cenozoic before that time, the area is only represented by marine sediments, such as limestones (made up of the hard parts of marine organisms) and underwater volcanics, suggesting that the Chathams were completely submerged.* Thus, we have indications of the ultimate in local mass extinctions: if there were any Gondwanan relict species on the islands before their inundation, afterward there would certainly have been none.

I'm reminded here that, when Tara and I were planning our trip to New Zealand, we briefly thought about taking the two-hour plane ride from Christchurch to the Chathams. That we decided against it was partly about time, partly about money, and, in a fairly direct way, partly about the fact that the whole place was probably underwater until recently. For me, the allure of the Chathams was their unique birds; for almost any serious birder, there's a weird thrill in seeing a species you know can only be found on some small and remote piece of ground. I imagined spotting a little songbird called the Black Robin on the Chathams, the only place on Earth where that bird can be seen. However, browsing the *Hand Guide to the Birds of New Zealand*, I was struck by the obvious similarity of all the

---

* The volcanic island that formed around 6 million years ago may have completely eroded away some 2 million years later, producing an even more recent submergence (Heenan et al. 2010). However, the evidence for this later submergence is less clear.

10.4 Not such a draw for the birdwatcher: the Black Robin of the Chathams (left) is a close relative of the similar North Island Robin of New Zealand (right). Photos by Frances Schmechel (Black Robin) and Tony Wills (North Island Robin).

endemic birds of the Chathams to New Zealand species. The Black Robin looks a lot like the New Zealand robins and the Tomtit, the Chatham Island Warbler like the Grey Warbler of New Zealand, and the Chatham Island Oystercatcher like New Zealand's Pied Oystercatcher (see Figure 10.4). The fact that none of the Chathams birds are as bizarrely unique as, say, a kiwi, or a kakapo, probably has everything to do with the islands' submergence; under that scenario, the ancestors of all the endemic birds must have arrived after the archipelago reemerged from the sea, and therefore they would have had little time to evolve into anything very distinctive. I'm not sure I could have resisted the lure of birds whose origins trace back tens of millions of years to the fragmentation of Gondwana, but species that hardly look different from their modern New Zealand relatives weren't such a draw.

The idea that species on the Chathams evolved only recently from ancestors that lived elsewhere has been confirmed by extensive molecular studies led by Steve Trewick, Adrian Paterson, and other scientists from New Zealand universities. These researchers have sequenced genes from a diverse collection of endemic Chathams plants and animals, including daisies, gentians, grasses, rails, parakeets, cockroaches, beetles, earwigs, and grasshopper relatives called wetas. They found exactly what one would expect for islands recently emerged from the sea: all the sampled Chathams species are very similar genetically to related species in New Zealand or elsewhere, supporting the idea that they arrived on the Chathams within the past few million years.

In short, there is no evidence whatsoever of Gondwanan relict species on the Chathams. Instead, the geological and biological evidence dovetails neatly to indicate that, although the Chathams once had a Gondwanan biota complete with southern beech trees and dinosaurs, the entire modern flora and fauna reached the islands only very recently and had to cross ocean barriers to do it. The slate was wiped clean, and a new story, with an entirely new cast of characters, was written on it.

In great contrast to the Chathams, New Caledonia, another piece of Zealandia, harbors an array of bizarre, seemingly ancient lineages that scream out for a Gondwanan interpretation. The most famous of these are perhaps *Amborella trichopoda*, a rainforest shrub that is likely the sister group to all other flowering plants, and the Kagu (*Rhynochetos jubatus*), a secretive forest bird that resembles a heron and is apparently most closely related to the Neotropical Sunbittern (which is not a bittern). New Caledonia also has the only known parasitic gymnosperm—a podocarp shrub that grows on the roots of another podocarp species—and, until people arrived and probably drove them to extinction, it was home to a large horned turtle and a small terrestrial crocodile that were the last surviving members of their respective genera. The island's deep history is also suggested by substantial endemic radiations of skinks, geckos, and *Araucaria* conifers, among other groups.

All of these lineages and others have been recognized as Gondwanan relicts. However, both geological and biological evidence suggest that, despite appearances, New Caledonia's history actually parallels that of the Chatham Islands. Geologists now think that New Caledonia was entirely underwater from about 37 to 70 million years ago, a period corresponding to a general subsidence of the area and, later, overthrusting of oceanic crust onto the original continental crust as New Caledonia collided with an island arc. Only marine strata are known from the period of inundation, including layers composed of a type of fine-grained chert that must have formed deep under the ocean.

As with the Chathams, molecular dating results are in line with this history of submergence and reemergence and, therefore, of mass extinction and recolonization. Many New Caledonian lineages, including the skinks and *Araucaria* conifers, are estimated to have split from relatives elsewhere only after the "drowning," indicating overwater colonization rather than persistence from Mesozoic times. Others, such as *Amborella*, diplodactylid geckos, and troglosironid harvestmen, diverged from relatives before

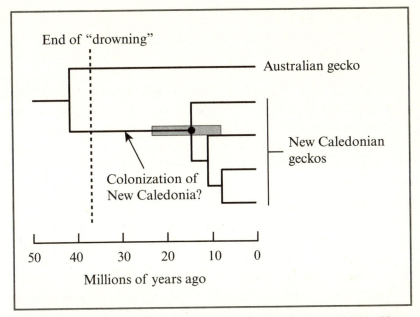

10.5 Timetree showing that the earliest branching point *within* New Caledonian geckos (marked by a black circle) occurred after the period of the island's assumed submergence. The shaded bar is the 95 percent confidence interval for the age of the branching point in question. Timetree redrawn and modified from Nielsen et al. (2011).

the submergence of New Caledonia, but, critically, there is no compelling evidence that these groups were actually *on* New Caledonia before the drowning. Such evidence could potentially come either from fossils or from molecular dating showing that branching points *within* an endemic New Caledonian radiation preceded submergence, but that evidence has not materialized. Within groups studied so far, none of the earliest branching points clearly occurred before the reemergence of land (see Figure 10.5). Thus, although some of these lineages are ancient in the sense of having separated from their nearest living relatives in the Mesozoic, there's little to show that they've been riding on a New Caledonian ark ever since then.[*] All could have arrived via overwater dispersal in the past 37 million years.

If the drowning scenario is true, it means that New Caledonia, like the Chathams, is biologically an oceanic island, despite its continental

---

[*] It has been claimed that the earliest branching point within the troglosironid harvestmen predates the reemergence of New Caledonia. However, the range of estimates for that branching point (28 to 102 million years ago) extends to after the hypothesized time of reemergence (37 million years ago).

roots. There is still a good deal of resistance to that idea because of those strange and demonstrably ancient lineages such as *Amborella* and the troglosironid harvestmen; those groups just *have* to be Gondwanan holdovers, the thought goes, even if that conclusion cannot be demonstrated by fossils or molecules. However, it is telling that similarly unique, ancient lineages inhabit islands that are unequivocally oceanic. For instance, the shrub *Lactoris* from the Juan Fernandez Islands is the sole living member of a family otherwise known only from Cretaceous to Miocene fossils from various continents; the Hawaiian perennial herb *Hillebrandia*, in the begonia family, apparently separated from all other living begonias more than 50 million years ago; and the Round Island boas of the family Bolyeriidae split from their nearest living relatives close to 70 million years ago. These ancient lineages must have colonized their young, volcanic island homes by overwater dispersal, subsequently becoming extinct in the source areas. Ancient lineages are often assumed to demonstrate the equally ancient persistence of an area, but the story of New Caledonia and the examples from oceanic islands tell us that, in fact, such an assumption is a leap of faith.

To complete our tour of Zealandian islands, we will briefly revisit our old friend New Zealand, where some of the first major cracks in the vicariance worldview materialized. We have already seen that the original Gondwanan plants of New Zealand have largely disappeared and, conversely, that almost the entire modern flora, from the southern beeches to the vegetable sheep, has descended from overwater colonists. The story for animals turns out to be similar, although perhaps not quite as extreme; in a recent compilation of molecular dating studies, some three-quarters of New Zealand animal lineages clearly fell into the ocean voyager category, while the rest were possibly, but not definitively, Gondwanan relicts.

Many of the cases of overwater colonization are not surprising, at least in hindsight; rails, for instance, which have dispersed to dozens of Pacific islands, made it to New Zealand several times, and Bob McDowall's galaxioid fishes colonized New Zealand at least twice, presumably as immature marine "whitebait." However, there is one example that, if it holds up, will rank among the most unexpected cases of long-distance, overwater dispersal yet known.

This case involves the ratite birds, possibly *the* most iconic example of Gondwanan vicariance. The ratites include living ostriches, rheas, emus, cassowaries, and kiwis, plus several extinct groups, such as the moas of New Zealand and the elephant birds of Madagascar. They are primarily a Southern Hemisphere group (although fossil ratites have been found in Eurasia and possibly North America) and, along with the tinamous, they are clearly the sister group to all other living birds, which suggests ancient origins. Those facts have pointed to Gondwanan breakup as the explanation for their wide distribution across the southern lands. And, of course, ratites are large, flightless birds, which, for many biologists, has closed the case: these birds couldn't possibly have crossed wide sea barriers, so their distribution must be the result of continental fragmentation.

This apparently clear example of Gondwanan vicariance, however, has started to unravel. A pure vicariance scenario for ratites now seems unlikely because of the lack of agreement between the branching order in the ratite evolutionary tree and the sequence of breakup of the Gondwanan fragments, and because molecular dating studies suggest that some splits in the ratite tree occurred after the separation of the associated landmasses. Furthermore, the case for dispersal is bolstered by another recent, rather astonishing finding that calls into question the whole notion that ratites cannot cross ocean barriers. The new result, convincingly shown in several different phylogenetic studies, is that tinamous, which are heavy-bodied but flying birds, are actually deeply embedded within the otherwise flightless ratite group. What this implies is that the common

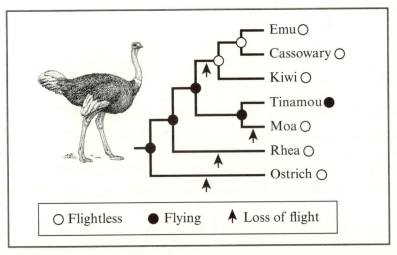

10.6 The placement of the flying tinamous within the ratite group suggests several losses of flight among the ratites. That, in turn, raises the possibility that some ratites flew across ocean barriers; for instance, moa ancestors could have flown to New Zealand. Evolutionary tree redrawn and modified from Phillips et al. (2010).

ancestor of all ratites could fly, and that the capacity for flight was lost several times within the group (see Figure 10.6). Loss of flight has been very common in birds, having occurred in eighteen living families, whereas the alternative in this case, that flightless ratites gave rise to flying tinamous, seems extremely unlikely. For New Zealand, the upshot is that the ancestors of the extinct moas might not have needed an improbable rafting voyage to reach New Zealand; they could have just flown there.

Even while evidence for the vicariant origins of "obvious" Gondwanan groups like the ratites and southern beeches has fallen apart, biologists still point to ancient lineages, such as the tuatara and the New Zealand wrens (which are likely the sister group to the enormous clade of all other perching birds), as if these were definitive Gondwanan relics. In general, though, there is no clear evidence from fossils or molecules that these taxa were in place in New Zealand before the birth of Zealandia some 80 million years ago. They were certainly around somewhere, but they could have arrived after Zealandia became an island. However, there are a few exceptions, involving two lineages of mite harvestmen—tiny, short-legged relatives of daddy longlegs—and two lineages of centipedes. All four of these groups have evolutionary branching points *within* New Zealand that have been dated to around 90 million years ago or earlier. Thus, it appears that members of these groups were resident on what would become Zealandia before it separated from Antarctica/Australia.* They really do seem to be Gondwanan relics.

Ironically, New Zealand, which has been so critical in the deconstruction of the vicariance worldview, might actually turn out to contain a greater proportion of relict lineages than almost any other Gondwanan island or archipelago (although these lineages are not necessarily the ones that people have often recognized as relicts). It might also be the only Zealandian island group that has *any* ancient holdovers. Nonetheless, these observations hardly come close to resurrecting "Moa's Ark." The message that came first out of studies of fossil plants still holds: in general, the biota of New Zealand is not Gondwanan.

---

* Some geologists and biologists have suggested that New Zealand might have been entirely submerged during the Oligocene, making its history parallel to those of the Chathams and New Caledonia. However, the divergence ages within New Zealand for the mite harvestmen, as well as for leiopelmatid frogs, refute this drowning hypothesis. Mike Pole has also argued that the fossil plant record does not indicate the kind of mass extinction required by this "drowning" hypothesis. For cogent arguments against the idea that New Zealand was completely submerged, see Pole et al. (2010) and Giribet and Boyer (2010).

Finally, we come to Madagascar, the largest and biologically the most diverse of the Gondwanan islands. Like New Zealand, it feels like a piece of some alternate Earth. It has no monkeys, but it is the only place in the world with native lemurs, almost a hundred species of them, from nocturnal, omnivorous, mouse-sized creatures to the diurnal, leaf-eating indri, which can weigh up to thirty pounds. Compared to nearby Africa, Madagascar has relatively few of the major groups of lizards and snakes, but among them are about half of the world's chameleon species, including some only as big as your thumb, and boas and iguana-like lizards whose closest relatives are in South America. It has three families of birds that are found nowhere else, as well as thirty of the world's thirty-three species of tenrecs, small mammals that have evolved into shrew-like, hedgehog-like, and even otter-like forms. It has nearly 10,000 endemic plant species. Like many places, it had a surprising number of large vertebrates that were almost certainly driven to extinction by humans. On Madagascar these were not big cats or mammoths, but, among others, a slew of giant lemurs (at least seventeen species) and elephant birds that topped out at over a thousand pounds (and laid the largest eggs of any bird ever known).

It's very clear that Madagascar harbors not just a huge number of unique species, but a large set of unique evolutionary radiations, such as the lemurs, tenrecs, and chameleons. These radiations must be fairly old—one-hundred-odd species of lemurs and thirty species of tenrecs do not evolve overnight, or even over 10 million nights—but it doesn't necessarily follow that they have been on Madagascar since its final separation from another Gondwanan fragment. That final split, when Madagascar detached from India, happened about 84 million years ago, and it wouldn't necessarily take 84 million years to generate those one hundred species of lemurs and thirty species of tenrecs. The lemurs, tenrecs, and others could have reached Madagascar by overwater dispersal 30 million or 50 million years ago and still had enough time to radiate into the diverse forms we see today.

In fact, most biologists did *not* interpret the biota of Madagascar as being predominantly Gondwanan, even during the height of the vicariance movement. This hesitation stemmed from two related observations. First, because Madagascar is so close to Africa, and has been for most of the island's history, overwater colonization has always seemed plausible, even for groups, like the lemurs, that are not very good at crossing ocean barriers. Second, several well-known Madagascan taxa, such as the tenrecs and the euplerids, a group of small carnivores related to mongooses, are members of African lineages, which is not what one would expect if those groups were Gondwanan relicts. Oddly enough, Africa and Madagascar, despite their current proximity, separated early on, at least 120 million years ago. This means that, under a Gondwanan vicariance explanation, one would not expect especially close ties between those two areas.

Recent studies show that the traditional, dispersalist view of the Madagascan biota is correct. In a key literature-review study published in 2006, two evolutionary biologists at Duke University—Anne Yoder, director of the Duke Lemur Center, and Mike Nowak, a graduate student interested in plant biogeography—compiled evidence on the evolutionary relationships of many Madagascan lineages, from lemurs and tenrecs to butterflies and baobab trees. They found that the closest relatives of Madagascan groups occurred all over the place; sometimes they were African, sometimes Indian, sometimes Southeast Asian, sometimes Australian, and so on. Crucially, though, Madagascan lineages were connected to African lineages about three times as often as to Indian ones. What Yoder and Nowak's survey suggests, then, is that the proximity of Africa has indeed allowed a great deal of dispersal from that continent across the Mozambique Channel to Madagascar; in effect, the original India-Madagascar connection has been overtaken by an Africa-Madagascar connection forged by overwater colonization.

Yoder and Nowak also looked at molecular divergence times between Madagascan lineages and their closest relatives elsewhere. Out of forty-five groups, only two had estimated divergence dates as old or older than the separation of Madagascar from India. These results confirmed that most groups originated by overwater colonization. That must be the case if, as the molecular results indicate, most of these lineages—including the lemurs, tenrecs, and mongoose relatives—didn't arrive on Madagascar until *after* it became isolated in the ocean.*

Subsequent studies have added a striking connection between colonization and ocean currents that reinforces Yoder and Nowak's conclusions. Models of ocean currents indicate that, from the time of the separation of Madagascar and India until the mid-Miocene, currents flowed eastward from Africa toward Madagascar, whereas since then the flow has been in the opposite direction. That shift in ocean currents, caused by Madagascar's northward movement into the equatorial gyre, would have made it far more difficult for land vertebrates to reach the island. When the molecular

---

* A proposed, but still controversial, Late Cretaceous connection between Madagascar and South America via Antarctica (Noonan and Chippindale 2006) would not change the general conclusions here, because there are relatively few connections of Madagascan to South American taxa.

10.7 Aye-ayes (*Daubentonia madagascariensis*), bizarre Madagascan primates that use their elongated middle fingers to extract insects from wood, making them ecological analogs of woodpeckers. Aye-ayes are descended from the same overwater colonist ancestor that gave rise to the lemurs. Drawing by Gustav Mützel.

dating results are examined in the light of this change in currents, a pattern jumps out: the frequency of arrival by land vertebrates, almost all coming from Africa, dropped significantly after the mid-Miocene. Part of that pattern, for instance, is that no land mammals have colonized the island since the reversal in the direction of the current. These results are an example of details validating the general idea that the biota originated mostly by overwater dispersal.

The fossil record also dovetails strikingly with Yoder and Nowak's work. The key studies have been headed by David Krause, a paleontologist at the State University of New York at Stony Brook. Krause and various colleagues (one of whom is Scott Sampson, a.k.a. "Dr. Scott the Paleontologist" on the PBS kids' show *Dinosaur Train*) have been studying vertebrate fossils on Madagascar since 1993, working in Late Cretaceous strata of the Mahajanga Basin on the northwest coast of the island. There they have unearthed an extraordinary array of forms—theropod and sauropod dinosaurs, bizarre insectivorous and herbivorous crocodiles, sparrow-sized and vulture-sized birds, marsupials and multituberculates (a diverse extinct group of rodent-like mammals), turtles, snakes, frogs, and freshwater fishes. Many of the specimens are wonderfully preserved;

for instance, a skull of the short-snouted theropod *Majungasaurus* is one of the most complete dinosaur skulls ever found. The Mahajanga Basin has turned out to be one of the world's great Mesozoic fossil sites.

At the time these animals died, Madagascar had only recently detached from India. Not surprisingly, the fossil fauna is clearly Gondwanan. That Cretaceous fauna, however, has almost no connection to the island's modern vertebrate fauna. With the possible exception of one of the turtles, none of the Cretaceous vertebrates are close relatives or potential ancestors of the modern species. In other words, during Madagascar's long journey through time, virtually all of those old lineages have disappeared. This inference from the fossils agrees with Yoder and Nowak's molecular divergence dates, which indicate that most of the sampled vertebrates arrived after the Cretaceous. For vertebrates, at least, the old Gondwanan fauna has been replaced by an almost entirely new set of creatures.

The biota of Madagascar, like that of the whole world, was undoubtedly battered by the end-Cretaceous extinction. Presumably the dinosaurs, which of course are not known to have survived anywhere after the Cretaceous, and many other lineages on Madagascar didn't make it through that catastrophe. Unfortunately, after the Late Cretaceous, the known fossil record of the island is pretty much blank until it picks up again in the Late Pleistocene, about 26,000 years ago. Thus, it's unclear to what extent the earlier biota disappeared in mass-extinction pulses, such as the end-Cretaceous event, versus more slowly, piece by piece over long spans of time. The important point, though, is that it did disappear.

To sum up, a general message from Madagascar and the other Gondwanan islands is that the loss of many lineages by extinction, on the one hand, and the arrival of many others by overwater colonization, on the other, have both been pervasive in all of these areas since their last separation from the other landmasses of the supercontinent. Whatever species they may have harbored at the time of fragmentation, extinction and oceanic dispersal have since put a massive, defining stamp on their character.

Earth and Life Evolve Together— Except When They Don't

On West Falkland Island, swarms of igneous dikes cut across the exposed geological strata, like dark walls running through the paler rock. These dikes were formed by magma pushing into older strata in the Early Jurassic, about 190 million years ago. Similar dikes are found in southern

Africa and are part of the argument that the Falklands microplate was once attached to that region. These rocks record a piece of the deep history of the place. They tell us that this is a fragment of old Gondwana.

Rocks are not living organisms, though. The volcanic dikes have persisted in many places while the plants and animals around them have disappeared, to be replaced by other, unrelated species. The rock of the Falklands may be ancient, but the biota, as a continuous entity, definitively is not. From a biological perspective, "Gondwanan" is a misnomer for the Falklands and other small fragments of the supercontinent.

Instead, these continental islands are converging on the history of oceanic islands like Hawaii and the Galápagos. In the case of the Chatham Islands (and perhaps the Falklands and New Caledonia), the transformation is complete; none of the original Gondwanan lineages have made it to the present. In others, some relict groups probably remain, intriguing but no longer part of the main theme: the leiopelmatid frogs and pettalid mite harvestmen on New Zealand, the caecilians and blob-like sooglossid frogs of the Seychelles (an archipelago I haven't discussed here that tells much the same story as the other islands*), and perhaps the boas and iguanian lizards of Madagascar, among others. You might think of these islands as the Gondwanan fragments that, because of their small size, have been especially prone to extinctions, perhaps especially local mass extinctions, such as those caused by the inundations of the Chathams and New Caledonia. Because of extinction, the histories of how their modern biotas came into being are mostly or entirely histories of overwater colonization and subsequent speciation. Biologists who study these islands, if they are not died-in-the-wool vicariance scientists, increasingly speak of them in these terms: "New Caledonia must be considered as a very old Darwinian [i.e., oceanic] island," one set of authors writes. "Extinction, colonization and speciation have yielded a biota in New Zealand which is, in most respects, more like that of an oceanic archipelago than a continent," says another group.

In hindsight, this whole business of the Gondwanan islands might seem obvious. These areas were isolated for 50 or 100 or 150 million

---

* The Seychelles, as part of a larger landmass called the Mascarene Plateau, detached from India about 65 million years ago (McLoughlin 2001). Several groups on the Seychelles—including the sooglossid frogs and caecilians—are most closely related to Indian lineages and are estimated to have diverged from the latter before the two landmasses separated (Zhang and Wake 2009; Biju and Bossuyt 2003). These findings are consistent with the notion that these groups have existed on the Seychelles since before that tectonic fragmentation event. However, most Seychellian species are in the same genera or even the same species as taxa found elsewhere, indicating that they arrived fairly recently by overwater dispersal (Stoddart 1984).

years. Maybe we should have expected them to get hammered by mass extinctions (both global and local) and to absorb an influx of species from overseas. Why did we ever believe that their floras and faunas would still strongly reflect their ancient geologic origins? The fact that many biogeographers did believe this is a reflection of the power of the myth of tectonic-driven vicariance in general and Gondwanan breakup in particular. The Gondwanan story is almost irresistibly simple and elegant—the wandering fragments of a former world, carrying their cargoes of species into the present. And there's something especially appealing about the myth applied to the Gondwanan *islands*, from Madagascar and the Seychelles to New Zealand and New Caledonia. These are remote, exotic places, where one can almost imagine (forgetting such things as the ubiquitous sheep pastures of New Zealand or the gargantuan nickel-mining operations on New Caledonia) that Jurassic dinosaurs still lurk in some dark and inaccessible corner. In our minds, at least, they are the lands that time forgot. I know that I believed the myth, without giving much thought to what the evidence actually revealed.

Bob McDowall was one of the few who never believed the myth, even when vicariance biogeography was at its height. From the beginning, however, he seems to have intuited that the story of Gondwana had a great power over others and that it needed to be countered head-on; the myth had to be identified and, if not shattered, at least drastically tempered. In the final paragraph of his Falklands paper, McDowall included a statement about the origins of the archipelago's biota that looks to me like an attempt at myth-breaking. He focused, in particular, on the iconic phrase coined by Léon Croizat that captured the essence of the vicariance worldview, a phrase that I imagine McDowall found doubly galling because of the vicious criticism he had absorbed from Croizat. "For me," McDowall wrote, in a voice that gains power from restraint, "the particularly interesting aspect of these patterns, and for biogeography generally, is the demonstration that the Croizatian dictum . . . that 'earth and life evolve together', does not have general application in the way that some believe it has." In other words, it isn't all about the fragmentation of the Earth generating a matching pattern in the living world. For the Falklands and other Gondwanan islands, the relevance of the Croizatian dictum had all but vanished, figuratively speaking, in the waves; in waves of extinction that erased the original inhabitants and in waves of immigrant species arriving from beyond the sea.

*In September 1995, hurricanes Luis and Marilyn swept through the Lesser Antilles, traveling northwest. In early October of that year, local fishermen watched a natural raft made up of logs and uprooted trees wash up in a bay on the eastern shore of the island of Anguilla, near the path of both hurricanes. On the raft were at least fifteen green iguanas (*Iguana iguana*), a species that did not previously occur on the island. From the tracks of the hurricanes as well as prevailing currents in the area, researchers conjectured that the iguanas probably came from Guadeloupe, 175 miles to the southeast. (Suggestively, a sign from the Parc National de Guadeloupe washed ashore on Anguilla at about the same time.) In the wake of the hurricanes, green iguanas also showed up on two other islands, Antigua and Barbuda, where they had not been seen before.*

As of 2011, a population of green iguanas remained on Anguilla, descendants of the rafting lizards, mixed with some escaped pets.

Section Four

TRANSFORMATIONS

Chapter Eleven

THE STRUCTURE OF BIOGEOGRAPHIC "REVOLUTIONS"

Hawaiian Anomalies

Steve Montgomery stopped on the muddy slope and pointed out a hard, thorny "house" constructed by a moth caterpillar. I had been scrambling to keep up, forcing aside tangles of branches and executing involuntary, spastic pirouettes in the rust-colored volcanic mud. I came up alongside him and briefly examined the insect's case, which I probably would have passed off as detritus. Steve identified it as being from a species in the genus *Hyposmocoma*, one of the many radiations of Hawaiian insects each resulting from a single colonization of the archipelago. *Hyposmocoma* is one of the biggest of these radiations, with several hundred species, and possibly the most bizarre, especially the larvae; some of the caterpillars are amphibious, living perfectly happily on land or in the water, and others eat snails after tying them down with silk strands, like a spider. They're some of the seemingly endless examples of weird island life forms.

We were in the Koolau Mountains on Oahu, in a forest that managed to look lush and scruffy at the same time, and unnaturally patchy, like many forests filled with nonnative plants. It was not the Hawaii of dreams. My longtime friends and current scientific collaborators, John Gatesy and Cheryl Hayashi, were a little ways behind, sliding around just like I was. In contrast, Steve, despite being the oldest of the four of us by at least a decade, moved quickly and seemingly effortlessly through the forest and kept up a running natural history commentary as he went. His reservoir of knowledge about the Hawaiian flora and fauna seemed bottomless. As we passed through a patch of ginger plants, with their verdant, footlong leaves, he showed me how to squeeze something like rosewater out of the flower buds, but said the rhizomes (underground stems) from this particular species weren't as good to eat as the usual marketplace ginger. He talked about how cattle had long ago trashed the native vegetation and, moving on both literally and figuratively, handed me some leaves that he said smelled like camphor when crushed. Finding introduced fiddlewoods growing next to native saplings, he turned into a one-man ecological restoration crew, taking out a small handsaw and cutting down the invaders to give the natives more light.

This was in August 2011, off the Pali Highway, which runs from Honolulu northeast to Oahu's Windward Coast. John, Cheryl, and I had met up with Steve at a place called the Nuuanu Pali ("Cool Heights Cliff") Wayside Park, where in 1795 the army of King Kamehameha I won a decisive battle for control of the island, driving hundreds of the men fighting for the Oahu chief, Kalanikupule, over the thousand-foot-high precipice. That shallow history wasn't on our minds, though. We had come to Hawaii to look for insects called jumping bristletails, and Steve—a freelance entomologist who, among many other things, provided the bugs for the TV show *Lost* and is an authority on another group of carnivorous Hawaiian caterpillars—was one of the very few people who knew where to find them. I had set up the rendezvous—look for an Asian guy (me), a big white guy (John), and a small Asian woman (Cheryl), I told him, and this abbreviated description apparently worked, because Steve found us as soon as he got out of his car at the Nuuanu Pali parking lot. Even without a description to go on, we probably would have picked him out, too; in jeans, T-shirt, and a baseball cap, with wild gray hair and beard, and glasses tilted askew, he stood out as the potential freelance field entomologist among the tourists.

After we introduced ourselves, Steve pulled out a glass vial and handed it to me. It was filled with bristletails preserved in alcohol that he had collected for us while visiting his sister in Indiana. Definitely a bug guy, I thought. Then he looked down at our feet, asked somewhat incredulously if we were going to wear those "tennis shoes" (actually we were all wearing lightweight hiking boots), and showed us his preferred footwear for Hawaiian fieldwork, sock-like Japanese *tabi* with an indentation between the big toe and the others, and metal spikes on the soles.

They looked like good mud shoes, so I borrowed an extra pair that he had (regretting it later because they pinched my feet). Then, after getting our gear together, the four of us headed roughly east from the parking lot into the forest, at first following a faint trail, but quickly veering off into the pathless mud and brush.

No more than half a mile from the parking lot, past the ginger patch and the *Hyposmocoma* caterpillar in its thorny house, we reached a kind of dell, noticeably darker than what we had come through. This was the spot where Steve had seen bristletails on the trunks of a native tree called *Pisonia*. He pointed out the *Pisonia* trees, with pale bark and whorls of large leaves at the ends of branches, and showed us what he thought would be a good collecting technique, brushing tree trunks and boulders over a white bedsheet placed on the ground.

11.1 Close-up of a jumping bristletail, *Neomachilis halophila*, from coastal California. Bristletails broke a biogeographic "rule" by crossing the ocean between North America and Hawaii. Photo by Merrill Peterson.

After setting our packs down and stuffing some plastic vials into our pockets, we fanned out, quickly immersing ourselves in the shared emotions of all collectors—the anticipation at each stone overturned, or trap checked, or, in this case, every sweep of a tree trunk or moss-covered boulder, and then the at-first-barely-perceptible-but-ever-increasing disappointment accompanying each small failure to find the thing looked for. But this time, we didn't have to wait long for success. Within minutes, Steve yelled that he'd found one, and we converged around him to peer at a small, dark form in a plastic vial.

Bristletails are more or less all of a piece: they all have a humped thorax (which they flex as part of the jumping mechanism); big compound eyes that meet at the midline like skiing goggles; long abdomens with short, spine-like appendages on each segment, which may or may not be vestiges of real legs; and three long bristles that stick out from their tail end. In gestalt, they look like tiny land shrimp. Once the basic form is learned, any bristletail is instantly recognizable as a bristletail, but, without a microscope, it's hard to tell one particular kind from another. The one in Steve's vial was, to ape Dr. Seuss's description of the Lorax, biggish and darkish, but beyond that it just looked like a bristletail.

There were a couple of things that had led John, Cheryl, and me to this first encounter with bristletails in Hawaii. One was that, although none of us are entomologists—John is an expert on mammal evolution and Cheryl on the evolution of spider silk—we had become intrigued by the deep history of these jumping, shrimp-like creatures that branched off from the rest of the evolutionary tree of insects in the Devonian or thereabouts, before insects evolved wings and took over the world.* (The bristletail order, Archaeognatha, with about five hundred known species worldwide, is, at best, a modest, albeit persistent, evolutionary success story in comparison to the winged insects.) Another was that some bristletails, including the Hawaiian ones, have unusually large *pseudogenes* (genes that have lost their function) that can be used in odd ways to decipher the history of populations. What interested us about the particular shrimp-like insect in Steve's vial, though, was that it was on a volcanic island in the middle of the ocean, some 2,400 miles from North America, and even farther from any other likely continental source area. Bristletails are supposed to be inept at crossing ocean barriers, because they're

---

* Some arboreal bristletails have directed aerial descent, meaning they glide, and it has been suggested that this represents an early stage in the evolution of insect flight. Apparently the long tail bristles are important for gliding; if they are removed, the animal can't maneuver in air nearly as well (Yanoviak et al. 2009).

flightless and delicate and have a tendency to jump in a completely random direction when disturbed, presumably not the best thing to do on a raft shaken by waves. However, their presence in Hawaii suggests that their voyaging capabilities have been underestimated, like those of frogs, monkeys, and other creatures. Not only had bristletails managed an improbable journey to some initial landing spot in the archipelago, they had also reached a total of at least six islands within it—Kauai, Oahu, Molokai, Lanai, Maui, and Hawaii—requiring at least three crossings of the channels between islands.**

A message that comes out of the new evidence for long-distance dispersal is that, although there are many useful generalizations about such events, one should not mistake these for laws that cannot be broken. Darwin himself made this mistake, discerning that mammals are not very good at crossing ocean barriers, but then going too far in declaring that "no terrestrial mammal can be transported across a wide space of sea." More generally, the message is that we should not be too wedded to our preconceptions. There is evidence about improbable events in deep history, and we need to follow it to wherever it leads. This brief interlude in Hawaii provides two final illustrations of such evidence leading to the unexpected. More importantly, these examples serve as a point of entry for discussing why views on chance dispersal have oscillated from Darwin's time on, and, especially, why the latest shift in thinking might be fundamentally different from earlier ones.

Regarding bristletails and their poor dispersal abilities, it turns out that there are native species on quite a few volcanic islands, such as the Azores, Lord Howe, and São Tomé (where they may fall prey to the ocean-voyaging amphibians studied by John Measey), islands that they must have reached by overwater dispersal. Also, one bristletail genus, *Neomachilellus*, suspiciously contains dozens of species in the New World tropics and just two in West Africa, a geographic pattern that may reflect a recent eastward crossing of the Atlantic. In short, bristletails have likely negotiated ocean barriers more often than one might have guessed based on their physical characteristics and behavior.

** Molokai, Lanai, and Maui have been connected by land at various times, together forming the prehistoric island of Maui Nui; thus, the presence of bristletails on those three islands could have been the result of a single overwater colonization (Ziegler 2002).

The evolutionary connections of the Hawaiian bristletails suggest that they have also failed to follow another generalization about dispersal and colonization. The generalization is that nonaerial animals—that is, ones that can neither fly well nor use silk to float by "ballooning"—can only reach Hawaii from the west, not from the New World. The thought here is that aerial animals, such as birds, butterflies, and ballooning spiders, might make the trip from the Americas in a matter of days, either under their own power or blown by storm winds, whereas nonaerial ones would require an almost impossibly long rafting voyage, because there are no islands between the Americas and Hawaii to use as waystations. However, there are many islands in the western Pacific, and these would allow nonaerial animals from, say, Asia or Polynesia to reach Hawaii by a series of shorter rafting "hops." A 2012 compilation of phylogenetic studies shows very clearly the pattern expected under this reasoning: out of twenty animal lineages that apparently arrived from the west, nine were aerial and eleven were nonaerial, whereas, out of thirty-seven lineages that came from the New World, all were aerial.

Bristletails might be the first exception to this generalization. In the early 1990s, a German entomologist named Helmut Sturm, who probably knows more about the bristletail order than anyone else in the world, examined the anatomical traits of the Hawaiian species—the shape of their second pair of eyes, the number of water-absorbing vesicles on their abdomens, and the like—and concluded that they were most closely related to a species called *Neomachilis halophila*, from the west coast of North America. Recent molecular evidence supports Sturm's conclusion: DNA sequences from our Hawaiian specimens and from various continental species strongly confirm the close connection of the island forms to *N. halophila*. The obvious interpretation of these results is that the Hawaiian bristletails originally came from North America, presumably by rafting.

A monumental rafting voyage might not be as implausible as it at first sounds for *Neomachilis*. To begin with, the mainland species, *N. halophila*, is found among rocks around the high tideline, where the chances of being swept out on a wave must be great. Furthermore, Sturm envisioned a mechanism that might get around the problem of delicate, erratically jumping insects surviving such a voyage: he suggested that they made the passage not as adults, but as eggs attached to driftwood. To bolster this argument, he noted that bristletail eggs in general take a long time to develop, and that *N. halophila* eggs, in particular, are resistant to salt water. (In fact, one investigator has claimed that the eggs of some bristletails are even resistant to chemicals used to preserve specimens.)

Sturm's "how-possibly" argument seems reasonable, but he might have gotten the direction of the journey backward. Specifically, the DNA sequences we've obtained so far suggest that the North American *N. halophila* might actually fall, in an evolutionary sense, *within* the Hawaiian bristletail group. We clearly need more data before we can say anything definitive, but if that result holds up, it would indicate, not a North America to Hawaii colonization, but the reverse, that is, island bristletails establishing themselves on the continent. Such an island-to-continent colonization would, of course, require a long unbroken journey across the eastern Pacific, just as the assumed continent-to-island colonization would. In addition, it would go against a very general and well-known idea about island organisms, an idea that is sometimes treated as a biogeographic rule.

The rule is this: Islands, especially small, remote islands like the Hawaiian chain, are evolutionary dead ends. Many groups colonize such islands and flourish on them, but once they have evolved into distinctive island forms, they do not go back and successfully establish themselves on continents. Islands are like evolutionary black holes; lineages go in, but they don't come out.

Two main arguments have been given to explain this islands-as-dead-ends rule. First, island species—especially ones on remote islands like Hawaii—are thought to suffer from a kind of evolutionary degeneracy, because, unlike continental organisms, they do not have to compete with a large array of other species. The notion here is that natural selection is relaxed on islands, producing an assemblage of "slacker" species that would never survive in the dog-eat-dog (or dog-outgrow-dog, or, ultimately, dog-outreproduce-dog) environments of continents. Island plants that have lost their defensive spines or thorns, and island animals that show no fear of terrestrial predators, are obvious examples, although "relaxed" features aren't always so blatant. The second argument is that oceanic islands, in particular, are tiny compared to continents, so they typically do not harbor very large populations from which "chance" colonists might be drawn. This effect probably explains, at least in part, why natural invasions recorded in recent times have usually been by species that are both widespread and abundant in their native areas; those species are the ones that contain especially large numbers of individuals.

These arguments against island-to-mainland colonizations sound reasonable if one believes that long-distance dispersal of any kind is exceedingly rare in general. However, if, through deep time, such dispersal events are fairly common, then maybe we should expect that at least some island lineages have broken the rules and succeeded in colonizing continents. In any case, we should let the evidence sort things out. For *Neomachilis*, the case for an "out of Hawaii" dispersal is tenuous, but there are many other Hawaiian groups, and some of them have been studied far longer and far more intensively than the bristletails.

In the scruffy forest by the Nuuanu Pali, between sweeping mossy boulders for bristletails, Steve Montgomery and I got to talking about ocean crossings. We were in Hawaii, after all, so it was a pretty obvious thing to bring up. Not surprisingly, Steve knew a fair amount about the subject, and he mentioned the case of Hawaiian fruit flies, a group that probably has been studied more than any other Hawaiian evolutionary radiation and perhaps more than any island radiation anywhere.

Steve was talking in particular about the work of a Berkeley entomologist named Patrick O'Grady (who also happened to be the person who had told me to contact Steve for advice on where to find bristletails in Hawaii). One of Patrick's scientific missions is to construct an evolutionary tree for all Hawaiian fruit flies, which might sound like a modest ambition, but in fact is a large and unending task, because the group includes close to six hundred described species, a backlog of several hundred more that are known to exist but haven't yet been formally described, and new forms being discovered every year. Beyond this, the latest research by Patrick and his colleagues indicates that piecing together the Hawaiian fruit-fly tree is actually a much larger and more complex undertaking than even the fly experts imagined.

The big complication is that not all Hawaiian fruit flies live in Hawaii. The fruit-fly evolutionary tree implies a single colonization of the archipelago leading to a large proliferation of flies—the hundreds of species that have been the focus of Patrick's research. However, it also indicates, unexpectedly, that several lineages of Hawaiian fruit flies in the genus *Scaptomyza* have "escaped" to other areas without the help of humans. Some of these travelers ended up on other warm-weather Pacific islands, such as the Marquesas, which maybe is not so surprising; after all, Hawaiian flies might be expected to do well on other islands with similar climates and a similar lack of a full continental array of species. A slacker species from Hawaii, the thought goes, ought to be able to hold its own among the comparable evolutionary degenerates on other oceanic islands.

The weird thing is that some of the Hawaiian escapees have given rise to flies that live not on islands, but on continents. They have been found in Australia and Africa, for instance, and, by the time Patrick and his colleagues are done, it is likely that some flies on every continent except Antarctica will be found to have Hawaiian ancestors. The meek species of islands are not supposed to inherit the Earth, but Hawaiian fruit flies have apparently taken a serious step toward populating much of the planet. In fact, the mainland *Scaptomyza* flies have become enough of a presence that they have also colonized numerous islands, replaying the original colonization of Hawaii by continental flies. Thus, a partial history of successful fruit-fly dispersal goes something like this: from some as yet unidentified continent to Hawaii, back to continents, then back to islands (many times), as well as a bunch of jumps between islands. When it comes to ocean crossings, the fruit flies are the insect equivalent of Matt Lavin's bean plants.

Why have *Scaptomyza* flies been able to cross ocean barriers so readily? Patrick thinks there are a couple of key factors. One is that members of this genus are resistant to desiccation compared with other fruit flies and with flies in general, a good property for surviving a transoceanic voyage, whether on a raft or on storm winds. Second, *Scaptomyza* flies tend to be generalist feeders, meaning they can subsist on many different kinds of flowers or fruits; thus, relative to more specialized species, they are more likely to find plants they can eat wherever they happen to end up. Finally, because they have short generations (even for flies), they can build up large populations quickly, perhaps making them less subject to the random "drunkard's walk" off the cliff of extinction.

In any case, however they managed to do it, these fruit flies add an especially striking example to the long list of "things that shouldn't have happened but did." Degenerate island lineages are not supposed to establish themselves on continents, but *Scaptomyza* flies have done exactly that, probably several times. In fact, they're apparently only one of several island groups that have managed this trick: DNA-based evolutionary trees also indicate, among other cases, that aquatic beetles in the genus *Rhantus* successfully dispersed from New Guinea to both Australia and Eurasia; that *Anolis* lizards went from the West Indies to Central or South America and, independently, to North America; and that monarch flycatchers traveled from Pacific islands to Australia.

These surprising events have been revealed by scientists like Patrick, who were gathering evidence and following where it led them instead of just believing the prevailing dogma. John, Cheryl, and I are attempting to do the same with the Hawaiian bristletails, as are many other researchers studying other groups, not only in Hawaii but all over the world. This is not to suggest that Patrick or any of us deserve a pat on the back; we're all just doing what scientists are supposed to do, namely, focusing on the evidence. That sounds like such a simple, not to mention essential, thing to do. However, in the field of historical biogeography, it often hasn't worked that way.

Beauty, Truth, and Evidence

Say that we could pluck four well-known biogeographers representing different time periods from the past 150 years, put them in a room together, and pose to them this question, "How is it that there are monkeys on both sides of the Atlantic Ocean?"

Our first biogeographer is none other than Charles Darwin, who hems and haws a bit because he doesn't think mammals have much chance of surviving a long ocean journey, but finally settles on the idea of monkeys crossing the Atlantic on an enormous raft. This "chance dispersal," though it appears unlikely, seems more plausible to Darwin than the obvious alternative, some sort of transatlantic land bridge. "It shocks my philosophy to create land," he says, "even more than to imagine an ocean-crossing primate."

Charles Schuchert, a Yale invertebrate paleontologist snatched from the early twentieth century, is the next to speak. Schuchert says that even invertebrates with larvae that drift in the open sea rarely traverse oceans, so the idea of a monkey making such a journey and establishing a new population is absurd, raft or no raft. Schuchert is a diehard land-bridge advocate, and his explanation of choice is a wide swath of continental crust stretching between the westward-projecting hump of Africa and the eastward-projecting hump of Brazil. In support of this hypothesis, he notes that the largely volcanic islands of Tristan da Cunha, St. Helena, and Ascension all contain some granite, which he claims is evidence of continental origins. "Those granitic pieces," Schuchert pronounces, "are the ancient remnants of the land bridge."

Then, from the middle of the twentieth century, comes George Gaylord Simpson, the prominent American Museum of Natural History mammal paleontologist and disciple of the great dispersalist William Diller Matthew. Simpson doesn't like Schuchert's hypothetical land bridge and seems to have less of a problem with overwater dispersal from Africa than Darwin did (although he wonders whether early monkeys colonized South America from North America across an ocean barrier not so wide as the Atlantic). He pulls out a pen and jots down some calculations to show that an event with a minuscule probability of occurring in any given year, such as monkeys rafting across 1,800-plus miles of ocean, might yet be reasonably likely given a long enough period of time. Monkeys, he notes, have been around for something like 50 million years, which makes a single, successful Atlantic crossing not so far-fetched. Darwin carefully follows Simpson's argument and nods approvingly; he's had the same thoughts, but couldn't do the math.

Our next biogeographer is Gary Nelson, the Gary Nelson of the 1970s at the American Museum, although, frankly, he might just as well be the Gary Nelson of today (who lives in Melbourne, Australia, and is as vehement as ever when it comes to cladistics and vicariance). Nelson drops a bombshell on the others: the geological evidence for continental drift. There's some rumbling over this, but when Nelson brings up the alternating magnetic stripes indicating seafloor spreading (after explaining the necessary background), the others grow quiet. (Schuchert, an outspoken critic of Wegener's theory, seems less than thrilled with the revelation.) Nelson goes on to sketch some crude maps that show the opening of the Atlantic Ocean through time and scribbles a cladogram of primates on top of the last map. (Darwin, who included an evolutionary tree as the only figure in *The Origin of Species*, seems to intuitively understand the importance of the cladogram more than Schuchert or even Simpson does.) "It's all about the fragmentation of ancestral biotas," Nelson says, "and not just for monkeys, for everything."

Simpson seems a bit discombobulated on seeing the evidence for continental drift, but he regains his composure and asks Nelson when the Atlantic first opened. On hearing that it was about 100 million years ago, he says that monkeys didn't even exist then, so the whole business of continental drift is irrelevant in this case. This Simpson from midcentury is meeting Nelson for the first time, but seems to instinctively distrust the brash young man. Meanwhile, Darwin is sitting back, silent, with one hand scratching through his beard. In *The Origin*, he spent a good deal of time arguing that one doesn't see much evidence of gradual evolution in the fossil record because that record has so many gaps, and now he's wondering if Simpson might be relying too much on fossils. "Maybe monkeys *were* in existence far earlier than we know," Darwin thinks to himself. "However, I greatly doubt it."

As Nelson and Simpson continue arguing, a door opens and a fifth biogeographer walks in, this one nameless and from the present day. "I think you people need to hear about relaxed molecular clocks," she says.

When it comes to explaining distributions broken up by oceans, biogeography has been as fickle as fashion. Back and forth the pendulum has swung, from the likes of Darwin to Schuchert to Simpson to Nelson to the present, from dispersal to vicariance to dispersal to vicariance and, finally, today again toward dispersal (but now including a healthy dose of vicariance). Given this history, why should we think that this most recent development has any more validity than the others? Why should we believe that we're finally getting it right?

The answer goes back to the Hawaiian fruit flies and bristletails, the many other examples I've described, and, most generally, to all the studies that have piled up over the past twenty years or so: it's all about following the evidence.

This may sound presumptuous. After all, weren't the scientists of previous generations following the evidence too? To an extent, of course, they were, but as the biogeographer John Briggs has said, too often the field has been "beset by attempts to make the facts fit the theory instead of consulting the evidence in the first place." People adopted biogeographic theories and then saw what they wanted to see, dismissing observations that were inconvenient for their pet hypotheses.

For example, although the land-bridge advocates were rightfully swayed by Paleozoic and Mesozoic fossils that indicated former land connections (which we now explain by continental drift), they tended to conjure up bridges on the flimsiest of geologic evidence, turning faint hints into the pillars of an argument. Charles Schuchert really did use the existence of granite on Tristan da Cunha, St. Helena, and Ascension to argue for an Atlantic land bridge, although there was in fact no reason to believe that granite could only form within continental crust. In the extreme, land-bridge enthusiasts used similarly weak reasoning to concoct vast land connections to the unlikeliest of places, including Hawaii, Samoa, and other remote volcanic islands. In those cases, facts no longer seemed to matter; everything became subservient to the theory.

On the surface, it might seem that the vicariance movement of the 1970s was, in contrast, deeply grounded in evidence, but that wasn't really the case. It is true that vicariance biogeography was founded in part on the conclusive evidence supporting the theory of plate tectonics. However, although plate tectonics provided a compelling mechanism for the fracturing of landmasses, other evidence was required to show that this fragmentation actually explains the piecemeal distributions of living things. To say that vicariance biogeography was based on evidence because it was tied to the validation of plate tectonics is like saying that oceanic dispersal is based on evidence because it's connected to the undeniable observation that rafts of vegetation can float on water. Identifying a way that something *could* have happened, while significant, is not the same as showing that is how it *did* happen; "how possibly" should not be equated with "how actually."

As described in Chapter Two, evidence that vicariance is the best explanation for disjunct distributions, including those broken up by oceans, was actually scant. There was Brundin's wonderful early example of the chironomid midges, and then . . . not much. The American Museum ornithologist Joel Cracraft used Gondwanan breakup to explain disjunctions in several groups, but none of these cases was as convincing as the midges (and some, such as the southern beeches and ratite birds, clearly haven't held up as pure examples of vicariance). It probably isn't coincidental that Donn Rosen's findings on Middle American swordtail fishes and their relatives were analyzed over and over again and became almost legendary among the vicariance crowd; Rosen's study was an excellent piece of work, to be sure, but one wonders if its importance got blown out of proportion because so few other good examples of vicariance materialized in the 1970s and 1980s.

If evidence for the ubiquity of vicariance was lacking, then what drove this scientific movement? As noted in Chapter Two, there seem to have been two influences at work: forceful personalities and the inherently attractive nature of the theory itself. The forceful personalities included Gary Nelson, the "hub" at the American Museum; Colin Patterson, the "voice of God" at the British Museum; and Léon Croizat, who was viewed by some as a sort of biogeographic messiah. Take away those three and a few others, erasing both their publications and personal influence, and perhaps there would have been no extreme vicariance movement.

Beyond this social aspect, however, vicariance may have been for many people an inherently seductive explanation, regardless of the evidence. In particular, it combined two traits that many scientists strive for in their theories: simplicity and generality. The basic idea of areas and their biotas fragmenting in concert was comprehensible to the average eight-year-old and, when coupled with plate tectonics, could be applied to the entire world.* Tectonic vicariance also had a feeling of almost Newtonian inevitability about it: start with lineages on a landmass, let rifting and seafloor spreading take their course, and the predictable end-product is vicariance. Long-distance dispersal, in comparison, was a messy proposition. By definition it involved random, unpredictable events—its other names were "chance" and "occasional" dispersal—not to mention a reliance on unverifiable assumptions about things such as the velocity of ocean currents in the Oligocene, or a beetle's ability to survive a tumultuous, freezing trip in the jet stream.

Finally, if vicariance was accepted as the dominant process, it led to the notion that the whole geography of life could be distilled into a series of fragmentation events, the sequence of Gondwanan breakup being the archetype. This was perhaps the grandest, cleanest idea of all, because it meant that one could reduce biogeographic history to cladograms, unambiguous branching diagrams that represented these area-splitting events. Hell, it was so clean, it was barely biological. The final product, the general area cladogram, didn't even have to mention living things, just the names of regions—southern South America, Australia, New Zealand, or whatever. This was the logical conclusion of the dictum that "Earth and life evolve together"; if that was indeed true in the way Croizat had envisioned, then, in a cladogram, areas of the Earth could stand in for the living groups within them. It was so simple, and so general, and so orderly.

If this interpretation is correct, the rise of vicariance biogeography had more to do with the beauty or elegance of the theory than with any flood of supporting observations. A pretty theory can, at least for a time and for some people, trump the evidence.** In great contrast, just the opposite was true of the next and latest swing of the pendulum, back toward dispersal.

---

* Panbiogeographers tended to complicate matters by also emphasizing the conglomeration of formerly separated landmasses, but the simpler view that focused primarily on fragmentation was more widespread.

** Physicists, in particular, often have chosen theories on the basis of beauty or elegance. For instance, Paul Dirac, known, among other things, for predicting the existence of antimatter, wrote of the general theory of relativity: "One has a great confidence in the theory arising from its great beauty, quite independent of its detailed successes" (McAllister 1998, 174). Similarly, some of Brian Greene's arguments for string theory—currently the most popular "theory of everything" but notoriously difficult to test—are based on the theory's apparent elegance (Greene 1999).

Evolutionary biologists who started their careers in the 1970s, 1980s, and 1990s, especially in English-speaking countries, were often trained to be suspicious of long-distance dispersal as an explanation and to view it as anything but elegant. Even if they weren't directly connected to the hardcore vicariance school, they had heard the rhetoric about dispersal hypotheses being the product of a muddy, storytelling mindset, that these hypotheses were unfalsifiable and therefore unscientific.

Then, something happened, namely, the polymerase chain reaction (PCR) created a giant pile of DNA sequence data, which in turn created a flood of molecular dating studies, putting ages on evolutionary branches. Initially, few scientists were thinking that this was going to shift biogeography away from vicariance and toward dispersal. In fact, some of the people doing these timetree studies were coming from the vicariance side and were anticipating that the molecular evidence would just clarify or confirm exactly which fragmentation events had been important for their chosen study organisms. For instance, Miguel Vences had aimed in his PhD project to clarify the vicariant origins of Malagasy frogs, and Matt Lavin had expected to confirm that an Early Tertiary land bridge and a subsequent cooling climate had left related woody legumes on both sides of the Atlantic. As we have seen, though, the clarification and confirmation of vicariance did not materialize. The ages estimated from the DNA sequences were almost always too young, turning both Vences and Lavin away from vicariant explanations and toward long-distance, overwater dispersal.

For other scientists, a loss of confidence in vicariance came before the molecular dating evidence arrived, but ultimately it was this evidence that shifted their thinking, not some prior faith that dispersal had to be the answer. Steve Trewick, who has done some of the key studies on the biogeography of New Zealand and, especially, the Chatham Islands, had what was probably a fairly common experience. "Early on I took the vicariance model as the default," he said, "and yes, I have to say when I first arrived [in New Zealand for graduate school] I probably did believe it. Why? Because that is what everyone talked about and I assumed that it was based on evidence. Quite quickly it became apparent that there is almost no direct evidence." Molecular dating studies—his own and others—and geological support for the partial or total drowning of landmasses eventually convinced him that most of the plants and animals of New Zealand and the entire biota of the Chathams has been derived from overseas colonists.

In short, the shift toward long-distance dispersal hypotheses was driven by evidence, not by any preconceived notions that dispersal was a particularly attractive, elegant theory. If anything, the unpredictable nature of long-distance dispersal and its disconnection from the grand events of Earth history made it an ugly duckling in comparison to vicariance. It was evidence, not theorizing, that turned it into a swan. And it is evidence—a mountain of evidence—that distinguishes this latest swing of the pendulum from the ones that have gone before.

Finally, a Paradigm?

In his seminal book, *The Structure of Scientific Revolutions*, the philosopher Thomas Kuhn describes a typical revolution beginning with some kind of anomaly (or anomalies), that is, an observation or experiment that isn't well explained by the current paradigm. For instance, under the view that the Sun, the other planets, and the stars revolve around the Earth, it was hard to explain the fact that some planets, when observed over a period of several days, sometimes appear to reverse their direction of movement (so called "retrograde motion"). Such anomalies, if they persist, can lead to the formulation of a new theory that better accounts for them. In this case, the theory that followed was Copernicus's heliocentrism, in which retrograde motion was explained by the idea that the Earth orbits the Sun faster than do the planets farther out, so that these outer planets seem to move backward as the Earth catches up with and then passes them.* Eventually, corroboration of the new theory piles up and a "paradigm shift" takes place, with the result that everyone who is considered a serious investigator in the field believes the new theory rather than the old one. The shift occurs partly because individuals are converted, but also because those who cling to the older view eventually die and are replaced by younger investigators who hold the new view.

To a degree, the shifting views about distributions broken up by oceans follow this model of anomaly leading to revolution. For instance, the frequent occurrence of very closely related fossil groups in widely separated regions—like the Triassic vertebrates of Africa and South America, or the *Glossopteris* flora on various Southern Hemisphere landmasses—was anomalous under the Darwinian view that distributions had come into being against a background of fixed continents and ocean basins. However, the various swings of the pendulum—from Darwinian dispersalism

---

* Retrograde motion was just one of many anomalies that had accumulated by the time Copernicus revolutionized astronomy.

to land bridges to dispersalism again (the New York School version) to vicariance biogeography—clearly do not fit Kuhn's view of typical revolutions in at least one way: none of them resulted in a complete or even nearly complete shift of the discipline from the old view to the new. In particular, "dispersalism" persisted through both the land bridge and more recent vicariance movements.

Kuhn's views, of course, should not be taken as gospel. It is possible, for instance, that his model of scientific revolutions applies much better to the physical sciences that were his focus than to evolutionary biology. However, what he had to say about situations in which multiple views have been tossed about, without any of them completely taking over the field, seems to describe extremely well the history of explanations of distributions broken up by oceans. If his interpretation of such situations is right, it makes sense of more than 150 years of confusion and disagreement among scientists interested in the geography of life.

The word to remember here is "pre-paradigm." Specifically, what Kuhn said is that the period before a discipline adopts its first paradigm is "regularly marked by frequent and deep debates over legitimate methods, problems, and standards of solution, though these [debates] serve rather to define schools than to produce agreement." This seems a perfect description of the history of biogeography, especially when it comes to explaining piecemeal distributions.** For instance, dispersalists criticized land-bridge advocates for erecting vast land connections on illusory geological evidence; in other words, the former questioned the latter's "standards of solution." Similarly, vicariance biogeographers thought that much of the evidence used by dispersalists, such as the fossil record and Darwin's seed-survival experiments, was basically worthless. Scientists in every discipline have disagreements, of course, but what is striking about biogeography is how deep the divisions ran, to the point that members of different schools of thought seemed, as Kuhn might have put it, to be living in different worlds. Thus, vicariance scientists thought that the approach of dispersalism was unscientific, because of faulty evidence and

---

** On a more general level, Darwin and Wallace established a paradigm for historical biogeography based on the ideas of common descent and deep time, as described in Chapter One. However, this evolutionary paradigm left open the specific explanations for disjunct distributions; the fact that Darwin and Wallace favored long-distance dispersal as an explanation did not make that explanation or the assumptions behind it part of the accepted paradigm. From the beginning of the Darwinian Revolution, some evolutionists preferred the land-bridge explanation, and by the time Wallace died in 1913, the land-bridge school was well established.

because dispersal hypotheses supposedly could not be falsified, and, further, that it was based on the erroneous notion that species and other taxa necessarily spread out from centers of origin. These were not the kinds of minor disagreements that could be resolved with a bit of new data; practitioners of vicariance biogeography were basically saying that dispersalists weren't even performing legitimate scientific research.

Kuhn also wrote that "Throughout the pre-paradigm period when there is a multiplicity of competing schools, evidence of progress, except within schools, is very hard to find. This is the period . . . during which individuals practice science, but in which the results of their enterprise do not add up to science as we know it." It doesn't add up to science because there is little that becomes an accepted foundation upon which the whole discipline can build. Instead, the different, conflicting schools develop independently, as if they aren't even dealing with the same subject.

With this in mind, recall the disdain that Croizat and Nelson had for Darwinian biogeography, describing it as "a world of make-believe and pretense"; clearly, they did not see Darwin's ideas as any sort of foundation for their own work. Similarly, in the dispersalist William Diller Matthew's criticism of land-bridge advocates, there was little sense of shared progress between the two schools, apart from the gathering of new information on distributions.

The disconnect between different biogeographic schools jumps out very baldly in communications I had with the New Zealand botanist Michael Heads and with John Briggs, an emeritus professor at the University of South Florida. Heads, as described earlier, is a staunch proponent of panbiogeography, and Briggs, although he has written books about the importance of plate tectonics in biogeography, has argued strongly for the significance of long-distance dispersal. When I asked Heads, "Who do you consider the most important contributors to the development of biogeography in the last fifty years?" he wrote, in an email: "Apart from Croizat I don't think there has been much development over the last 50 years. I think biogeography has gone downhill. In many ways the 19th century writers were better informed than the moderns and I enjoy reading them much more. I study lots of molecular phylogenies but don't usually read the text of the papers as they're so predictable." In other words, from Heads's perspective, the biogeography of the past half-century (except for the work of Croizat and his followers) hasn't gone anywhere. There hasn't been any progress to speak of.

In great contrast, when I asked Briggs about the significance of Croizat's work, he replied, in no uncertain terms, "I don't think he made any lasting contribution." Similarly, Briggs had this to say about cladistic vicariance biogeography: "Cladistics by itself has been a useful systematic procedure but the vicariance inclusion was a distraction that led to a lot of argument and wasted time. Except for stimulating the interest of some who were not acquainted with biogeography, I don't think the vicariance movement had any lasting benefit." Thus, again, an entire segment of the discipline—a rather large segment, at that—had been banished to irrelevance.

In short, Heads thinks that Croizat made the only significant contributions to biogeography in the past fifty years, Briggs thinks that Croizat made no contribution at all, and neither of them seem to have much use for the cladistic vicariance biogeography advocated by Nelson, Rosen, and others. Heads, with his panbiogeography, and Briggs, with his "balanced" view (more or less the view taken in this book, although I don't agree with his statement about the irrelevance of cladistic vicariance biogeography), are technically researchers in the same field, but they seem to share little common ground from which to move forward.

According to Kuhn, such fundamental disagreements occur not only in the period before a discipline's first paradigm has been established, but also when an accepted paradigm falls apart and different views are jostling to take its place, that is, during the first stages of revolution. However, the history of biogeography points to the pre-paradigm explanation: the disagreements have been deep and persistent, basically going back in one form or another to Darwin's criticism of land bridges, and that kind of lasting discord signifies not a scientific revolution in the making but a field searching for its initial consensus. If this view is correct, historical biogeography has been spinning its wheels for the past 150 years in a pre-paradigm state, an immature condition in which the most consistent agreement has been that the "other guys"—dispersalists, land-bridgers, cladistic vicariance scientists—were fundamentally wrong in their approach to the science and, therefore, also in their conclusions about the world.

Optimistically, this immature phase may finally be ending; historical biogeography may at last be developing a full-fledged paradigm. In this view, several steps within the past half-century have been critical to the science's maturation. The first and probably the most obvious one was the acceptance of plate tectonics, which produced a dynamic view of Earth history that, at least in rough outline, almost everyone interested in biogeography agrees upon. In the face of all the evidence for tectonics, returning to a belief in fixed continents and ocean basins would be absurd, like an astronomer regressing to a geocentric view of the universe, or a physicist trying to study subatomic particles using Newtonian mechanics. Even if one doesn't believe that "Earth and life evolve together" in the extreme sense envisioned by Croizat, one has to place life's history within the context of the changing configurations of continents and oceans.

The second step, following closely on the heels of the first, was the rise of cladistic thinking, the replacement of the muddy "river delta" approach to evolutionary relationships with methods for building those precise, unambiguous branching diagrams called cladograms. Like plate tectonics, cladistics was a major step forward; knowledge of relationships is critical to interpreting where lineages came from, and precise knowledge is better than muddiness. Beginning in the 1960s, there have been heated debates over exactly how to figure out such relationships, with statistical methods largely taking over from the nonstatistical approaches of the die-hard cladists (a few of whom are still unhappy about this change). However, for biogeography, these debates are just a sidelight, and almost everyone agrees that building evolutionary trees is a necessary part of inferring the geographic history of life. After reading the summary of a paper on historical biogeography, what is the first thing that most biogeographers do? More than likely, they look for the tree diagrams.

As we have seen, though, the melding of plate tectonics and cladistics did not by itself produce a movement that took over the field. The vicariance biogeography that came out of this marriage of ideas was influential and was adopted by many, but, as noted above, it never came close to being universally embraced. In that sense, the vicariance "revolution" wasn't like the rise of Copernican heliocentrism or Einsteinian relativity, or even, within geology, plate tectonics itself. It was not, in the Kuhnian sense, the rise of a paradigm.

The problem was that something was missing, something fundamental to understanding history. The opponents of vicariance biogeography saw this and, as a result, could not be fully converted to the new view, while the proponents of vicariance biogeography tried to ignore or gloss over the omission. The missing element was the river in which all of history flows. The missing element was time.

The addition of time, in the form of molecular dating results, is, in my view, the final step that might finally produce a paradigm in historical biogeography. Of course, timelines for some evolutionary events have existed since before publication of *The Origin of Species*, but, because these were based entirely on fossils, the information was missing for many groups and could be easily dismissed for others. In the 1940s, it was possible for a late-lingering land-bridger, Carl Skottsberg, to argue for the great age of many Hawaiian lineages without sounding like a complete lunatic. Today it is much harder to make that kind of claim (although a few have done so) because molecular evidence indicates that nearly all Hawaiian lineages studied thus far are relatively young. Basically, molecular dating has greatly extended the reach and precision of information on the age of evolutionary events. This cumulative evidence of timelines—what I called the "forest" in Chapter Six—has for many evolutionary biologists now passed a tipping point, such that it strains credibility to simply dismiss it as some sort of misguided intellectual fashion.

The question of the moment is whether molecular dating will progress to the point where virtually everyone who studies historical biogeography accepts this approach as valid (even if they don't believe every single age estimate). There are signs that the field is heading in that direction. Certainly, many mainstream scientists now see the rejection of molecular dating as irrational and view the extreme vicariance school as intellectually stagnant. For instance, Anne Yoder, the lemur biologist who has focused on the biogeography of Madagascar, refers to vicariance scientists who refuse to accept the dispersal origins of Madagascan mammals as being "impervious to evidence." Along the same lines, when I asked Steve Trewick why the panbiogeographers, once prominent in New Zealand, had been "exiled," he replied, "They were seen for what they are, a group of fundamentalists who have refused to engage with other thinkers or other evidence." And here's how the botanist Michael Donoghue, in typically informal fashion, described his reactions to a research talk by a cladistic vicariance biogeographer: "I thought, man, I'm having some really weird flashback to some era that I can even barely remember. . . . And she's livin' that, she's livin' it. It's bizarre and, like, no movement forward as far as I can tell."

It is true that some biogeographers continue to argue vehemently against the timetree approach, referring to the dispersalist conclusions drawn from it as "artefactual," "reactionary," and "ignoring basic biogeographic realities." Increasingly, however, those criticisms sound like the desperate salvos of an endangered group. The extent to which the field is shifting away from these vicariance scientists can be seen in a simple graph showing the parallel increases, over the past twenty years, of studies that use molecular dating and studies that discuss long-distance dispersal (see Figure 11.2). Those rapidly ascending curves reflect real shifts in the

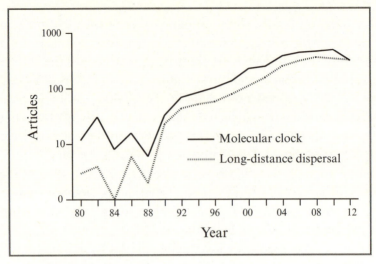

11.2 Graphing an intellectual sea change: scientific studies using or discussing molecular clock analyses have taken off in the past twenty years, almost certainly driving the parallel rise in studies that use the term "long-distance dispersal." Note the log scale. Redrawn and modified from an unpublished graph made by Susanne Renner from a search of the Web of Science database on November 21, 2012.

methods biogeographers are using, and in what they believe about the history of life. The curves illustrate how timetrees have shifted thought away from vicariance and toward long-distance dispersal.

None of this is to imply that vicariance as a process is insignificant; in fact, molecular dating studies support the role of Gondwanan breakup and other fragmentation events in quite a few cases. However, the graph, and the strong opinions of many mainstream scientists, suggest that the vicariance *worldview* may be on its way out. Despite the past contributions of vicariance biogeography—especially the clear use of evolutionary trees and the search for patterns common to many different taxonomic groups—that school of thought now appears to be an intellectual branch that is drying up and may eventually fall off the tree for lack of sustenance. If and when the vicariance approach, in its extreme form, fades away, and the evidence of time is at the core of all historical biogeography, the discipline might at last achieve its elusive paradigm.

On or around October 12, 1988, fishermen near the island of Trinidad, near Venezuela, came across thousands of dead locusts floating on the sea. Soon reports of live locusts were cropping up from scattered sites in the West Indies and northern South America—from Grenada, Guadeloupe, Jamaica, Antigua, Guyana, and Suriname, among other places. The locusts were memorable for their large size, pinkish color, and, in some places, their vast numbers; a supertanker nine hundred miles southeast of St. Lucia became a landing spot for so many of these insects that the entire deck was awash in pink, while on Dominica a single swarm (one of six reported there) was estimated to contain some 10 million to 20 million individuals. The locusts appeared to be weather-beaten and starving, as if they had endured a long and harrowing journey, which indeed they had: they were desert locusts, Schistocerca gregaria, *an Old World species, and they had just flown across the Atlantic Ocean.*

In North Africa, plagues of locusts have devastated crops throughout history—swarms are referred to as the "teeth of the wind"—and farmers and government officials in the Caribbean imagined such a disastrous outcome from the 1988 invasion. Fortunately, that scenario never materialized. The damage caused by the desert locusts was minimal, and the species quickly disappeared and has not been reported again in the Americas. However, recent DNA studies indicate that desert locust ancestors or their close relatives were more successful in the deeper past; within the past several million years, they successfully dispersed across the Atlantic from Africa, perhaps several times, and very rapidly gave rise to some fifty New World species.

Locust swarm in North Africa. Photo by Eugenio Morales Agacino.

Chapter Twelve

A WORLD SHAPED BY MIRACLES

Potatoes and the Homogenocene

This book began with the great conundrum of biogeography, the observation that certain closely related lineages occur on land areas separated by wide stretches of ocean, that is, by barriers that seem insurmountable for many organisms. By now it should be clear that, for a large number of these cases, the primary explanation of the vicariance biogeographers—that is, that such distributions came about through the movement of drifting tectonic plates—was the wrong explanation. Instead, many different kinds of plants and animals have crossed such ocean barriers and then have successfully established populations in their new lands. Chance, long-distance voyages of this sort have taken place many times across every sea and ocean in the world. Furthermore, plants and animals that seem to have no business making such long ocean journeys have done so, from monkeys and southern beech trees to frogs, freshwater snails, and probably even dinosaurs.

These examples have a kind of parlor game interest. *What flightless arthropods have been extraordinarily successful at crossing ocean barriers?* Spiders, especially ones that can "fly" by ballooning. *Where can one find dinosaur fossils—Gondwanan relics—alongside a modern biota descended only from overwater colonists?* The Chatham Islands, east of New Zealand. *What do crocodiles, geckos, skinks, amphisbaenians (worm lizards), and thread snakes have in common?* All of them are reptiles that crossed the Atlantic.

However, these cases of long-distance dispersal, collectively, have deeper meanings as well. Specifically, they help to answer a fundamental question that people have been asking for millennia, namely, "Why is our world the way that it is?" Of course, there is a multitude of ways to answer that question—an infinite number, actually—even if we leave Divine purposes out of it. I suggest, however, that the answers that emerge from the biogeographic evidence are at once exceptionally broad and deep in their application. They are answers that can change the way we look at the world.

In this final chapter, I first address this fundamental question in a relatively straightforward way, arguing that the biota of the Earth—specifically, the identities and locations of living species—has been profoundly influenced by natural ocean crossings. This influence has been so great, I will suggest, that if one could somehow go back in time and eliminate all of these overwater colonizations, and then jump back to the present, the living world as we know it would be transformed into an alien one. This conclusion that ocean crossings have had an enormous impact then leads to another, more conceptual, kind of answer to the question, an answer that reveals a general property of the history of life. As will be discussed below, some (probably most) biologists believe that evolution is unpredictable; they recognize that even small events can have unforeseen and profound consequences. Take out one species in the Cambrian, so the story goes, and the subsequent course of life on Earth would be radically altered. However, the exact nature of these history-changing events has typically been an intangible, a subject only of speculation. The evidence for ocean crossings, I will argue, makes the nature of such events far more concrete and indicates that the diversification of life has frequently been deflected by purely chance occurrences. This deduction, in turn, implies a fundamental unpredictability to the large-scale history of living things, to the way in which evolution has unfolded through deep time.

As a launching point for thinking about the general importance of natural ocean crossings, we will consider first an unnatural one, an organism transported by Europeans from the Americas to the Old World. The point of this detour is to introduce the subject of historical consequences, of paths taken and not taken, by considering a species that is both familiar and has had an undeniably large impact on humanity. From this recent and readily comprehended history of dissemination, we will then segue to the deeper and less apparent past that encompasses the multitude of natural ocean crossings.

Let us consider the potato, *Solanum tuberosum*. Native people in what is now Peru first cultivated potato plants from wild progenitors more than 4,000 years ago, an event that would change the history first of South America and then of the entire world. Potatoes became the staple crop in the Andean altiplano region and, rendered into a freeze-dried form that could be kept in centralized storehouses, were the fuel for the labor gangs and armies of the Inca Empire. When the Spaniard Francisco Pizarro met the Inca in 1526, he was encountering a civilization built upon the potato.

As with cases of natural dispersal, no one knows exactly when the Spanish first brought potatoes to the Old World, but it probably happened in the 1550s, on ships coming from the Pacific side of South America, and the potatoes likely were on board as provisions for the voyage rather than as an intended import. Rumors dogged the plant at first. The tubers were thought to be poisonous. They were said to cause leprosy. Ultimately, however, need and convenience won out. People found that potatoes were not only edible and highly nutritious, but also could be grown on land that was marginal for most other crops. As a result, by 1800, potato farming had spread all across northern Europe and into Asia.

In Ireland, the impetus for growing potatoes was especially strong, because the climate was too soggy for productive wheat farming, and the new crop was adopted early on, in the mid-1600s. For poor Irish, in particular, that changed everything, first for apparent good, then great ill. Families that had barely been able to grow enough food to sustain themselves found they could produce an excess of nutritious spuds; thus, a region that had been held back by strong limitations on food production was released from restraint. Young Irish men and women could start families at an earlier age; children who would have succumbed to starvation or associated diseases survived. The predictable result was a population explosion. In the early 1600s, pre-potato, there were about 1.5 million people in Ireland; two centuries later, the number had jumped to over 8 million. Other factors probably were involved, but in large part the increase seems to have been driven by potato farming.

After that dramatic, exponential rise came the crash. In 1844, near the border of France and Belgium, a botanist noticed dark spots on the leaves of some potato plants. This was the potato blight, a disease that started in the leaves and then moved down the stem to the tubers and roots, quickly causing the whole plant to disintegrate into a putrid, black mess. In the summer of 1845, the blight spread rapidly outward from its epicenter in

Belgium, reaching Ireland by mid-September. The culprit, discovered in 1861, was a microorganism called an *oomycete*, fungus-like in many of its properties but actually more closely related to brown algae. This particular kind of oomycete, *Phytophthora infestans*, was, not surprisingly, an import from the New World, just like the potato itself. It had evolved with *S. tuberosum* in the Americas, and perhaps had been carried across the Atlantic to Belgium in a shipment of seed potatoes from the United States.

In 1846 the blight reached its peak in Ireland, destroying some three-quarters of the potato crop, and with it the main source of nutrition for nearly all poor Irish. The toll on the population was devastating: it is estimated that, in three years, a million people died of starvation or, in a weakened state due to malnutrition, succumbed to typhus, cholera, and other diseases. At the same time, thousands of the living, deprived of essential nutrients, went blind or insane. Within a decade, the Great Famine had pushed some 2 million people out of Ireland, including more than half a million to the United States, radically changing the history of that country as well; if you're an American with Irish blood, you probably owe your existence in a very direct way to the introduction of the potato and the potato blight oomycete to the Old World. In Ireland itself, the famine was such a cultural and political watershed that historians often divide the country's history into prefamine and postfamine periods.

Similar history-changing stories could be told about the ocean-crossing potato in other countries where it eventually became a staple crop—Russia, Germany, China, and India, for example. One can make a strong case, for instance, that potatoes fueled the ascendance of Russia and Germany as world powers in the nineteenth century. More generally, it has been estimated that about a quarter of the rise in population that took place in the Old World from 1700 to 1900, from roughly 600 million to about 1.5 billion people, can be attributed to the adoption of potato farming.* In other words, by 1900, the introduction of this one species had added some 200 million people to the Old World. Potatoes, in short, propelled world history into a radically different course from what would otherwise have been. Population explosions, famine, mass immigration, and the rise of world powers were all brought about because *S. tuberosum* came across the ocean in Spanish ships.

The potato is obviously somewhat extreme in its influence, but countless other introduced species that humans have purposefully or

---

* Nathan Nunn and Nancy Qian generated these figures in a 2011 study by comparing population growth for regions where potatoes became an important crop to those for regions where it did not.

inadvertently transported across oceans have had massive effects as well, from tomatoes, rubber trees, and cheatgrass to rats, pigs, and gypsy moths to the microorganisms that cause malaria, smallpox, and yellow fever. These human introductions have been so numerous, and their consequences so world-altering, that the era beginning with Columbus's discovery of the New World has been called the "Homogenocene," referring specifically to the fact that movements of species blend formerly distinct biotas, tending to homogenize them into a single, uniform global biota. Whatever we want to call this era, it seems abundantly clear that one can hardly understand human history and the nature of modern environments without recognizing the influence of the flood of species that we have moved across oceans.

A strong message of the biogeographic studies presented in this book is that a parallel statement holds for the deeper past: one can hardly understand the history of living things on Earth and the nature of the modern world without recognizing the influence of species that have crossed oceans *naturally*, without the aid of humans. Quantifying the full effects of these chance ocean crossings with any accuracy isn't possible, any more than it's possible to know, for instance, all the historical ramifications of the introduction of potatoes to the Old World. History is simply too complex for such an undertaking. Nonetheless, just as we can infer, at a minimum, that the introduction of potatoes had a huge effect, we can also infer that the influence of natural ocean crossings has been enormous. This influence has been so great, in fact, that I can only hint at its scope.

Changes in Shallow Time

Imagine an apparently unremarkable day in the life. In the morning, your wife says she's stopping at the grocery store on the way home from work and asks if there's anything she should pick up. "Can you get some nutmeg?" you ask. "I need some for a cake." It's summer and you dress for the heat, which means more or less entirely in cotton—cotton T-shirt, cotton shorts, cotton underwear, cotton socks, and a cotton baseball cap when you go outside. After breakfast, you putter around with your daughter and son in the vegetable garden to see what's ripe. The kids are picky eaters but good pickers, and they twist off a couple of eggplants and some oversized zucchinis. A watermelon is getting big, but it isn't quite ready yet. The kids take the vegetables inside and then head off to school, and you stay home and write all morning and then drive to a Mexican fast-food place for lunch, where you order a burrito and are asked if you want it with black or pinto beans. "Pinto," you say, "and I'd like the corn salsa." There's a large TV in the restaurant and, as you eat, you watch the news, a report of a storage compartment, washed out to sea by the Japanese tsunami, drifting all the way across the Pacific to a beach in British Columbia. The compartment is open, but, amazingly, some of its contents still remain inside, including a Harley Davidson motorcycle.

After lunch you head home and try to keep writing, but the ideas are elusive, flitting around like butterflies, so you take a break and think about an upcoming family trip to the Yucatán. In a big book on Mayan ruins, two photos of blocky statues hewn out of the pale, volcanic rock jump out, one of a crocodile and one of a monkey god. The book also mentions a tame spider monkey that lives among the ruins of Copán. (You remember that there was a tame spider monkey at the Smithsonian field station on Barro Colorado Island in Panama years ago and wonder if this species has some special affinity for humans.) After that, more writing, or at least an attempt at writing, and then, almost as soon as the kids are home, pandemonium breaks out: your daughter sets the pet guinea pig down on the living room floor and seems shocked when, as usual, it scurries off and lodges itself under the sofa. The kids' running and screaming has the pair of lovebirds chattering, too, but the kaleidoscopically colored panther chameleon seems nonplussed, only scanning with one turreted eye as it perches motionless on a branch in its terrarium.

How would this ordinary day be different if living things had never made chance ocean crossings? Pretty much in every way: the evidence to date, mostly from molecular timetrees and the fossil record, suggests that all of the organisms mentioned had ancestors that dispersed naturally across ocean barriers. The nutmeg tree's ancestors colonized the volcanic Banda Islands, near New Guinea, presumably from other islands in the area; the eggplant's ancestors went from South America to India, probably via Africa; the zucchini's from Africa to South America; the watermelon's from Asia to an island Africa; the common bean's from Africa to the New World; the corn's from Africa to North America (although the evidence in this case is a bit iffy); the crocodile's, spider monkey's, and guinea pig's from Africa to the New World; the lovebird's from Australasia to Africa via Madagascar; and the chameleon's from Africa to Madagascar. The oddest story belongs to cotton. Most of the cotton we wear comes from *Gossypium hirsutum*, part of a group of New World species that originated when one cotton lineage hybridized with another. The strange thing is that these two lineages appear to have independently colonized the New World from Africa by crossing the Atlantic. In other words, *G. hirsutum* is simultaneously the product of two separate chance, long-distance dispersal events, like a miracle squared.

The point is that none of these lineages would even exist if their ancestors had not crossed oceans. Keeping our focus on human history for the moment, we can easily see that without these and other ocean-crossing taxa, many small details of the world as it exists today would necessarily be altered. For instance, all the zucchini breads and eggplant parmigianas would disappear. No more guinea pigs for genetic experiments or for Andean people to roast on sticks. No more nutmeg in the eggnog. However, the changes would go far deeper than that. Consider, as just one of a multitude of possible examples, that of the great triumvirate of pre-Columbian crops in the New World, beans and squash almost certainly evolved from ancestors that crossed oceans naturally, and the third, maize, probably did as well. Make all three of those plants disappear, and history would lurch into some unknown but fundamentally different space. The nature of prehistoric agriculture and its associated cultures would have been radically shifted almost everywhere from tropical Mesoamerica through the desert American Southwest to the eastern seaboard of temperate North America. The Toltecs and Mayans and Aztecs, the cliff-dwelling Anasazi, the mound builders of the Mississippi drainage and Great Lakes, the eastern tribes that encountered the Pilgrims—in this alternate history, none of these cultures would have existed in anything like their known form. In post-Columbian times, billions of tons of beans, squash, and corn would be erased from agricultural ledgers all over the world. No corn-fed hogs, no squash in the cellar or sacks of dried beans to get pioneers through hard winters, no Mexican food as we know it, no high-fructose corn syrup.

Finally, there are indications that all of these details might be moot. In particular, some primatologists have argued from the fossil record that an ancestor of anthropoids, the group that includes monkeys and apes, must have crossed the Tethys Sea to colonize Africa from Asia. This conclusion is based on placing certain Asian and African fossils in specific positions within the primate evolutionary tree, and those placements remain controversial (as they often are with fossils). Let us assume for the moment, however, that this crossing of the Tethys really happened, and then imagine the consequences of erasing that unlikely journey from deep history. Without that ancestor, the whole African anthropoid tree would disappear. Colobus monkeys, baboons, and macaques would all wink out of history, along with the monkeys that crossed the Atlantic to populate the New World. There would be no gibbons, orangutans, gorillas, chimpanzees, *Australopithecus*, or *Homo erectus*. There would be no us. Darkness would descend, literally—the night would be lit once again only by the stars and moon.

Contemplating the elimination of that ancient anthropoid ancestor and, with it, a large segment of the primate tree, leads into our next subject. In considering cases such as the civilizations founded on beans, corn, and squash, we've been taking a "shallow-time" view, pushing only slightly further into the past than the history of potatoes in the Old World. However, such examples represent just the thin surface of what would have to be different in a world without chance, oceanic dispersal. To more fully appreciate the ramifications of natural ocean crossings, one has to recognize that most of these events are far more ancient than any human introduction, which means that their immediate effects have had much more time to cascade into others, to expand their spheres of influence through deep time.

As an illustration of that influence, we will turn, appropriately, to South America, the continent that, more than any other, put the seeds of evolutionary thought into Darwin's mind, the same continent where Alfred Russel Wallace, noticing that the boundaries of species or varieties often coincide with the courses of great rivers, began thinking deeply about barriers to dispersal.

The Branches That Became Trees

Not so long ago, just a thin section on the geologic timescale, South America was an island continent. From roughly 30 million years ago, when a fairly tenuous link to Antarctica was broken, until the emergence of the Isthmus of Panama about 3 million years ago, South America was disconnected from all other large landmasses (see Figure 12.1). One has to go much further back, to 50 million years ago or so, to find the next most recent possible connection to North America ("possible" because some geologists doubt this connection existed), and then much further still, to roughly 100 million years ago, to see the final attachment between South America and Africa prior to the opening of the Atlantic.

That geologic history is the setting for one of the great stories in biogeography, focusing on the origins and fates of South American land mammals. And that story begins with the continent's long period of isolation.

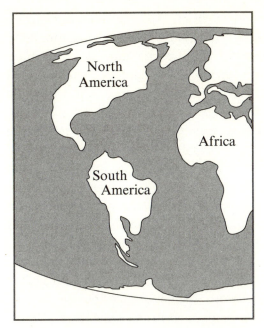

12.1 South America as an island continent. The reconstruction pictured is for the Middle Miocene, roughly 10 million years ago. Redrawn and modified from Schaefer et al. (2009).

During and because of that isolation, a diverse and distinctive mammal fauna developed, including anteaters, sloths, armadillos, opossums, and a great assortment of entirely extinct groups, such as hyena-like and sabertooth-cat-like forms that were closely related to marsupials, as well as a variety of hoofed mammals with unfamiliar names like Pyrotheria, Astrapotheria, Litopterna, and Xenungulata. In that same period, mammals from other continents generally did not enter South America, kept out by ocean barriers. Then the Panamanian Isthmus emerged, providing a corridor for the passage of land mammals (and other organisms); whereas before, a natural raft was needed for mammals to pass between the continents, now they could make their way by walking, hopping, or running. The result was a rapid and dramatic mixing of faunas: during a relatively short period—the past 3 million years—deer, cats, dogs, squirrels, peccaries, camels, and ten other North American mammal families colonized South America, while opossums, armadillos, anteaters, marmosets, porcupines, and thirteen other South American families moved in the other direction, colonizing North America. This mixing of faunas, dubbed the "Great American Interchange," was a classic case of "geodispersal," that is, the movement of many species into a new area after a barrier has been eliminated, in this case, the ocean barrier that had separated North and South America. As a prominent example in historical biogeography, it rivals Gondwanan vicariance.

The colonizations of the Great American Interchange presumably happened by normal dispersal over the newly emerged land bridge, that is, by the expected "garden-variety" movement of organisms into suitable habitat. However, somewhat lost in this story is the fact that two North American groups showed up in South America shortly *before* the emergence of the isthmus. Specifically, sigmodontine rodent fossils have been found in Argentina in strata 4 to 6 million years old, and procyonid carnivores (in the same family as raccoons and coatimundis) in strata dated to about 7 million years old, also in Argentina. For the sigmodontines, molecular dating corroborates the notion that the dispersal event happened before the existence of the land bridge. Members of those two groups apparently arrived in South America by way of an archipelago in what is now Central America, probably using a series of islands as stepping stones.

These island-hopping journeys by sigmodontines and procyonids don't rank among the truly startling long-distance colonizations. However, given that terrestrial mammals are not adept at crossing substantial water barriers—even dispersalists like Charles Darwin and George Gaylord Simpson admitted that fact—these events certainly qualify as unexpected colonizations. Thus, although these two southward dispersals are sometimes treated as "heralds" of the Great American Interchange, they can also be placed in a different category: they are examples of land mammals that arrived in South America by chance, oceanic dispersal.

We have already run across two much better known examples that fall into this category, the monkeys and caviomorph rodents (guinea pigs and relatives), both of which likely crossed the Atlantic from Africa (Chapter Nine) no later than about 26 million and 41 million years ago, respectively. Beyond these, there are a few less convincing cases. Thus, there are exactly four well-established examples of overwater colonization by land mammals in a period of some 50 million years, dating from the disappearance of the previous North America–South America land connection. Four may not sound like many, and certainly doesn't compare with the tide of mammal groups that has passed over the Isthmus of Panama in the past 3 million years. However, that small number is misleading when it comes to thinking about the importance of overwater colonists in the history of the continent. It's misleading for the simple reason that, during the course of deep time, evolutionary lineages can ramify. Single branches can turn into large evolutionary trees.

That, as it turns out, is exactly what has happened for three of the four groups of oceanic dispersers—the monkeys, caviomorphs, and sigmodontines.* In all likelihood, the ancestors of each of those groups arrived in the form of just a few individuals of a single species carried on a natural raft. From that tenuous beginning, however, each of those ancestral species evolved into a great range of forms. The diversity of New World monkeys, from tiny, acrobatic marmosets and tamarins to lanky spider monkeys and robust, deep-voiced howlers is well known, but the radiations of caviomorphs and sigmodontines are arguably even more impressive. The caviomorphs include, among many others, guinea pigs and chinchillas; a dozen species of prehensile-tailed porcupines; some sixty kinds of burrowing, gopher-like tuco-tucos; many species of arboreal spiny rats; the semiaquatic nutria (a common introduced pest outside of South America); tail-less, long-legged agoutis; mountain viscachas, all with big ears like a rabbit's and a bushy tail like a squirrel's; and the plains viscacha, which lives in communal burrows, looks like a giant guinea pig, and can both run rapidly and make prodigious leaps (see Figure 12.2). The world's largest living rodent, the capybara, is a caviomorph, as was the world's largest known extinct rodent, *Josephoartigasia monesi*, which had a skull twenty inches long and may have weighed as much as a rhino. Sigmodontine rodents are anatomically less strikingly varied—most of them look like mice or rats—but they occupy almost every kind of environment in South America, from deserts to alpine tundra to rainforests. Like the caviomorphs, the sigmodontines include terrestrial, arboreal, semiaquatic, and burrowing species. Most sigmodontines are herbivores of one sort or another, but some semiaquatic species eat arthropods and, occasionally, fish.

Beyond their diversity of form and habits, the success of the immigrant rodents and monkeys shows up in a more subtle form also: these groups became so ubiquitous, forming such a large pool of potential dispersers, that they managed to make quite a few successful long-distance

---

* The procyonid that colonized South America before emergence of the isthmus is thought to have given rise to several species, including one that was the size of a bear, but all of these lineages apparently went extinct by the end of the Pliocene (Koepfli et al. 2007).

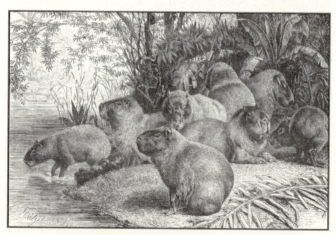

12.2 A hint of the great diversity of caviomorph rodents. From top to bottom: Brazilian porcupine (*Coendou prehensilis*), plains viscacha (*Lagostomus maximus*), capybara (*Hydrochoerus hydrochaeris*). Drawings by Gustav Mützel.

voyages *from* South America. In particular, all three lineages reached the West Indies, and the sigmodontines also traveled west in the Pacific to colonize the Galápagos at least three separate times, and east in the Atlantic to Fernando de Noronha (where, before going extinct, the immigrant rodent lived alongside the *Mabuya* skink from Africa).

Numbers cannot tell the whole story of biological influence, but here are some that must indicate a massive impact: In South America today, there are 124 described species of monkeys, 219 species of caviomorphs, and 330 species of sigmodontines, a total of 673 species. Together, these three groups make up 73 percent of all living, native South American mammal species, excluding bats and purely aquatic species. In terms of individual animals the percentage is undoubtedly much higher, because the caviomorphs and sigmodontines include many of the most abundant species on the continent. These numbers, in turn, imply countless others—all the fruits and leaves and insects eaten by monkeys; the seeds destroyed or dispersed by agoutis; the square miles of ground undercut by the burrows of tuco tucos;* the furry meals for various species of weasels, cats, dogs, snakes, hawks, and other predators; and the monkey and rodent bodies scavenged by birds and mammals and decomposed by various invertebrates, bacteria, and fungi.

It is hard to say *exactly* what effects these and other ecological interactions have had on the evolutionary history of the South American biota, but those effects must be vast. A very clear impact comes from the observation that the monkey, caviomorph, and sigmodontine radiations essentially encompass the evolution of many other species, namely, parasites that depend on these vertebrates as hosts. As an especially striking example, consider a group of nematodes, related to human pinworms, that are intestinal parasites of New World monkeys (and presumably give monkeys itchy butts just as pinworms do in humans). It turns out that the evolutionary tree of these nematodes largely mirrors the evolutionary tree of the monkeys; nematode lineages tend to branch in concert with the branching of their monkey host lineages. That pattern of one tree matching another indicates that the evolutionary history of the parasites has been dictated by the history of the hosts; to a degree, the parasite acts as if it were simply part of the host. For our purposes, what this phenomenon of trees mirroring trees means is that, if there were no monkeys in the New

---

* In *The Voyage of the Beagle*, Darwin wrote, "Considerable tracts of country are so completely undermined by these animals [tuco tucos], that horses, in passing over, sink above their fetlocks" (Darwin 1839, 79).

World, the entire associated nematode tree also would not exist. More generally, even for parasites that do not evolve in such lock-step with the species they infect, the hosts are still critical to the evolution of the parasites. Thus, any South American parasite that requires a monkey, a caviomorph, or a sigmodontine as a host—and there are many such parasites, from protists and fungi to tapeworms, chewing lice, and botflies—owes its existence to those host lineages and, therefore, to the fact that the ancestors of these mammals were able to colonize the continent in the first place. In addition, many of these parasites have complex life histories that require a mammal host for part of the life cycle and another, completely unrelated animal for another part of the cycle. And this means that the influence of the mammals expands outward through the parasites to other kinds of animals, all part of the cascade of ecological effects.

The cases of monkeys and rodents are striking, but the importance of overwater dispersal in the history of the South American biota is not just about land mammals. Similar histories of colonization and subsequent radiation—a single branch becoming a sizable evolutionary tree—apply to other vertebrate taxa as well (see Figure 12.3). From my own somewhat serpent-centric point of view, an especially memorable example involves a group of snakes called the Xenodontinae, whose common ancestor apparently rafted from North America to South America sometime between about 12 and 28 million years ago, when those two continents were still separated by an expanse of ocean hundreds of miles wide. That single branch persisted and eventually gave rise to a great diversity of forms, including terrestrial and arboreal species that kill prey using a combination of constriction and envenomation; small snakes that specialize in eating reptile eggs; large species, called false water cobras, that search for fish by probing with their tails in aquatic vegetation; and hog-nosed snakes with upturned snouts and with color patterns that mimic the warning coloration of highly venomous coral snakes. When one thinks of South America, the snakes that come to mind are boa constrictors and anacondas, lineages that have been on the continent since it broke away from the other fragments of Gondwana. However, the Xenodontinae, with nearly three hundred living South American species, is actually the largest evolutionary radiation of snakes on the continent, much larger than any group of Gondwanan snakes.

Group	Source	So. Amer. Species
Rhinella toads	No. America	85
Phyllodactylid geckos	Africa	45
Mabuya skinks	Africa	18–20
Amphisbaenid lizards	Africa	81
Xenodontine snakes	No. America	~300
Turdus thrushes	No. America	25

12.3 More single branches leading to substantial trees: some nonmammalian vertebrate groups that reached South America by overwater dispersal, as indicated by molecular timetrees. In the list below, if only one reference is given for a taxon, that reference includes the molecular dating analysis and a tally of the number of species in the group. If two references are given, the first is the molecular dating study and the second provides the number of species. *Rhinella* toads, Pramuk et al. (2008), Frost (2011); phyllodactylid geckos, Gamble et al. (2011); *Mabuya* skinks, Whiting et al. (2006); Amphisbaenidae, Vidal et al. (2008), Gans (2005); *Turdus* thrushes, Voelker et al. (2009). Pramuk et al. (2008) did not conclude that *Rhinella* reached South America by oceanic dispersal; however, the authors estimated that its separation from North American relatives took place between about 30 million and 50 million years ago, which would indicate overwater colonization. For references on the xenodontine snakes, see the endnotes for Chapter Twelve. Photo of *Rhinella alata* by Brian Gratwicke.

The xenodontines also are instructive in that they make up part of an ecological intersection of colonizing lineages in the New World. Specifically, many of these snakes eat lizards, small mammals, and/or toads (the hog-nosed coral-snake mimics are toad specialists), and this must mean that often a snake descended from an overwater colonist ends up eating another vertebrate that is also descended from an overwater colonist. That kind of colonist-meets-colonist interaction underscores the depth of the influence of oceanic dispersal on the South American biota. And if we add plants to the discussion, such encounters become truly rampant. Although the geographic origins of most South American plants have not been worked out, there are certainly hundreds, and probably thousands, of species derived from overseas immigrants, and these plants must have ecological connections with tens of thousands of other species, including many of the vertebrate overseas colonists. For instance, sigmodontine rodents of the genus *Oecomys* eat the fruits of *Renealmia alpinia* (Zingiberaceae), part of a group of some sixty species of gingers whose common ancestor came from Africa in the Miocene or Pliocene; at least five kinds of South American monkeys feed on the flowers of the rainforest tree *Symphonia globulifera* (Clusiaceae), a recent immigrant from Africa; and *Turdus* thrushes eat the fruits (and presumably disperse the seeds) of *Ocotea* (Lauraceae) and *Miconia* (Melastomataceae), plant lineages descended from species that arrived over water from North America within the past 40 million years.

All of this is to say that there is nothing subtle about the effect of natural, overwater colonization on the history of life in South America. It is clear that such colonists have given rise to a large part of the continent's biota and that this must have entailed an enormous cascade of impacts on other species. Ripples, domino effects, the fire that transforms the landscape, all of those metaphors and others apply. The cumulative effects are hard to even imagine, but it is safe to say that, without those colonists, the modern biota would feel very alien, both in what it would have and what it would lack. And if we can extrapolate from South America to other regions, it must be that the living history of the entire planet has been deeply influenced by ocean crossings and other long-distance colonizations. That is an inescapable conclusion, a tangible and general message, from the recent flood of biogeographic case histories.

Pikaia and the Nature of Deep History

If we accept that chance colonizations have had clear and profound effects, as it seems we must, what larger meaning can we take from this? What does this conclusion say about the nature of the history of life in general? What does it tell us, on a higher, conceptual level, about how the world came to be the way it is?

To answer these questions, we need to think about the linked ideas of contingency and unpredictability as they apply to history. In this context, *contingency* refers to the property of events being dependent on prior events: D happened because C happened; C happened because B happened; and B happened because A happened. To put it in the form of the usual sort of historical example: in 1972, South Dakota senator George McGovern ran for president on an antiwar platform (event D), in response to the continuing Vietnam War (C), which was itself caused by (among other things) escalating fear within the United States of Soviet communist influence (B), which was a result of (among other things) the occupation of much of Eastern Europe by the Soviet Union during the Second World War (A). (Incidentally, for this particular case, if you keep going back in time, you'll eventually reach Spanish ships bringing potatoes across the Atlantic, leading to the rise of Russia.) Eliminate A or B or C, and D never happens; thus, D is contingent upon A and B and C.

The observation that history has this property of contingency is a truism (how could it *not* have this property?). Similarly, at least for human history, it is uncontroversial that at least some events in any sequence are unpredictable, which means that the course of history in general must be unpredictable; destiny might pivot on the mood of a head of state at some critical juncture (do the negotiations continue or do we launch the missiles?), or even on whether a child runs into the street at exactly 2:13 on a Tuesday afternoon, causing a car to swerve and crash into a telephone pole, making the driver miss an engagement where she would have met her future husband, etc., etc., etc. That sort of unpredictability is the basis for an entire cottage industry of "what if?" novels, short stories, TV episodes, and movies that depend on the notion that things might have turned out very differently (for Peggy Sue or Jean-Luc Picard or the entire human race) given some small and entirely plausible alteration of events. Even among many academic historians, such "what if?" scenarios are considered legitimate exercises, an acknowledgment that there truly is very little separating what actually has happened from an infinite number of radically different alternative histories.

This notion that apparently minor events can cause profound changes down the line, although widely accepted for human history, has not been so obvious for evolution, particularly when it comes to the large-scale history of life. Could some seemingly insignificant event hundreds of millions of years ago—the survival or demise of a few individuals of a single species, for instance—lead through long chains of cause and effect to great evolutionary success for some groups and the extinction of other groups, that is, to massive consequences for the overall form of biological diversity? Or, alternatively, is that large-scale history more like water running through deep, narrow channels, such that minor events have little effect on the long-term outcome, analogous to pebbles that fall into those channels without diverting the flow?

The late paleontologist and evolutionary theorist Stephen Jay Gould argued vehemently for the former, particularly in *Wonderful Life*, his 1989 book on the Cambrian fossils of the famous Burgess Shale Formation in the Canadian Rockies. That book still stands as the most detailed and widely read explication of contingency and unpredictability in evolution, so it is worth considering what Gould said and, in particular, how he reached the conclusions he did.

Life's history, Gould claimed, is dominated by events that have produced massive effects that could not have been anticipated. Importantly, many of these events would have appeared small in magnitude and thus insignificant when they happened. Thus, evolutionary history on the grand scale is not only contingent, but thoroughly unpredictable, because of its sensitivity to tiny perturbations. Replay the tape of life over and over, Gould suggested, "altered by an apparently insignificant jot or tittle at the outset," and the result would be entirely different each time.

In *Wonderful Life*, Gould used the Burgess Shale fossils, representing the rapid early diversification of animals known as the Cambrian Explosion, as his centerpiece example of the contingent and unpredictable nature of history. He followed Harry Whittington, Simon Conway Morris, and Derek Briggs—experts on the Burgess specimens—in suggesting that the Cambrian Explosion gave rise to a great array of very distinct anatomical types, most of which disappeared shortly thereafter. According to Gould, the decimation of most of these early forms dictated the course of all subsequent animal evolution or, to put it another way, the history of animals was contingent on exactly which forms survived that early winnowing.

Gould then went on to argue that the set of surviving lineages was unpredictable. For instance, he asks us to consider a fairly unassuming Burgess fossil called *Pikaia*, once considered a polychaete worm, but recognized by Conway Morris and subsequent researchers as an early member of our own phylum, the chordates. Gould notes that there was nothing special about *Pikaia* to indicate that it would be one of the survivors of the early decimation. Replay the tape of life again, he suggests, making some slight change in initial conditions, and perhaps *Pikaia* is one of the losers this time around. And if *Pikaia* was a direct ancestor of all subsequent chordates, then a substantial portion of the tree of life, including ourselves and all other vertebrates, would never come into being in the replay.

My gut reaction on reading *Wonderful Life* years ago was that Gould was probably right. Perhaps I was thinking in human terms—how the course of a life could depend on that child running into the street or a thousand other events that, if the tape were replayed, might easily turn out differently. However, at the same time I was dissatisfied with Gould's examples of the unpredictable. So if *Pikaia* went extinct without producing any descendants, then the course of life on Earth would be radically changed. That I could easily accept (having no religious or other beliefs to dictate to me that humans are either necessary or inevitable). But what kind of perturbation, what "apparently insignificant jot or tittle at the outset," would suffice to make it so? Would an underwater avalanche that buried a few of these worm-like chordates push the species toward an early exit from history, or would some much larger deviation be necessary? Perhaps there was, in fact, something special about *Pikaia*, some working of its muscles or gills or rudimentary brain that earmarked it for survival, even if, 500 million years later, nothing of the kind is apparent to us. (After all, most of what went on 500 million years ago is not apparent to us.) If that were the case, maybe *Pikaia* was destined to make it through the period of decimation in *any* replay of the tape.

The point here is not to suggest that Gould was wrong about the unpredictability of life's history. The problem, as I see it, was in how he reached this conclusion. Gould claimed that the early decimation of most of the incredible early diversity of animal life, on the one hand, and the survival of a few lineages, including *Pikaia*, on the other hand, is a supreme example of the contingency and unpredictability of history, but actually it isn't, or, at least, it's not a convincing one. We really have no idea why some lineages survived the decimation and others did not, so it doesn't follow from this case that the set of survivors was unpredictable and that a hypothetical replay of the tape would produce an entirely different outcome.* Gould brought up many other cases, too, from the extinction of the bizarre set of early multicellular organisms known as the Ediacaran fauna to the origins of *Homo sapiens*, but they were similarly unsatisfying. *Wonderful Life* is in many ways a brilliant book—it's deep, erudite, and entertaining, all trademarks of Gould's writing in general. But his argument in this instance could have used some better examples.

Archetypes of the Unexpected

Biogeography, especially now, with the evidence from molecular time-trees, can provide those better examples of the unpredictable. In fact, if I'm even close to correct about the frequency of chance ocean crossings and other long-distance colonizations, we are up to our eyeballs in cases that Gould could have used in *Wonderful Life*. The transoceanic journeys of monkeys, rodents, lizards, crocodiles, sundews, araucarian conifers, southern beeches, and countless others were all rare, chance events, the

---

* Some paleontologists also have argued against Gould's view of the Burgess fauna as representing an exceptional breadth of distinct animal forms, most of which are unrelated to later lineages. Interestingly, one of these paleontologists is Simon Conway Morris (1998), who initially agreed with Gould's view and whose work formed the basis for much of what Gould concluded about the early evolution of animals.

kinds of events that, for any particular time and place, one would *not* expect to happen. Replay the tape of life again, altering some "apparently insignificant jot or tittle," and none of those colonizations occur (but others, absent from the history we know, do take place).

For instance, consider a reasonable scenario for monkeys colonizing South America from Africa. Start with a group of African monkeys lounging in a tree by a major river, a river similar to the modern-day Congo or Niger. Heavy rains have been falling for days, and the resulting floodwaters undercut a large chunk of the bank, including the monkeys' tree, and the whole piece falls into the water and is swept away. This natural raft is carried many miles downriver, eventually reaching the ocean, where it gets caught up in a westward-flowing current. On the ocean, the monkeys eat anything edible they can find on their small, floating island. When it rains, they drink the water that briefly pools up. Weeks later, the raft, now waterlogged and barely buoyant, is grounded on the South American coast, and the few surviving monkeys, scrawny and dehydrated, splash onto the beach and disappear into the adjacent forest. They stay together, feeding on unfamiliar but edible fruits and insects, eventually mating, giving birth, and raising young. A thousand years later, their descendants form a substantial population that will ultimately give rise to the entire radiation of New World monkeys.

This scenario is rife with points at which the successful colonization could have been derailed. For starters, the monkeys might easily have chosen a different tree in which to rest. On its journey downriver, their raft of earth and vegetation might have meandered along some other course in the current, and thus perhaps could have ended up caught in a tangle of flood debris. Reaching the delta, the raft could have run aground on a sandbar, as many such rafts do today at the mouths of great rivers. The exact path across the ocean might have been different, depending on exactly when and where the raft picked up the westward current, and some of those alternate routes might have ended with the monkeys dying of thirst or starvation before ever making landfall. Finally, even if the monkeys reached South America, their tiny population could easily have disappeared, as extremely small populations often do, from simple bad luck: the only female in the group, for example, might have succumbed to disease, or been picked off by a predator. In short, the success of the colonization likely depended on a whole series of occurrences, any one of which could easily have gone a different way. And similar fortuitous sequences probably apply to nearly all cases of long-distance, chance dispersal—that is what makes them "chance" events—from seeds hooked onto a bird's feathers to moths caught up in the wind to crocodiles riding ocean currents. In essence, these colonizations and their subsequent impacts hinge on events akin to the child running into the street at exactly 2:13 p.m., not ten seconds earlier or ten seconds later.

If one is still skeptical of the chance nature of these events, consider, finally, that many ocean crossings probably depend to a large extent on that archetype of unpredictable phenomena, the weather. In particular, many transoceanic colonizations likely occurred through the agency of severe storms, which, as we all know, are impossible to accurately forecast even a few days before they happen. Furthermore, current thinking about how weather is generated suggests that long-range forecasting is not only impossible given our current means of measuring the important variables (the distributions over the landscape of atmospheric pressure, wind velocity, temperature, and so forth), but is likely impossible in principle. The key finding here, from quantitative weather models, is that the generation of storms and other weather phenomena shows "sensitive dependence on initial conditions." In other words, tiny differences in starting values (of pressure in one particular spot, for instance) can produce huge changes at some later time. This is the famous "butterfly effect": a butterfly flapping its wings in Brazil, it is said, might set off a tornado in Texas. For our purposes, what this finding suggests is that the development of any particular storm is the epitome of a chance event, the kind of event that could be erased by a tiny perturbation in a replay of history. In fact, a "jot or tittle" in the Jurassic atmosphere could change the details of all subsequent weather.* Those changes, in turn, would mean massive alterations in the history of long-distance dispersal.

In short, the means by which ocean crossings and other long-distance colonization events take place suggest that they are fundamentally unpredictable. Forget about *Pikaia* and the other creatures of the Burgess Shale; the shaping of life's history by oceanic dispersal is a much more compelling illustration of the chance nature of that history. Run the tape again (starting, let's say, from 50 million years ago), and South America would have no monkeys, caviomorph rodents, xenodontine snakes, or *Renealmia* ginger plants, but it *would* contain many lineages that never made it there in the history that we know. Similar changes would apply over every other part of the globe.

None of this is to say that chance dispersals are the only or even the main unpredictable events that have shaped the history of life. In fact, chaos theory—a research area that focuses on all kinds of systems that

---

* This is not to imply that such changes would affect climate, the long-term pattern of weather for an area.

show "sensitive dependence on initial conditions"—suggests that such sensitivity is a property of many biological phenomena. For example, fluctuations in population size and the rate of spread of an infectious disease should be affected by small perturbations. Such phenomena almost certainly influence the long-term history of life. For instance, a small, chance perturbation that ultimately caused a population to go extinct (a fluctuation to a population size of zero) might have major ramifications. Similarly, a particular mutation (that is, a particular base-pair change at a particular site in a particular gene at a particular time in a particular individual's germ line), the quintessential sort of chance event in biology, no doubt sometimes influences the grand sweep of evolution. For example, a mutation that produced a new advantageous trait could dictate the future course of evolution for a species, which could, in turn, influence the fates of many other species.

Nonetheless, among all these possible small, unpredictable perturbations with large effects, ocean crossings and other chance colonizations stand out. What sets this class of events apart is that we can identify individual instances of them and at least the most obvious of their far-reaching consequences. Strange though it may seem, this is not true even for mutations. Most kinds of mutations are actually common enough that, in a replay of the tape in which a particular mutation (that is, in a particular individual organism) did not occur, the exact same kind of mutant likely would already be present in the population or would soon turn up. Thus, at best we usually can only point to a *kind* of mutation—one that changes an alanine to a threonine at amino acid site 269 in an opsin gene, for instance—that has had a major impact. In that sense, mutations are more akin to "normal" rather than "chance" dispersal. They are analogous to, say, a common land-bird species colonizing an island very close to the mainland; replay the tape and the colonization still occurs, even if the exact same individuals are not involved this time around.* Cases like monkeys or caviomorph rodents crossing the Atlantic are immune to that sort of argument because of their extreme rarity and improbability; rerun history and, almost without a doubt, monkeys and caviomorphs do *not*

---

* The effects of single, rare mutations have been demonstrated in experiments in which bacteria that were initially genetically identical were propagated as separate populations, and only some of the populations evolved specific genetic adaptations for using a food source. Presumably such mutations have had major effects on the large-scale history of life, but it would be difficult to point to an actual case, that is, to identify a single mutational event that occurred millions of years ago and dramatically changed the course of evolution. See Blount et al. (2008).

colonize the New World *at all*. This means that we can point to these events as the particular ones that have changed history. Furthermore, we don't need to strain to imagine the consequences of these ocean crossings. Traveling through Latin America, the results are plain to see (and hear) in the form of species that would not exist without these colonizations—a troop of squirrel monkeys, noisily chattering as they move through a rainforest in Amazonia; the bell-like calls rising from the burrows of tuco-tucos on the Argentine pampas; a herd of capybara splashing through a flooded field in the Brazilian Pantanal. As testimonies to life's unpredictability on a grand scale, there can hardly be examples more telling. They are the embodiment of Gould's notion of history's sensitivity to tiny perturbations.

For some, the view of life described here may seem disheartening. We discovered the geographic distributions of all kinds of organisms, a rich record of past life in the rocks, the changing configurations of continents and oceans, the means to reconstruct evolutionary trees and to use molecular data to turn them into timetrees, and what did it all tell us? That the course of life on Earth has been frequently buffeted, deflected into new pathways, by the random and improbable. That many of the key events in this history are essentially inexplicable in the sense that they cannot be derived from any natural laws (such as "Earth and life evolve together"). That these events are impossible to fully reconstruct. We may infer, for instance, that monkeys crossed the Atlantic sometime between 26 and 51 million years ago, but we will never know exactly how they did it.

This is not to imply that there are no generalizations that can be derived from studying chance dispersal. In fact, we have run across several of these, such as the inference that vertebrates have crossed the Atlantic and the Mozambique Channel in the directions predicted by ocean currents, and the deduction that nonaerial animals have much more trouble colonizing Hawaii from the Americas than from the Indo-Pacific. In fact, one of the obvious consequences of the renewed interest in long-distance dispersal is that many scientists are trying to discover such tendencies, retrieving patterns from the seemingly random. However, this doesn't change the fact that individual instances of successful dispersal and establishment are fundamentally unpredictable. Even if, after the fact, it supports a general tendency, the monkey's voyage was a fluke.

Nonetheless, there is wonder in this view when we consider the immensity of nature it implies. Ocean crossings and other chance colonizations are outcomes of a game of almost unimaginably large numbers. Seen in isolation, they may seem mysterious and miraculous, but, taken in proper context, that mystery disappears and is replaced by understanding. These occurrences are like the man struck twice by lightning, in his own eyes an act of God or the Devil, but an event with high probability when the whole human population is taken into account. Similarly, the many seemingly implausible colonizations remind us that we are living in a tiny slice of a deep history, a history acted out over many millions of years, with a vast array of living organisms as the players, moved by their own powers, by inexorable ocean currents, by storms beyond count. We may be surprised by the notion that monkeys once crossed an ocean; we should be in awe of a history in which events of that sort (although not that *particular* event) are inevitable. The large number of these colonizations tells us that, in the long history of this living world, the miraculous has become the expected.

EPILOGUE: THE DRIFTWOOD COAST

It is a misty July morning on the Oregon coast. There is no horizon, just a blending of dull sea and dull sky. Inland lies a dark forest of Sitka spruce, western redcedar, and Douglas fir. Tara and I and our kids, two-year-old Eiji and almost-five-year-old Hana, are in the middle of a long road trip from Reno across California and up the coast to northern Washington, a journey that has become a lazy exploration of beach after beach—long, serene days spent flying kites, building and destroying sand castles and fairy houses, chasing gulls, chasing the waves, tending to scraped knees and small disappointments.

Scattered on the beaches is a profusion of organic debris, a disorganized museum of the recently or not-quite deceased. On this trip we have seen or will see giant strands of orange kelp, a pelican skull, the mysterious mustard-colored and grub-like husk of some unknown crustacean, disembodied gull wings, mussel shells, clam shells, abalone shells, bleached sand-crab skeletons gathered into detrital ribbons by the waves, dried-out thistle stalks, the desiccated corpse of a fledgling murre.

By far the most obvious debris is driftwood. It's everywhere, from finger-sized pieces that Hana uses in her fairy houses to the trunks of large trees. To the kids' delight, people have built driftwood lean-tos, driftwood tepees, driftwood forts, whole driftwood villages. In places, the ocean has tossed up great chaotic piles of the stuff, arrayed along the beach like mangled pike fences, as if the continent were defending itself against invasion from the Pacific. I'm amazed by the sheer volume of driftwood, but I later learn that these massive accumulations only hint at the amounts that washed up on these beaches before Europeans came, before people began in earnest to log the forests and clear the rivers for navigation.

On the beach at Carl G. Washburne Memorial State Park, I'm following the line of high water, looking, not quite idly, under any debris light enough to lift. I take hold of one four-foot-long piece of driftwood, charred from a campfire, and flip it over. On the underside are half a dozen mottled, grayish brown insects, ensconced in crevices and hollows in the blackened wood. I recognize them immediately (I was looking for them): they're jumping bristletails, and I'm fairly sure they're *Neomachilis halophila*, the mainland relative of the Hawaiian bristletails that John, Cheryl, and I have been studying. (Weeks later, with a dissecting microscope, I find that these Oregon specimens have only a single pair of water-absorbing vesicles on the underside of each abdominal segment, confirming their identity.) Helmut Sturm's notion that bristletails could have reached Hawaii as eggs attached to driftwood jumps into my head. Even though I haven't found the eggs, Sturm's idea is suddenly transformed from the abstraction of words on a page to tangible form: here are his bristletails actually clinging to driftwood. And there is certainly no shortage of this kind of debris. I keep looking, turning over other pieces, and find several more *N. halophila*.

Later, Eiji joins me, not quite helping to find insects, but adding his toddler's enthusiasm to the search. He stomps. He dances. He squeals about earwigs and bristletails (he calls them "briss-oh-tayuhs"). Already, he seems to have developed a visceral appreciation for small, creeping things. Half of my mind is with him, excited to see whatever turns up in our informal biological survey of the beach at Washburne State Park. The other half, though, is thinking about those *Neomachilis* drifting on the ocean, about the wondrous means by which the history of life has unfolded.

A little ways south of us, rocky headlands rise out of the mist. To the north, the nearly deserted beach stretches on for miles, making the two children appear even smaller than they are. From the west, the endless ranks of gray waves curl and crash to foam then slide hurriedly in thin sheets over the sand.

Beyond the waves, the horizonless ocean seems infinite, an impossibly wide barrier, yet I know that, in the fullness of time, many living things have crossed it.

ACKNOWLEDGMENTS

It's a great pleasure to thank the large crowd of people who have helped me complete this book, including friends, family, research collaborators and other colleagues, and quite a few scientists whom I know only through phone conversations and email correspondence.

For critiquing chapters—an enormously important part of the process—I'm grateful to Merrill Peterson, Carol Yoon, Peter Wimberger, Chris Feldman, Marjorie Matocq, Peter Murphy, Angela Hornsby, Brandi Coyner, Mitchell Gritts, John Measey, Mike Pole, Matt Lavin, Mott Greene, Susanne Renner, Patrick O'Grady, Karen de Queiroz, Sean de Queiroz, Jade Keehn, Sarah Hegg, Jason Malaney, Guanyang Zhang, Heather Heinz, Joseph Collette, Eric Gordon, David Haisten, Derek O'Meara, Eric Stiner, and Rebecca Swab. I owe a special debt to my wife, Tara de Queiroz, and my friend and research collaborator John Gatesy. Both gave constructive and supportive feedback on the entire manuscript—Tara in a relatively calm way, John in his usual biting, provocative style.

I interviewed or more informally chatted with many people, mostly scientists, to gain their insights, to check facts, and to provide more of a human touch to the narrative. For extensive telephone or email interviews I'm grateful to Anne Yoder, Dennis McCarthy, John Briggs, Michael Donoghue, Michael Heads, Gary Nelson, Matt Lavin, Susanne Renner, the late Robert McDowall, John Measey, Miguel Vences, Bob Drewes, Mike Pole, Dallas Mildenhall, Isabel Sanmartín, Steve Trewick, and Jorge Crisci. In this context, I especially want to thank Michael Heads, Dennis McCarthy, and Gary Nelson, who all knew that I disagreed mightily with their scientific views yet were unfailingly cooperative in answering my questions. For shorter correspondence or conversations I thank Norm Platnick, Andreas Fleischmann, Hamish Campbell, Ellen Censky, David Krause, Rob Meredith, Sean de Queiroz, Kevin de Queiroz, Jerome Salador, Mike Crisp, Nicolas Vidal, Patrick O'Grady, Peter Wimberger, Chris Feldman, Ben Normark, Rudolf Scheffrahn, Greger Larson, Graham Wallis, Jimmy McGuire, Cheryl Hayashi, and Steve Montgomery.

Any book about science builds on an enormous amount of previous work and, in that sense, the reference list can be viewed as part of the acknowledgments. However, I also want to single out five books on which I leaned especially heavily for information and/or inspiration: Janet Browne's *The Secular Ark*; David Quammen's *The Song of the Dodo*; the late David Hull's *Science as a Process*; Mark Lomolino, Dov Sax, and James Brown's edited volume *Foundations of Biogeography*; and Blair Hedges and Sudhir Kumar's edited volume *The Timetree of Life*. This book would have suffered if those five others had not existed. I'm also indebted to Andreas Fleischmann, John Gatesy, Patrick O'Grady, Mike Crisp, Michael Heads, Gary Nelson, Jason Ali, Susanne Renner, and the late Robert McDowall for providing important unpublished manuscripts or results.

For help obtaining or modifying photos or other images I thank Bob Drewes, Isabel Sanmartín, Andreas Fleischmann, John Measey, Susanne Renner, Gary Nelson, Miguel Vences, Kenneth Miller, Chris Burridge, Don Jellyman, Chris Feldman, Hernán Sosa, Nathan Muchhala, Gary Nafis, and Martin Meyers (sorry Martin, I couldn't quite find a place for the bedraggled nuthatch stranded on the boat). I obtained virtually all of the references I needed through the Mathewson-IGT Knowledge Center and the DeLaMare (Earth Sciences) Library, both at the University of Nevada at Reno.

As I described in the Introduction, a research project on garter snakes is what originally drew me into the subject of biogeography in general and oceanic dispersal in particular. That project and, thus, this book, would not have happened without my herpetological collaborators, Robin Lawson and Julio Lemos-Espinal. My interest in and knowledge of biogeography has been deepened by another research project, an ongoing set of studies of jumping bristletails in collaboration with John Gatesy, Cheryl Hayashi, Laura Baldo, Marshal Hedin, Rob Meredith, Eric Stiner, and Jim Liebherr. Several mud-splattered bristletail trips to Hawaii with my friends John and Cheryl, in particular, gave me a greater appreciation for and knowledge of the colonization of oceanic islands.

My longtime friends Carol Yoon and Merrill Peterson did me a great favor by providing detailed and extremely constructive feedback on the proposal that gave rise to this book. I'd also like to thank my nononsense literary agent, Russ Galen, for efficiently selling the project to a publisher. T. J. Kelleher, my editor at Basic Books, is very knowledgeable about science and has a great sense of the big picture; I'm grateful to him for improving this book in ways small and large. For helping turn the manuscript into an actual book, I'd also like to thank other members of the team at Basic Books, especially Tisse Takagi, Collin Tracy, Kathy Streckfus, and Jack Lenzo.

My immediate and extended family provided moral and material support, and, more importantly, often helped take my mind off of writing. In particular, I want to thank Kristine de Queiroz; Richard and Chizuko de Queiroz; Sean and Karen de Queiroz; Janice Harvey; Jim Forbis; Robyn, Hernán, Gabriela, and Natalia Sosa; Tess, Matt, and Sammy Gallegos; Jerome Salador; Eric Saijo, Mari Rose Taruc, Sidhartha Taruc, and Kawayan Saijo-Taruc; Rani Saijo; and Lee Saijo.

My wife, Tara—fellow naturalist, gardener, lover of the Great Basin, and writer—deserves special thanks. She was there at the beginning (in Baja California, collecting garter snakes with surprising goodwill), and her love and support (and, at times, salary) has bolstered me through this entire, sometimes seemingly interminable project. I cannot thank her enough; she made this book possible. She is also a great mother to our children, Hana and Eiji, and our family journey in these past few years has been a life-changing experience. It was especially gratifying that, toward the end, Hana, a bookworm almost from birth, could appreciate what I was doing, even if she seemed to have doubts about the subject matter compared to a project I had put aside. Finding me at my desk once, she asked earnestly, "Dad, are you ever going to write that book about eyes?"

FIGURE CREDITS

*F*or images obtained from Wikimedia Commons, the title of the image is given, followed by a URL that provides the image and a link to the associated license.

Geologic timescale: Ages are from the International Chronostratigraphic Chart (version of January, 2013) constructed by the International Commission on Stratigraphy (www.stratigraphy.org).
I.1: CaliforniaHerps.com (www.californiaherps.com)
3.3: White-faced Heron, http://commons.wikimedia.org/wiki/File:White-faced_Heron.jpg
5.1: Kary Mullis at TED, http://commons.wikimedia.org/wiki/File:Kary_Mullis_at_TED.jpg
9.1: Mabuia Noronha Skink, http://commons.wikimedia.org/wiki/File:Mabuia_Noronha_Skink.jpg
9.3: Amphisbaena alba03, http://commons.wikimedia.org/wiki/File:Amphisbaena_alba03.jpg
9.5: Lémur catta, http://commons.wikimedia.org/wiki/File:Lémur_catta.jpg; Tarsier Tarsius sp., http://commons.wikimedia.org/wiki/File:Tarsier_Tarsius_sp._.jpg; Golden-headed-Lion-Tamarins, http://commons.wikimedia.org/wiki/File:Golden-headed-Lion-Tamarins.jpg; Papio ursinus 2, http://commons.wikimedia.org/wiki/File:Papio_ursinus_2.jpg
10.4: Black Robin on Rangatira Island, http://commons.wikimedia.org/wiki/File:Black_Robin_on_Rangatira_Island.jpg; NZ North Island Robin-3, http://commons.wikimedia.org/wiki/File:NZ_North_Island_Robin-3.jpg
Figure after Chapter 11: Nube de langostas en el Sáhara Occidental (1944), http://commons.wikimedia.org/wiki/File:Nube_de_langostas_en_el_Sáhara_Occidental_(1944).jpg
12.3: Leaflitter toad Rhinella alata, http://commons.wikimedia.org/wiki/File:Leaflitter_toad_Rhinella_alata.jpg

NOTES

Introduction: Of Garter Snakes and Gondwana

1 *"Science must begin with myths . . . ":* Popper (1965), 66.
3 *The enormous supercontinent of Gondwana:* For reviews of the sequence of Gondwanan breakup, see McLoughlin (2001); Sanmartín and Ronquist (2004). For Gondwanan breakup as the iconic tale of historical biogeography, see Raven and Axelrod (1972); Hallam (1994); Gibbs (2006); and McCarthy (2009), among many others.
6 *Its geologic history is reminiscent:* For the geologic history of Baja California, Carreño and Helenes (2002); de Queiroz and Lawson (2008); and references in both.
7 *We collected* T. validus *specimens:* For the Baja California garter-snake study, de Queiroz and Lawson (2008).
9–12 *Continental Southeast Asia . . . was inundated:* Inger and Voris (2001).
13 *The rise of the Isthmus of Panama:* Lessios (2008).
14 *Often has been called a scientific revolution:* Funk (2004); de Queiroz (2005). This topic is discussed in detail in Chapters Two and Three.
14 *The papers arguing for ocean crossings:* For tortoises, Caccone et al. (1999); for plants between Tasmania and New Zealand, Jordan (2001); for southern beeches, Cook and Crisp (2005), Knapp et al. (2005); for baobabs, Baum et al. (1998); for rodents, Poux et al. (2006).
16 *"The nearest approach to life on another planet":* Gibbs (2006), 7.
17 *Field guide to the trees of New Zealand:* Metcalf (2002).
19 *Aldabra giant tortoise:* Gerlach et al. (2006).

Chapter One: From Noah's Ark to New York: The Roots of the Story

23 *Croizat, an Italian botanist:* For background on Croizat and panbiogeography, see the notes for Chapter Three.
24 *Darwin was aware of the other natural:* For Darwin's distrust of land-bridge explanations, Browne (1983).
25 *Darwin did believe that lands had risen and fallen:* Darwin (1839); Browne (1983).
25 *"It shocks my philosophy to create land . . . ":* Darwin letter to J. D. Hooker, June 5, 1855, from Burkhardt and Smith (1989), vol. 5, 344.
25 *Hooker . . . was equally skeptical:* Browne (1983).
25 *A whole series of experiments:* Darwin (1859), in the chapters on geographical distribution.
26 *"It is quite surprising that the Radishes . . . ":* Darwin letter to J. D. Hooker, April 19, 1855, from Burkhardt and Smith (1989), vol. 5, 308.
26 *"When I wrote last . . . ":* Darwin letter to J. D. Hooker, April 13, 1855, from Burkhardt and Smith (1989), vol. 5, 305.
26 *"I am more reconciled to Iceberg transport . . . ":* J. D. Hooker letter to Darwin, November 9, 1856, from the website of the Darwin Correspondence Project, www.darwinproject.ac.uk/home.
27 *Noah's Ark at face value:* Browne (1983).
27 *Darwin had come to accept as fact:* Browne (1995).
28 *"Deep time":* John McPhee coined the term "deep time" in *Basin and Range* (1981).
28 *Alfred Russel Wallace:* For general biographical information on Wallace, Shermer

(2002); Quammen (1996).
29 *"Towards solving the problem of the origin of species":* Shermer (2002), 58.
29 *Writing a theoretical paper:* For Wallace's Sarawak and Ternate (survival of the fittest) papers, Wallace (1891).
30 *"Nothing very new"* and *"It all seems creation with him":* Shermer (2002), 89.
30 *Sent the manuscript to . . . Charles Darwin:* Quammen (2006); Shermer (2002).
30 *Were read at a meeting of the Linnean Society:* For the reactions of Lyell and Darwin to Wallace's Sarawak paper, Browne (1983); Shermer (2002); Quammen (2006).
31 *Take the work of Edward Forbes:* For general biographical information on Forbes, Browne (1983).
31 *"An acute and subtle thinker":* Browne (1983), 114.
31 *Finding molluscs in the Aegean Sea:* Browne (1983).
32 *The landmark for this new biogeography:* Darwin (1859).
33 *Wallace would write a two-volume work:* For Wallace's later biogeographic work, Shermer (2002).
33 *Another belief in common:* For the belief in the fixity of continents by Darwin and Wallace, Darwin (1859); Shermer (2002).
34 *Flemish cartographer Abraham Ortelius:* Lawrence (2002).
34 *French geographer . . . Antonio Snider-Pelligrini:* Lawrence (2002); McCoy (2006).
34 *Frank Bursley Taylor:* Lawrence (2002); McCoy (2006).
34 *Intense man named Alfred Wegener:* Most of the biographical information on Wegener is from McCoy (2006). When I used other sources, those are indicated below.
35 *"Fine features and penetrating blue-gray eyes":* German geologist Hans Cloos, quoted in Hughes (1994).
35 *Seven-hundred-mile trek:* Lawrence (2002).
36 *"Please look at a map . . . ":* Lawrence (2002), 34.
36 *"If it turns out that sense and meaning . . . ":* Yount (2009), 34.
36 *An obvious initial goal was to figure out:* The description of Wegener's theory is based on Wegener (1924), which is the third edition of his book on continental movement and the first one translated into English, and on the summaries in Lawrence (2002) and McCoy (2006).
39 *"Wegener's hypothesis in general is of the foot-loose type . . . ":* Chamberlin (1928), 87.
39 *"A beautiful dream, the dream of a poet":* P. Termier, quoted in Schuchert (1928), 140.
39 *"Ending in a state of auto-intoxication . . . ":* E. Berry, quoted in Oreskes (1988), 336.
39 *Very inconsistent from place to place:* Oreskes (1988); McCoy (2006).
39 *Rejection and ridicule:* Oreskes (1988).
41 *"If we are to believe Wegener's hypothesis . . . ":* Oreskes (1988), 336.
42 *"A great loss to geophysical science":* Oreskes (1988), 332.
42 *"The expression on his face was calm . . . ":* McCoy (2006), 133.
42 *Adamantly used Wegener's theory:* du Toit (1937). Frankel (1981) discussed others, mostly paleobotanists, who supported Wegener's theory.
42 *In the terminology of . . . Thomas Kuhn:* Kuhn (1970).
43 *Had the upper hand:* On the dominance of the land-bridge "school," Mayr (1982). Gadow (1913) includes a series of maps depicting extensive land bridges at various times starting from the Lower Triassic. Schuchert (1932) reviews evidence for Gondwanan land bridges.
44 *William Diller Matthew:* For Matthew's influence and the development of the New York School, Nelson and Ladiges (2001).
45 *Take the small number of natural rafts:* For the specifics about Matthew's views on oceanic dispersal, including the raft calculations and the nature of the mammal fauna of Madagascar, Matthew (1915).
45 *Matthew's "disciples":* Karl Schmidt, a herpetologist, was the scientist who referred to himself and others as "disciples" of Matthew and called "Climate and Evolution" a "Holy Writ." He is quoted in Nelson (1973), 313.
45 *"New York School of Zoogeography":* Nelson and Ladiges (2001).
46 *"On several occasions, when the vessel . . . ":* Darwin (1839), 148.

Chapter Two: The Fragmented World

47 *In his book* The Tipping Point: Gladwell (2000).
48 *"A lanky youngster . . . ":* Hull (1988), 144. For biographical information about Nelson, I referred to Hull (1988) and emails from Nelson, April 20 and October 22, 2010.
48 *"I'm not about to let Mayr . . . ":* Hull (1988), 145.
48 *Picked up a hefty new publication:* Hull (1988); email from Nelson to the author, October 22, 2010. The Brundin monograph was Brundin (1966).
49 *Brundin had been studying:* Brundin (1966).
49 *German entomologist Willi Hennig:* Hennig (1966). The development of cladism and the "wars" between the cladists and more traditional systematists are described in detail in Hull (1988).
50 *River delta diagrams:* For other examples of these diagrams, see Simpson (1953), 261; Young (1962), 239.
54 *Arthur Holmes, a British geologist:* For Holmes's radiometric dating work, Lewis (2002); Frankel (1978). For his mantle convection theory, Gohau (1990); Frankel (1978).
55 *Holmes first published his theory:* Holmes (1928, 1944). For lack of acceptance of Holmes's theory, Frankel (1978); Lawrence (2002).
55 *"I have never succeeded in freeing myself . . . ":* Quoted in Bryson (2003), 177.
56 *Hess, a Princeton geologist:* For biographical information, James (1973); Lawrence (2002). Hess's discovery of guyots is from those references and Hess (1946).
56 *"Have been able to get about a dozen . . . ":* Lawrence (2002), 157.
56 *An explosion of new information:* McCoy (2006); Lawrence (2002).
57 *Hess had put all the facts together:* Hess (1962).
57 *"I shall consider this paper an essay . . . ":* Hess (1962), 599.
57 *"If it . . . were accepted . . . ":* Hess (1962), 607.
58 *Explained Hess's guyots:* For the importance of guyots to Hess's theory, Hess (1962); Lawrence (2002).
59 *His theory was mostly just ignored:* Bryson (2003).
59 *The textbook version:* This version of the influence of the magnetic anomaly work of Vine and Matthews and of Morley is given in, for example, Gohau (1990); Macdougall (1996); Bryson (2003).
59 *He saw Hess give a talk:* Lawrence (2002). For description of the Vine-Matthews-Morley hypothesis, Vine and Matthews (1963); Lawrence (2002).
60 *"It was the classic lead balloon . . . ":* Lawrence (2002), 212.
60 *If people didn't believe:* Lawrence (2002).
60 *The conversion was not far off:* For additional evidence that accumulated after the Vine and Matthews paper and the description of the 1966 symposium in New York, McCoy (2006); Lawrence (2002).
62 *A heavy drinker and smoker:* Lawrence (2002).
63 *"Not the slightest evidence of chance dispersal . . . ":* Brundin (1966), 451.
65 *"No one denies . . . ":* Brundin (1966), 439.
65 *"Several troubled biogeographers . . . ":* This and the following two quotations are from Brundin (1966), 51.
66 *"Biogeography has a great future . . . ":* Brundin (1966), 5.
66 *Nelson had a chance to meet him:* Hull (1988); emails from Nelson to the author, October 22 and 23, 2010. Nelson discussed Brundin's lasting influence on him in emails to the author on April 20 and 21, 2010; October 22, 2010.
66 *His first stop:* For Nelson's conversion of scientists at the British Museum of Natural History and the American Museum of Natural History, Hull (1988); Nelson, "Cladistics at an Earlier Time" (unpublished manuscript). Where these accounts are in conflict, I have followed Nelson's version. For indications of the prior, noncladistic views of Greenwood and Patterson, see Patterson (1997); Nelson (2000).
67 *"The Hennigian methodology . . . ":* Mayr (1982), 620.
67 *"I have therefore become a Wegenerian . . . ":* Giller et al. (2004), 274.
68 *Looking at two groups of these fishes:* Rosen (1978, 1979).

69 *A young ornithologist:* For the examples of ratites and *Nothofagus*, Cracraft (1974) and Cracraft (1975), respectively.
70 *"We were sitting at a place . . .":* Donoghue, telephone conversation with the author, September 3, 2010.
70 *Bold and bullying:* The attitude of the cladists is described in detail in Hull (1988) and more briefly and engagingly in Yoon (2009).
70 *Stevie Wonder lyric:* Nelson (1978b).
70 *"Central attitude is one of self-conscious superiority":* Darlington (1970), 16.
70 *"Shouting at meetings . . .":* Email from John Briggs to the author, August 25, 2010.
72 *Anoles, the common arboreal lizards:* The distribution of *Norops sagrei* is from Williams (1969); Norval et al. (2002); and Tan and Lim (2012). The anole experiment is from Schoener and Schoener (1984).

Chapter Three: Over the Edge of Reason

73 *Submarine volcano, named Loʻihi:* Ziegler (2002); Tilling et al. (2010).
74 *Essentially a mythological process:* For Heads's arguments against long-distance dispersal, Heads (2009, 2011).
74 *"The taxa that colonized Hawaii . . .":* Email from Heads to the author, August 5, 2010.
74 *The hotspot that created the current islands:* Sharp and Clague (2006).
75 *Molecular evidence indicates:* Price and Clague (2002); Baldwin and Wagner (2010).
75 *Subspecies of more widespread species:* On birds, Hawaiian Audubon Society (2005). On the hoary bat, Jacobs (1994). On the beach strawberry and oval-leaf clustervine, Wagner et al. (1999). On the Boston swordfern, Palmer (2003).
77 *Then there was Léon Croizat:* For biographical information about Croizat, Croizat (1982); Hull (1988, 2009); Colacino and Grehan (2003).
78 *"Gone forever are the days . . .":* Croizat, vol. 1 (1958), xii–xiii.
79 *"Centres of origin"* and *"means of dispersal":* Darwin (1859); Matthew (1915).
80 *Looking for facts that supported it:* Croizat (1962), 637.
80 *Baconian view:* Croizat (1962).
80 *Distributional line that he called a* track: For succinct descriptions of Croizat's method, Croizat et al. (1974); Craw et al. (1999).
81 *"Patterns of geographic distributions . . .":* Croizat (1962), 712.
81 *Even to the ancient Greeks:* Mayr (1982), who noted that Hippocrates, Aristotle, Theophrastus, and others attributed some discontinuous distributions, such as elephants occurring in India and Africa, to former connections.
82 *Not even worth studying:* Croizat (1962), 213, for example.
82 *"Earth and life evolve together":* Croizat (1962), 604.
82 *The tracks . . . run all over the place:* For New Zealand, Page (1989); Jordan (2001); Winkworth et al. (2002); Wallis and Trewick (2009); Trewick and Gibb (2010). For Hawaii, Baldwin and Wagner (2010); Gillespie et al. (2012). Of these studies, only Page (1989) actually used the track method; however, the other studies imply the existence of diverse tracks for the two areas mentioned.
82 *New Caledonia is an amalgamation:* Heads (2008).
82 *To collectively produce a clear pattern:* The notion that chance dispersal can give rise to general patterns and, therefore, that tracks shared by different taxa are not evidence of a single (presumably vicariant) event is given in Ball (1975).
83 *Had hardly been cited at all:* For the fact that Croizat was widely known among botanists but not zoologists until the mid-1970s, Schmid (1986).
83 *"Conspiracy of silence":* Colacino and Grehan (2003), 10.
83 *"Totally unscientific style and methodology"* and *"a member of the lunatic fringe":* Nelson (1977), 452 and 451, respectively.
83 *"Pondered for a time":* Email from Nelson to the author, April 20, 2010.
83 *"Not wholly sound"* and *"blazing sermon":* Brundin (1966), 61.
84 *An overlooked visionary:* Hull (1988) describes Nelson's championing of Croizat and the interactions between the two at length. The paper in which Nelson described Croizat's work enthusiastically is Nelson (1973).

84 *Should collaborate on a paper:* The history of the Croizat, Nelson, and Rosen collaboration is discussed in Hull (1988). The resulting paper is Croizat et al. (1974).
85 *"Is to be understood first . . .":* Croizat et al. (1974), 269.
85 *"Having failed to dissect these concepts . . .":* Croizat et al. (1974), 276.
85 *"A world of make-believe and pretense"* and *"No one well informed . . .":* Croizat et al. (1974), 277.
85 *A "hardening" of vicariance biogeography:* Gould (1986).
86 *"A science of the improbable . . .":* Nelson (1978a), 289.
86 *Cannot be refuted:* The notion that dispersal hypotheses are untestable and/or unfalsifiable is found in Ball (1975); Rosen (1978); Craw (1979); Nelson and Platnick (1981).
86 *We know long-distance dispersal happens:* McDowall (1978).
87 *Simply too young to have been influenced:* For Mayr's claim that most living groups of mammals, birds, and flowering plants are too young to have been affected by the opening of the Atlantic, Mayr (1952).
88 *Assumed a strict molecular clock:* The molecular clock and its use in biogeography are more fully discussed in Chapters Five and Six. An early demonstration of the inconstant nature of the clock is found in Goodman et al. (1971).
88 *Vicariance scientists used this . . . to completely discount:* Heads (2005); Nelson and Ladiges (2009); Parenti and Ebach (2009, 2013).
90 *"Yep":* Email from Nelson to the author, October 23, 2010. Nelson also intimated that *Homo sapiens* might be much older than generally believed in Nelson (1978c).
91 *"Offends a critical mind":* Brundin (1966), 51.
91 *"Will have powerful influence . . .":* Croizat (1962), 708.
91 *"To formulate explicit methods . . .":* Croizat et al. (1974), 277.
91 *"Non-equilibrium thermodynamics theory":* Brooks and Wiley (1988).
93 *Upland Sandpipers:* McAtee (1914). The quotation, "I used to count the number . . ." is from p. 404 of that article.

Chapter Four: New Zealand Stirrings

96 *That tuatara's name is Henry:* Marks (2009).
97 *Contains just the one species:* Tuatara from North Brother Island have been considered a separate species, *Sphenodon guntheri*, but a recent genetic study suggests these animals should be lumped with *S. punctatus*. See Hay et al. (2010).
97 *Split from each other some 250 million years ago:* This estimate is from Shedlock and Edwards (2009).
97 *Tuatara . . . are anatomically distinct:* For tuatara anatomy, Pough et al. (1998).
97 *Sphenodontids were never a large group:* For the fossil record of tuatara and other sphenodontids, Jones et al. (2009).
97 *By about 83 million years ago:* For the geologic history of New Zealand, Campbell and Hutching (2007).
99 *Some seven hundred years ago:* Higham et al. (1999).
99 *The rats ate tuatara eggs:* For the impact of rat and other mammal introductions on tuatara, Towns and Daugherty (1994).
99 *Left some* Sphenodon *isolated:* Hay et al. (2010).
99 *Tuatara eating the giant, cricket-like weta:* Angier (2010).
100 *"With regard to general problems . . .":* Nelson (1975) 494.
100 *Joseph Hooker placed it:* Gibbs (2006). For Darwin, Matthew, and Darlington arguing for dispersal origins of New Zealand's biota, Darwin (1859); Matthew (1915); Darlington (1965). Pole (1994) gives a brief summary of the history of ideas about the origins of New Zealand's biota.
100 *"Stranded upon the shores of New Zealand . . .":* Nelson (1975), 494.
101 *I asked several New Zealand biologists:* Email correspondence with the author from Mike Pole, November 1, 2010; Dallas Mildenhall, December 12, 2010; Steve Trewick, December 13, 2010.
101 *"By the middle of the Cretaceous . . .":* Pole (1994), 625, quoting Enting and Molloy (1982).
102 *The moa record peters out:* For the fossil records of moa and kiwi, Tennyson et al.

(2010); Tennyson (2010). On possible kiwi fossil footprint, Fleming (1975). For Miocene tuatara, Jones et al. (2009). For New Zealand vertebrate fossil record in general, Tennyson (2010).

102 *A handful of dinosaurs:* For New Zealand vertebrate fossils from the Late Cretaceous, around the time of the separation of Zealandia from Antarctica/Australia, Tennyson (2010).

102 *"Much of the present lowland flora . . .":* Raven and Axelrod (1972), 1382.

103 *Pollen is tough stuff:* For general information on pollen and its analysis, Moore et al. (1991).

103 *A history of taxa in constant flux:* Fleming (1975); Mildenhall (1980); Pole (1993, 1994); Lee et al. (2001).

103 *A 2001 survey:* Lee et al. (2001).

104 *A Field Guide to the Alpine Plants of New Zealand:* Salmon (1992). For fossil records of plants we identified at Arthur's Pass, Lee et al. (2001), except for *Drosera*, which is from Raine et al. (2008).

106 *A paleontologist named Charles Fleming:* Fleming (1962, 1975).

106 *"There was a disconnect between paleontologists and biologists . . .":* Email from Dallas Mildenhall to the author, December 12, 2010.

106 *He published a paper making that point:* Mildenhall (1980). For the citation history of Mildenhall's 1980 paper, Web of Science citation database, accessed in March 2010 (http://thomsonreuters.com/web-of-science/).

106 *"Do Araucarias have double trunks?":* Email from Pole to the author, November 1, 2010. For biographical information about Mike Pole, emails from Pole to the author, November 1, 2010; December 3, 2010; March 23, 2011.

107 *That botanist was Michael Heads:* Email from Heads to the author, April 5, 2011.

107 *Pole compared the past floras:* Pole (1993).

108 *Pole fleshed out the case:* Pole (1994).

109 *Mike Pole spends much of his time:* Email from Pole to the author, November 1, 2010.

110 *He believes that explaining the origins:* Email from Pole to the author, November 1, 2010.

111 *Pole mentioned two sets of studies:* For molecular clock studies of southern beeches and ratites, respectively, Martin and Dowd (1988); Sibley and Ahlquist (1987); and references in both.

112 *A natural floating island:* Powers (1911).

Chapter Five: The DNA Explosion

115 *Estimates of the age of the universe:* Simanek and Holden (2001); Overbye (2013). The Simanek and Holden book is "science humor," but the age estimates it provides are from real studies.

116 *Selection for increased brain size:* Shultz et al. (2012).

117 *Whose focus was on cladograms:* For the focus on cladograms or tracks and lack of use of evidence for the ages of groups, Croizat et al. (1974); Cracraft (1975); Rosen (1978); Nelson and Platnick (1981); Wiley (1988); and Humphries and Parenti (1989), among many others. Croizat et al. (1974) and Nelson and Platnick (1981) explicitly criticized the use of fossils to place ages on groups.

117 *Both Hallam and Briggs wrote books:* Hallam (1973, 1994); Briggs (1987, 1995). The ichthyologist John Lundberg, although a diehard cladist, was (and is) similar to Hallam and Briggs in that he believed that the fossil record was very informative. In an important paper, Lundberg (1993) used fossil evidence to challenge continental drift as a general explanation for freshwater fish distributions.

118 *"Chicken scratchings":* Donoghue, telephone conversation with the author, September 3, 2010.

119 *"Degree of divergence is a guide neither . . .":* Heads (2005), 71.

119 *"Does not solve biogeographical problems . . .":* Heads (2005), 72.

119 *"Molecular dating game":* Nelson and Ladiges (2009).

119 *The idea of the molecular clock itself:* The first presentation of the molecular clock is in Zuckerkandl and Pauling (1962). For the early history of the molecular clock idea,

see Morgan (1998).
- 120 *Darwin became a confirmed believer:* The development of Darwin's belief in evolution is from Browne (1995) and Quammen (2006), although those authors would not necessarily agree with my interpretation.
- 120 *It was an unseasonably warm night:* For Mullis's invention of PCR and his other exploits, Mullis (1998); Wade (1998).
- 124 *In 1985 the description of PCR had been published:* The original paper describing PCR, with Mullis as the fourth of seven authors is Saiki et al. (1985). A later, more complete description of the method is in Mullis and Faloona (1987).
- 126 *Examining that tree now:* The first, fairly pathetic garter-snake sequencing study was de Queiroz and Lawson (1994). The later, much better one is de Queiroz et al. (2002). Incidentally, this later paper also presents a scenario for the vicariant origins of some highland garter snakes through fragmentation of woodland environments because of a drying climate.
- 126 *The clock didn't tick at a constant rate:* Some relatively early studies on the inconstancy of the molecular clock are Goodman et al. (1971); Britten (1986); Vawter and Brown (1986).
- 127 *A house made almost entirely of old bottles:* For the history of the bottle house, Gildart and Gildart (2005).
- 128 *Mathematical models to describe evolution:* For a summary of many models of DNA change including pre-PCR examples such as the Jukes-Cantor and Kimura two-parameter models, Felsenstein (2003).
- 129 *The estimates are getting better and better:* For a recent validation of complex over very simple models of evolution, Huelsenbeck et al. (2011). Also see Felsenstein (2003).
- 131 *Rizal Shahputra* and *Melawati:* For Rizal Shahputra, Fernandez (2005); Jones (2005). For Melawati, Associated Press (2005).

Chapter Six: Believe the Forest

- 133 *A molecular clock analysis using tapeworm DNA:* Hoberg et al. (2001).
- 134 *"I've been involved in a bunch of dating . . .":* This and subsequent quotes from Michael Donoghue, Donoghue, telephone conversation with the author, September 3, 2010.
- 135 *Their dismissal of this evidence:* For pointed critiques of molecular dating from vicariance scientists, Heads (2005); Parenti (2006); Nelson and Ladiges (2009).
- 136 *Certain apes:* Benton et al. (2009).
- 136 *The actual age will almost always be older:* For the problem of first fossil occurrences being misleading for calibrating molecular clocks, Heads (2005); Nelson and Ladiges (2009).
- 136 *The first fossil hummingbirds:* Mayr (2003).
- 137 *Use only especially good calibrations:* For the notion of using especially good fossil calibration points and justification for many specific points, including the bird-crocodile split, Benton et al. (2009).
- 137 *Incorporating uncertainty:* For specific methods for taking into account uncertainty in calibration ages, Kumar et al. (2005); Drummond et al. (2006). Ho and Phillips (2009) and Hedges and Kumar (2009b) give reviews of the subject of uncertainty in calibrations.
- 138 *Which ones were out of synch:* For comparisons among multiple calibration points, Douzery et al. (2003); Near and Sanderson (2004).
- 139 *The clock is pretty variable:* Bousquet et al. (1992); Martin and Palumbi (1993); Rodríguez-Trelles et al. (2004); and Thomas et al. (2006), among many others. Mammal and shark rates are from Marshall et al. (1994); Martin and Palumbi (1993).
- 139 *Reasons for the inconstancy of the clock:* Bromham (2009).
- 140 Relaxed clock *methods:* Welch and Bromham (2005); Hedges and Kumar (2009b).
- 140 *Tested using computer programs:* Kishino et al. (2001); Kumar et al. (2005); and Battistuzzi et al. (2010), among others.
- 140 *A 2011 study:* Meredith et al. (2011b). John Gatesy's statement about the drastic

effects of using calibrations from different groups is from a telephone conversation between Gatesy and the author, August 2011.

141 *The two most convincing studies:* Vidal et al. (2009); Pyron and Burbrink (2011). Many comparisons of timetree studies for individual taxa are in Hedges and Kumar (2009a), the book resulting from the Timetree of Life project.

144 *Molecular age estimates match the fossil-based ones:* Hedges and Kumar (2009b).

144 *Molecular estimates are consistently older:* For discussion of the discrepancies between molecular and fossil-based ages for branching points over 400 million years old and for the early radiations of birds and mammals, Cooper and Fortey (1999); Smith and Peterson (2002). Note, however, that the discrepancies between the fossil record and molecular age estimates may be lessening for the bird and mammal cases with the discovery of new fossils (Clarke et al. 2005) and the use of more sophisticated molecular dating approaches (Meredith et al. 2011b).

148 *Three estuarine crocodiles:* Campbell et al. (2010). On fossils indicating multiple crocodile oceanic dispersals, Brochu (2001). For the molecular genetic study indicating crocodile transatlantic dispersal, Meredith et al. (2011a).

Box: The Molecular Clock and the Pitfalls of Extremism

142 *Fossil-calibrated clock of cichlid fishes:* Azuma et al. (2008).

142 *A 2004 study on amphisbaenians:* Macey et al. (2004). On amphisbaenians not being an early branch in the lizard evolutionary tree, Vidal and Hedges (2009). On the oldest lizard fossils being less than 200 million years old, Evans (2003).

143 *Fragmentation events should be used as calibration points:* Heads (2005, 2011). On the tenrec example using assumed vicariance for calibration, Heads (2005). On the oldest placental mammal fossils, O'Leary et al. (2013). Heads has suggested equally implausible calibration points for several other groups. One involving primates (Heads 2010) would push the origin of that order back to 180 million years ago, some 115 million years before the earliest known placental mammal fossil. For a rebuttal of this primate case, see Goswami and Upchurch (2010).

Chapter Seven: The Green Web

151 *Monolithic remnants of a sandstone plateau:* On the geological history of tepuis, Briceño et al. (1990). On the environment on tepui summits and the possibility of being stranded on them, Jackson (1985).

152 *Areas populated by relicts:* Malatesta (1996).

152 *"Remnants of an ancient biota . . .":* Funk and McDiarmid (1988), 48.

152 *"They are like a place where time stopped":* Handwerk (2004).

152 *Carnivorous plants . . . are a dime a dozen:* On the connection between carnivorous tepui plants and nutrient-poor soil, Butschi (1989). On the protozoan-trapping plant, Barthlott et al. (1998).

153 *Their seeds don't float:* For the facts that *Drosera* seeds do not float and are killed by saltwater and the suggestion that *Drosera* might have dispersed from Australia to South America as seeds stuck to the feet of a bird, email from Andreas Fleischmann to the author, January 19, 2010.

153 *Related species tend to occur near each other:* For the "geographic structure" for *Drosera* and the placement of some North American species with a South American group, Rivadavia et al. (2003).

153 *The weirdest exception to the rule:* For the study of *Drosera meristocaulis*, including both the DNA and anatomical work, Rivadavia et al. (2012).

155 *"I grew up all over the West . . . "* and Matt Lavin's personal and professional background: Lavin, telephone conversation with the author, September 16, 2009.

157 *For a group called the* dalbergioid *legumes:* Lavin et al. (2000). The quotation "relatively few instances where over-water . . . " is from Lavin et al. (2000), 461.

157 *"Oh definitely, definitely"* and *"Man these things have to be old . . . ":* Lavin, telephone conversation with the author, September 16, 2009.

157, 158 *"We tried to bias 'em . . . "* and *" . . . We could hardly make them old . . . ":* Lavin, telephone conversation with the author, September 16, 2009.

158 *An extensive molecular dating study:* Lavin et al. (2004).
158 *"A long slow death":* Lavin, telephone conversation with the author, September 16, 2009. Lavin also discussed the informal talk on Araucariaceae in this conversation.
159 *A massive flood of evidence:* Some recent papers (mostly review articles) collectively indicating hundreds of instances of oceanic dispersal by plants are Renner (2004a, b); Pennington et al. (2004); Pennington and Dick (2004); Yoder and Nowak (2006); Sanmartín et al. (2007); Crisp et al. (2009); Wallis and Trewick (2009); Christenhusz and Chase (2013). For the Cucurbitaceae study, Schaefer et al. (2009).
161 *Graham Wallis and Steve Trewick . . . compiled those exact lists:* Wallis and Trewick (2009).
162 *Many* entire *plant families . . . are too young:* Families of plants native to New Zealand are from the Flora of New Zealand Series web page, http://floraseries.land careresearch.co.nz. The age estimates for plant families that appear to be too young to have been present at the birth of Zealandia are from Forest and Chase (2009a, b, c, d); Bremer (2009). Theoretically, some of these families could have originated in Zealandia and colonized other landmasses by overwater dispersal. However, I'm not aware of studies that indicate the origins of any of these families in Zealandia. In any case, whether they originated in Zealandia or elsewhere, their distributions must be explained by oceanic dispersal. Information on the numbers of genera and species of native New Zealand sunflowers (family Asteraceae) is from the Flora of New Zealand Series web page.
163 *Even these iconic Gondwanan trees:* For age estimates from molecular dating that support oceanic dispersal of southern beeches (*Nothofagus*) to New Zealand, Cook and Crisp (2005); Knapp et al. (2005). In the case of the kauri (*Agathis australis*), some recent age estimates based on molecular dating are consistent with the Gondwanan breakup hypothesis (Knapp et al. 2007), but these rely on calibrations using very fragmentary fossils that may be incorrectly identified (Biffin et al. 2010). In any case, most recent estimates are too young to support the Gondwanan breakup explanation (Knapp et al. 2007; Biffin et al. 2010).
164 *Produced the biggest shudder of all:* Sanmartín and Ronquist (2004).
166 *"The last great gasp . . .":* Email from Michael Crisp to the author, January 24, 2012.
166 *Review of the flora of Australia:* Crisp and Cook (2013). Note that the Gondwanan connection of Australian to South American plants does not contradict Sanmartín and Ronquist's results. The latter study showed that *most* Southern Hemisphere plant distributions cannot be explained by Gondwanan breakup, whereas Crisp and Cook's study indicates that the timetrees for a minority of Australian plant groups do indicate such a history. The fact that the Australia–South America separation happened relatively recently might be part of the reason why the signal of vicariance is still strong for that subset of plants.
167 *"I don't see the formation . . ."* and *"Plants seem to get around . . .":* Lavin, telephone conversation with the author, September 16, 2009. The view of the distribution of woody legumes being limited by suitable environments rather than by dispersal is in Lavin et al. (2004). Along similar lines, Crisp et al. (2009) provided evidence for the importance of suitable environments for the establishment of overwater colonists for Southern Hemisphere plants. Specifically, they found that colonizing lineages had a strong tendency to occupy the same biomes in their new areas (for instance, wet forest or temperate grassland) as they had in their original locations.
168 *"I must now say a few words . . . "* and *"I shall here confine myself to plants":* Darwin (1859), 358.
169 *Seeds are essentially in suspended animation:* For a technical discussion that relates to the idea that seeds, because of their dormant state, can be dispersed in a multitude of ways, see Nathan et al. (2008). These researchers emphasized that long-distance dispersal of seeds often occurs by nonstandard means. For instance, a hooked seed adapted for dispersal by mammals might be dispersed on a raft or on strong storm winds. The authors suggest that such nonstandard mechanisms often make long-distance dispersal much more likely than standard mechanisms would indicate.
169 *Colonized the remote Hawaiian Islands:* Ziegler (2002).

169 *Many flowering plant groups . . . are too young:* For the argument that fossil evidence indicates many flowering plant groups are too young to have been affected by Gondwanan break-up, Thorne (1973); Smith (1973).
169 *Perceived importance of long-distance dispersal:* A key work emphasizing long-distance dispersal during the period dominated by land-bridge thinking is Guppy (1906). The volume *Biological Relationships Between Africa and South America* is Goldblatt (1993).
170 *"Dogmatic and a bit silly":* This quotation and information on Susanne Renner's career and opinions are from an email from Renner to the author, January 13, 2010.
170 *They also examined animal groups:* Sanmartín and Ronquist (2004).
172 *This distinction holds for the Northern Hemisphere:* Donoghue and Smith (2004).
173 *Fossils of elephants:* Johnson (1980). The quotes "wildly in every direction . . . " and "elephants swimming in circles . . . " are from p. 398 of that article.

Chapter Eight: A Frog's Tale

175 *John Measey was looking for the cobra bobo:* For how Measey came to study amphibians on São Tomé and Príncipe, emails to the author from Measey, March 5 and 6, 2009, and from Bob Drewes, March 18, 2009.
175 *A massive project on islands in the Gulf of Guinea:* Drewes's blog at http://islandbiodiversityrace.wildlifedirect.org/.
176 *"Mission from God":* Email from Drewes to the author, March 18, 2009.
177 *"Had eyes for nothing else":* Email from Measey to the author, March 5, 2009.
178 *Deep history and geography of the Gulf of Guinea islands:* Measey et al. (2007).
178 *"Batrachians (frogs, toads, newts) . . . ":* Darwin (1859), 393.
178 *Rejected the story of Noah's Ark:* Browne (1983).
179 *"These animals and their spawn . . . ":* Darwin (1859), 393.
180 *Amphibian species that are apparently native:* Measey et al. (2007).
181 *Measey was standing in the lunch line:* For Measey's conversation with Alain Morlière and the lack of response from scientists studying ocean currents, email from Measey to the author, March 5, 2009.
182 *Miguel Vences was studying frogs:* For biographical information on Vences and his original vicariance-oriented thinking about frogs on Indian Ocean islands, email from Vences to the author, February 27, 2009.
182 *Vences's most straightforward study:* Vences et al. (2003).
184 *Its mating call was especially different:* The descriptions of the calls of the *Mantidactylus* on Mayotte compared to the "same" species on Madagascar are based on sound recordings sent by Vences to the author, May 19, 2009.
186 *Discovered two other instances:* Vences et al. (2003).
188 *The old story is that alpine species moved:* Brown (1971); Grayson (1993).
188 *"No endemic amphibian species . . . ":* Vences et al. (2003), 2435.
189 *Measey and Drewes briefly considered:* Emails to the author from Measey, March 5, 2009; Drewes, March 19, 2009; and Vences, February 27, 2009.
189 *Frog called* Ptychadena newtoni: For reasons that Measey and Drewes chose to study *P. newtoni*, emails to the author from Measey, March 5, 2009, and Drewes, March 19, 2009.
190 *After getting more DNA sequences:* The *Ptychadena newtoni* study is Measey et al. (2007).
191 *This eastern connection:* For the East African connection of four of the five amphibian lineages on São Tomé and Príncipe, Measey et al. (2007).
192 *Measey finally called up Bernard Bourles:* Email from Measey to the author, March 5, 2009.
193 *Showing arcs of lowered salinity:* Measey et al. (2007).
193 *Let us now imagine the whole story:* The description of the dispersal scenario is based on Measey et al. (2007); Drewes's blog post of November 18, 2008, at http://islandbiodiversityrace.wildlifedirect.org/.
195 *"We cannot prove this happened":* Drewes's blog post of November 18, 2008, at http://islandbiodiversityrace.wildlifedirect.org/.

196 *Dray called this kind of account:* Dray (1957).
196 *The famous "origin of life" experiments:* Miller (1953).
197 *Robert O'Hara pointed out:* For *The Origin of Species* as a series of how-possibly arguments, O'Hara (1988).
197 *"How possibly could evolution have occurred . . . "* and *"How possibly could species isolated on islands . . . ":* O'Hara (1988), 148.
198 *A long line of how-possibly arguments:* For Wallace's description of natural rafts, Wallace (1880). For Muir's speculation about seeds carried on pumice, Thiel and Gutow (2005). For Yoder's speculation about lemurs surviving the trip to Madagascar by going into torpor, Yoder et al. (2003). On garter snakes being resistant to the desiccating effect of salt water, de Queiroz and Lawson (2008).
199 *Measey has been contemplating a trip:* Email from Measey to the author, March 5, 2009.
199 *"We vs they":* Email from Drewes to the author, March 27, 2009.
199 *Other phylogenetic studies extend that list:* On toads, Pramuk et al. (2008); on fanged frogs, Evans et al. (2003); on slender salamanders, Jockusch and Wake (2002); on Caribbean frogs, Hedges (2006), Heinicke et al. (2007), Fouquet et al. (2013).
200 *The caecilians and some of the frogs on the Seychelles:* Zhang and Wake (2009); Biju and Bossuyt (2003).
201 *"That insects can be transported . . . ":* Guppy (1906), 509–510.

Chapter Nine: The Monkey's Voyage

203 *To a small island called Rapta:* Branner (1888).
204 *Both species are . . . closely related to South American taxa:* For relationships of the worm lizard (*Amphisbaena ridleyi*) and the rat (*Noronhomys vespucci*), respectively, Laguna et al. (2010); Carleton and Olson (1999).
204 *"Offensively familiar":* Branner (1888), 867.
205 M. atlantica's *African roots have been verified:* For the anatomical and genetic evidence for the close relationship of *Mabuya atlantica* to African species and the evidence for transatlantic crossings by the ancestors of this species and, separately, by the ancestors of all other New World (South American and Caribbean) *Mabuya*, Mausfeld et al. (2002); Carranza and Arnold (2003); Whiting et al. (2006).
205 *At least 1,800 miles over water:* Mausfeld et al. (2002).
206 *Other ocean crossings by lizards and snakes:* For skinks (*Cryptoblepharus* and *Leiolopisma*) crossing the Indian Ocean, Arnold (2000); Austin and Arnold (2006); Rocha et al. (2006). For geckos (*Phelsuma* and *Nactus*) crossing the Indian Ocean, Arnold (2000); Austin et al. (2004). For geckos (*Tarentola*, two lineages of *Hemidactylus*, and *Lygodactylus*) crossing the Atlantic, Carranza et al. (2000); Gamble et al. (2011). For worm lizards (Amphisbaenidae) crossing the Atlantic, Vidal et al. (2008). For blindsnakes (*Typhlops*) crossing the Atlantic, Vidal et al. (2010b). For threadsnakes (Epictini) crossing the Atlantic, Adalsteinsson et al. (2009). This list only includes cases for which molecular dating or other evidence refutes alternative hypotheses, including vicariance by continental drift, movement over land bridges, and human introduction. There are several cases not included here that can be explained plausibly by either transatlantic dispersal or movement over a North Atlantic land bridge. Conspicuously missing from the list are the iguanas (*Brachylophus*) of Fiji and Tonga, whose ancestors may have reached those islands by crossing the Pacific from South America, a journey of some 5,000 miles. Authors of a recent study (Noonan and Sites 2010) argued plausibly, although far from conclusively, that iguanas reached Fiji and Tonga overland or by relatively short-distance island-hopping from Asia. The view that burrowing reptiles are especially poor overwater dispersers is presented (but refuted) by Vidal et al. (2008).
209 *The history of primate dispersal:* The record of primate oceanic dispersal is reviewed in McCarthy (2005); Rossie and Seiffert (2006); and Fleagle and Gilbert (2006). A discussion of the dispersal of lemur ancestors to Madagascar is found in Yoder et al. (2003) and Springer et al. (2012), with the age of 50 million years ago coming from the latter. Regarding the overwater colonization of the Greater Antilles by monkeys:

according to geologic reconstructions by Iturralde-Vinent (2006), there has been no land connection between the Greater Antilles and South America since about 30 million years ago. The earliest monkey fossils in the Caribbean (from Cuba) probably are about 18 million years old (MacPhee et al. 2003), and the oldest New World monkey fossils of any kind are about 26 million years old (Takai et al. 2000), suggesting overwater colonization of the Caribbean. However, Iturralde-Vinent and MacPhee (1999) have argued for a vicariant origin for Greater Antillean monkeys. The dispersal of macaques to Southeast Asian islands is reviewed in Abegg and Thierry (2002). The case for dispersal by *Homo floresiensis* between islands is from Morwood and Jungers (2009). Paleontologists also have inferred crossings of the Tethys Sea by primates early in their history (Beard 1998; Kay et al. 2004; Chaimanee et al. 2012). For Tethys crossings, see also Chapter Twelve.

210 *One branching point in the tree:* For strong support for the sister-group relationship of platyrrhines and catarrhines, Fleagle (1999); Steiper and Young (2009); Springer et al. (2012).

210 *Easy to recognize a New World monkey:* For the general characteristics of platyrrhines and catarrhines, Fleagle (1999).

210–211 *50 million years older than the earliest known primate fossils:* For the earliest fossils of primates and placental mammals, O'Leary et al. (2013). The fossil ages are for so-called "crown group" primates and placental mammals, that is, the earliest fossils that fall within the clades defined as the most recent common ancestor of living primates and all its descendants and the most recent common ancestor of living placental mammals and all its descendants. Heads's advocacy for the opening of the Atlantic as the explanation for the platyrrhine-catarrhine split is found in Heads (2010). For rebuttal of Heads's arguments, Goswami and Upchurch (2010).

211 *South America was either an island continent:* For South America as an island continent or connected only to Antarctica/Australia, Iturralde-Vinent (2006); Brown et al. (2006); Barker et al. (2007); Mann et al. (2007).

211 *Quite a few fossils in the Old World:* Fleagle (1999); Seiffert et al. (2005); Heesy et al. (2006); Ni et al. (2013; supplementary information). On the first fossil monkey in South America, Takai et al. (2000).

212 *The studies estimated the age:* Molecular divergence date studies for primates in general and for the platyrrhine-catarrhine split in particular are reviewed in Steiper and Young (2009) and Poux et al. (2006).

212 *The study that especially stood out:* Arnason et al. (2000).

212 *Made their age estimate far too old:* For criticism of calibration points used by Arnason et al. (2000), Raaum et al. (2005), in which the authors present several relatively good primate fossil calibration points. On problems with the use of mtDNA as applied to primates, Glazko and Nei (2003).

213 *All yield estimates of 51 million years or younger:* The date of 51 million years for the platyrrhine-catarrhine split is from Bininda-Emonds et al. (2007). These authors initially gave a date of 54 million years, but, after correcting errors in their analyses, revised that date to 51 million years.

213 *The single estimate I consider the best:* Springer et al. (2012).

214 *Crossing from Asia over the Bering Land Bridge:* The North American origin hypothesis for New World Monkeys is described and refuted in Fleagle (1999).

216 *Share several obvious features:* For traits linking Old World and New World monkeys, the disbelief in this relationship by many primatologists, and the alternate hypothesis of convergent evolution of these groups from prosimian ancestors, Fleagle (1999).

216 *Alain Houle built up a how-possibly argument:* Houle (1999).

217 *Used a model that takes subsidence into account:* Bandoni de Oliveira et al. (2009).

219 *At least eleven such cases:* The lizard and snake references are given on page 321 connected to the phrase, *Other ocean crossings by lizards and snakes*. For the caviomorph Atlantic crossing, Poux et al. (2006) and Rowe et al. (2010). For the past and present configuration of currents in the Atlantic, Renner (2004a). I have not included the case of crocodiles crossing the Atlantic (Meredith et al. 2011a) because crocodiles might be considered more aquatic than terrestrial.

219 *Primates are probably second only to rodents:* Rossie and Seiffert (2006).
219 *"Why, it may be asked . . . ":* Darwin (1859), 394.
220 *In his 2007 book* The Black Swan*:* Taleb (2010).
223 *Stephen Jay Gould and others have argued:* For Gould's argument about the importance of unpredictable events in evolution, Gould (1989).
223 *In the case of worm lizards:* For the number of amphisbaenid species in the New World, Gans (2005).
223 *Produced some 130 modern species:* For the number of platyrrhine species (128 to be precise), Wilson and Reeder (2005).
224 Cerithideopsis *is a genus of small snails:* Miura et al. (2012). On snails emerging alive from Willet pellets, Sousa (1993).

Box: Dinosaurs Too?
221 *Has been dubbed* Tethyshadros insularis: For this example of dispersal, Dalla Vecchia (2009).
222 *The collection of other fossils:* For other vertebrates at the *Tethyshadros* site, Delfino et al. (2008).
222 Ajkaceratops kozmai: For this example of dispersal, Osi et al. (2010).

Chapter Ten: The Long, Strange History of the Gondwanan Islands

226 *By all accounts it is a harsh place:* For general description of the Falklands, Darwin (1839); Strange (1972).
227 *"Wretched place":* Darwin letter to E. C. (Catherine) Darwin, April 6, 1834, from the Darwin Correspondence Project website, www.darwinproject.ac.uk/home.
227 *Bob McDowall visited the Falklands:* For McDowall's background and trip to the Falklands, emails from McDowall to the author, October 26, 2009, and August 15, 2010, and an unpublished manuscript by McDowall sent to the author on October 15, 2009. Some additional information is from McDowall (2005) and McDowall et al. (2005).
227 *"Blast him out of the water"* and *"execute":* Hull (1988), 173. For the quotation, "destroy this person, . . . " McDowall, unpublished manuscript sent to the author on October 15, 2009.
229 *Small tectonic plate that contained the Falklands:* For the geological history of the Falklands, Curtis and Hyam (1998); Storey et al. (1999).
231 *The case, as it turned out, was very clear:* McDowall (2005).
231 *Recent DNA analyses using museum specimens:* Austin et al. (2013). For Darwin's idea of the wolf reaching the islands on icebergs, Darwin (1859).
233 *"The Falkland Isld. flora seems to combine . . . ":* Darwin letter to J. D. Hooker, November 28, 1843, from the Darwin Correspondence Project website, www.darwinproject.ac.uk/home. For Croizat linking the Falklands biota to Patagonia, McDowall (2005).
235 *A species' chance of going extinct:* For a general review of the effect of area on extinction, Rosenzweig (2002).
237 *Graphs plotting the number of extinctions:* A good recent study of origination and extinction rates for taxa in the marine fossil record for the past 500 million years is Alroy (2008).
237 *"The Great Dying":* Gould (1977). The figure of more than 90 percent extinction for the end-Permian event is in Jablonski (1994).
238 *"A severe sorting of the Falklands biota":* McDowall (2005), 59.
238 *The ranges of many species . . . shifted to the south:* Grayson (1993).
239 *The endemic lineages consist of just one or a few species:* McDowall (2005).
239 *The Chathams . . . are a small archipelago:* For the appearance and human history of the Chathams, Harding et al. (2002).
240 *Formed part of Zealandia:* For the Chathams as part of Zealandia and Gondwana, Campbell and Hutching (2007).
240 *The fossil remains of a typical Gondwanan flora:* For Late Cretaceous plants and dinosaurs on the Chathams, Stillwell et al. (2006); Campbell and Hutching (2007).

240 *The Chathams were completely submerged:* For the Cenozoic geological history of the Chathams, including submergence, Campbell et al. (1988); Campbell and Hutching (2007); Stillwell and Consoli (2012).

241 *Endemic birds of the Chathams:* For the similarity of Chatham Islands and New Zealand birds, Robertson and Heather (2005). A study showing the genetic similarity of the Black Robin of the Chathams to related New Zealand species is Miller and Lambert (2006).

241 *Confirmed by extensive molecular studies:* Molecular studies of Chatham Islands species are reviewed in Paterson et al. (2006); Goldberg et al. (2008); and Heenan et al. (2010). An apparent anomaly is *Myosotidium hortensium*, a Chathams plant in the Boraginaceae family, which is estimated to have split from its closest relatives 3.6 to 22.4 million years ago (Goldberg et al. 2008). However, even if the older age limit is correct, this example does not refute the submergence hypothesis, since the divergence did not necessarily occur on the Chathams (see the discussion of New Caledonia later in this chapter). Also, the investigators in this case suggest that this potentially old divergence could be the result of researchers not yet having sampled the closest relatives outside of the Chathams, which seems especially probable, since the nearest known relative is from the Mediterranean region.

242 *Amborella trichopoda . . . and the Kagu:* For the evolutionary positions of *Amborella* and the kagu, respectively, Soltis et al. (2008); Fain and Houde (2004). For an entertaining description of the flora and fauna of New Caledonia, see Flannery (1994).

242 *New Caledonia was entirely underwater:* For geological evidence for the submergence of New Caledonia, Pelletier (2006); Grandcolas et al. (2008).

242 *Many New Caledonian lineages, including the skinks:* Lygosomine skinks are estimated to have diverged from relatives elsewhere 12.7 to 40.7 million years ago (Smith et al. 2007), which slightly overlaps with the drowning period, but Smith et al. argue that the younger limit is probably more accurate. *Araucaria* diverged from relatives elsewhere 10.8 to 38.2 million years ago, which also slightly overlaps with the drowning period (Pillon 2012). Other taxa with estimated divergence ages that fall after the drowning include many other plants; *Angustonicus* cockroaches (formerly considered an ancient relict group); galaxiid fishes; and *Paratya* freshwater shrimp, among others (Pillon 2012; Grandcolas et al. 2008).

242 *Others, such as* Amborella: Divergence ages for earliest branching points within ancient New Caledonian groups include the following: diplodactylid geckos, 9 to 21.6 million years (Nielsen et al. 2011); troglosironid harvestmen, 28 to 49 million years (Boyer et al. 2007), 49 million years (Sharma and Giribet 2009), 52 to 102 million years (Giribet et al. 2010), and 40 to 73 million years (Giribet et al. 2012). *Amborella trichopoda* and the kagu are single species and therefore are unlikely to show ancient divergences within New Caledonia.

244 *Similarly unique, ancient lineages:* Lactoris, Gamerro and Barreda (2008); *Hillebrandia*, Clement et al. (2004); Bolyeriidae, Pyron and Burbrink (2011).

244 *Compilation of molecular dating studies:* Wallis and Trewick (2009).

244 *Rails . . . have dispersed:* Steadman (1995).

245 *Pure vicariance scenario for ratites:* For the branching order in the ratite tree, Harshman et al. (2008); Phillips et al. (2010). For molecular dating studies of ratites, Baker and Pereira (2009), which gives estimates from seven studies, and Phillips et al. (2010).

246 *The common ancestor of all ratites could fly:* Harshman et al. (2008); Phillips et al. (2010). For loss of flight in eighteen living families of birds, Harshman et al. (2008).

246 *Two lineages of mite harvestmen:* Giribet et al. (2012).

246 *Two lineages of centipedes:* Murienne et al. (2010).

246 *Ironically, New Zealand:* For a discussion of the origins of New Zealand's terrestrial invertebrate fauna, see Giribet and Boyer (2010).

247 *We come to Madagascar:* For description of Madagascar's biota, Yoder and Nowak (2006); Goodman and Benstead (2004).

247 *When Madagascar detached from India:* Krause (2003).
247 *Most biologists did* not *interpret the biota:* For origins of Madagascar's biota by long-distance dispersal, Briggs (1987); Schatz (1996). Others did tend to interpret the biota as a vicariant, relict one. See, for instance, Wickens (1982); Gillespie and Roderick (2002).
248 *A key literature-review study:* Yoder and Nowak (2006).
248 *Colonization and ocean currents:* Ali and Huber (2010); Samonds et al. (2012).
249 *The key studies have been headed by David Krause:* Krause (2003); Krause et al. (2006). The 2003 paper gives the argument for the lack of a connection between the Cretaceous and modern vertebrate faunas.
250 *Madagascar and the other Gondwanan islands:* In this tour of Gondwanan islands, I have left out Sri Lanka, New Guinea, and Socotra, because those islands have been connected to continents relatively recently.
250 *Swarms of igneous dikes:* Storey et al. (1999).
251 *An archipelago I haven't discussed:* For specific examples in which molecular dating supports overwater colonization of the Seychelles, Austin et al. (2004) on daygeckos; Daniels (2011) on freshwater crabs; and Guo et al. (2012) on wolf snakes.
251 *"New Caledonia must be considered . . . ":* Grandcolas et al. (2008), 3309.
251 *"Extinction, colonization and speciation . . . ":* Goldberg et al. (2008), 3319.
252 *"For me, . . . the particularly interesting aspect of these patterns . . . ":* McDowall (2005), 59.
253 *In September 1995, hurricanes Luis and Marilyn:* On the story of the green iguanas arrival by rafting, Censky et al. (1998). The status of the iguanas in 2011 is from an email from Ellen Censky to the author, January 10, 2011.

Chapter Eleven: The Structure of Biogeographic "Revolutions"

257 *Some of the caterpillars are amphibious:* For amphibious caterpillars and snail-eating caterpillars, Rubinoff and Schmitz (2010) and Rubinoff and Haines (2005), respectively.
260 *Unusually large* pseudogenes: Baldo et al. (2011).
261 *Their presence in Hawaii:* For the distribution of bristletails in Hawaii, Sturm (1993).
261 *Quite a few volcanic islands:* For the distributions of bristletails in general, including that of *Neomachilellus*, Sturm and Machida (2001).
262 *A 2012 compilation:* Gillespie et al. (2012).
262 *Examined the anatomical traits:* Sturm (1993). The DNA sequences from Hawaiian specimens are from unpublished work by Robert Meredith, John Gatesy, Cheryl Hayashi, Eric Stiner, and myself.
262 *Eggs attached to driftwood:* Sturm (1993). For the observation that some bristletail eggs are resistant to chemicals, Larink (1972), cited in Sturm and Bach de Roca (1988).
263 *Islands-as-dead-ends rule:* Arguments against island-to-mainland colonizations are in Bellemain and Ricklefs (2008).
263 *Natural invasions recorded in recent times:* Levin (2006).
264 *Group includes close to six hundred:* O'Grady et al. (2010).
265 *Hawaiian escapees have given rise:* Published DNA-based phylogenetic evidence for at least one Hawaii-to-mainland dispersal by *Scaptomyza*, and speculation about why these flies are such good dispersers are in O'Grady and DeSalle (2008). For evidence for other out-of-Hawaii dispersals by *Scaptomyza*, email from O'Grady to the author, May 23, 2012.
265 *Only one of several:* For *Rhantus* beetles, Balke et al. (2009); for *Anolis* lizards, Nicholson et al. (2005), Glor et al. (2005); for monarchid flycatchers, Filardi and Moyle (2005).
266 *A diehard land-bridge advocate:* Bowler (1996); Schuchert (1932). The latter includes Schuchert's argument about granitic rocks on Atlantic islands.
267 *Jots down some calculations:* Simpson (1952) gives actual calculations of this sort.
268 *"Beset by attempts to make the facts fit the theory . . . ":* Email from Briggs to the author, August 25, 2010.

268 *Existence of granite on Tristan da Cunha . . . and Ascension:* The granitic rock on Ascension Island has been dated as at most a few million years old (Kar et al. 1998), so it clearly has nothing to do with Mesozoic land bridges.

268 *Land connections to . . . Hawaii:* Skottsberg (1925).

270 *The final product, the general area cladogram:* For the view that the goal of historical biogeography is to infer relationships among areas, Ebach and Humphries (2002); Ebach (2003); Parenti and Ebach (2009).

271 *"Early on I took the vicariance model as the default . . . ":* Email from Trewick to the author, December 13, 2010.

272 *It was hard to explain the fact:* For anomalies leading to the Copernican revolution, Kuhn (1970).

273 *"Regularly marked by frequent and deep debates":* Kuhn (1970), 47–48.

274 *"Throughout the pre-paradigm period . . . ":* Kuhn (1970), 163.

274 *Matthew's criticism of land-bridge advocates:* Matthew (1915).

274 *"Apart from Croizat . . . ":* Email from Heads to the author, August 6, 2010.

274–275 *"I don't think he made . . . " and "Cladistics by itself . . . ":* Email from Briggs to the author, August 25, 2010.

275 *If this view is correct:* Mayr (1982, 857), while not dismissing the notion of a scientific revolution, wrote, "I cannot think of a single case in biology where there was a drastic replacement of paradigms between two periods of "normal science." Mayr was suggesting that Kuhn's views do not apply well to biology, but his observation can be taken in a different way. Specifically, it may be that some of the great "revolutions" in biology represent the initial emergence of a paradigm that has yet to be replaced, rather than shifts from one accepted paradigm to another. The development of genetics, of cladistic/phylogenetic thinking, and, most conspicuously, of Darwinian evolutionary thinking in general, might be seen in that way.

276 *The missing element was time:* The importance of placing ages on evolutionary branching points has been emphasized by Donoghue and Moore (2003); de Queiroz (2005); Renner (2005); and Yoder and Nowak (2006), among many others.

277 *The great age of many Hawaiian lineages:* Skottsberg (1941).

277 *"Impervious to evidence":* Yoder, telephone conversation with the author, March 10, 2009.

277 *"They were seen for what they are . . . ":* Email from Trewick to the author, December 13, 2010.

277 *"I thought, man, I'm having some really weird flashback . . . ":* Donoghue, telephone conversation with the author, September 3, 2010.

277 *"Artefactual":* Nelson and Ladiges (2001), 389.

277 *"Reactionary":* Santos (2007), 1471.

277 *"Ignoring basic biogeographic realities":* McCarthy (2005), 3.

279 *On or around October 12, 1988:* The description of the locust invasion is in Richardson and Nemeth (1991). The DNA studies indicating that African locusts colonized the New World are Lovejoy et al. (2006); Song et al. (2013).

Chapter Twelve: A World Shaped by Miracles

283 *Let us consider the potato:* For the initial cultivation and early South American history of potatoes, Spooner et al. (2005); Mann (2011); McNeill (1999).

283 *Ships coming from the Pacific side:* McNeill (1999).

283 *Rumors dogged the plant:* Nunn and Qian (2011).

283 *Spread all across northern Europe:* McNeill (1999); Nunn and Qian (2011).

283 *In Ireland, the impetus:* Mann (2011); Pollan (2001).

283 *This was the potato blight:* Mann (2011); Reader (2008, 2009).

284 *The blight reached its peak:* Kennedy et al. (2000); Pollan (2001); Mann (2011).

284 *Some 2 million people out of Ireland:* Mann (2011); Reader (2008).

284 *Russia, Germany:* McNeill (1999).

284 *Rise in population:* Nunn and Qian (2011). The population of 600 million in 1700 is from Nunn and Qian, but the figure of 1.5 billion for 1900 subtracts the population of the New World (roughly 100 million) from their figure of 1.6 billion.

285 *"Homogenocene":* Charles Mann's 2011 book *1493* provides an engaging and detailed look at many of the consequences of the intentional and inadvertent introduction of species all over the world since the discovery of the Americas by Europeans. The term "Homogenocene" was coined by Samways (1999).

285 *Unremarkable day in the life:* This account draws from the following: nutmeg, Joseph (1980); eggplant, Olmstead and Palmer (1997); zucchini and watermelon, Schaefer et al. (2009); common bean, Lavin et al. (2004); corn, Bouchenak-Khelladi et al. (2010); crocodiles, Meredith et al. (2011a); monkeys, Fleagle (1999) and Poux et al. (2006); guinea pigs, Poux et al. (2006) and Rowe et al. (2010); lovebirds, Schweizer et al. (2010); chameleons, Tolley et al. (2013); cotton, Wendell et al. (2010).

287 *Ancestor of anthropoids:* Chaimanee et al. (2012). All placements of the fossils within the primate evolutionary tree indicate a crossing of the Tethys Sea, but whether the crossing was to or from Africa is unclear.

288 *South America was an island continent:* McLoughlin (2001); Iturralde-Vinent (2006); Brown et al. (2006); Mann et al. (2007).

290 *"Great American Interchange":* Webb (2006). The history of South American mammals through the Cenozoic (the past 66 million years), including the Great American Interchange, was described most famously by George Gaylord Simpson in his 1980 book, *Splendid Isolation*, but the basic story of evolution in isolation, followed by mixing, had been suggested much earlier, particularly by W. B. Scott in 1932.

290 *Island-hopping journeys:* For the earliest procyonid fossils in South America, Webb (2006). For the earliest sigmodontine fossils in South America, Verzi and Montalvo (2008). The exact age of the sigmodontine fossils has been questioned by Prevosti and Pardiñas (2009), but these authors agree that the fossils predate the emergence of the Isthmus of Panama. Overwater dispersal to South America by procyonids and sigmodontines is also covered by Koepfli et al. (2007) and Steppan et al. (2004), respectively. The radiation of South American sigmodontines described here refers only to the descendants of the presumed single oceanic dispersal event, that is, Oryzomyalia plus *Neusticomys* (Ichthyomyini). *Neusticomys* is placed in this group by Parada et al. (2013). Other lineages of sigmodontines, namely, the tribe Sigmodontini and some Ichthyomyini, may have reached South America separately and perhaps by land over the Isthmus of Panama (Steppan et al. 2004).

290 *Two much better known examples that fall into this category:* The latest possible ages for colonizations by monkeys and caviomorphs are based on the earliest New World fossils for these groups (Takai et al. 2000; Antoine et al. 2012).

291 *A great range of forms:* The diversity of caviomorphs and sigmodontines is described in Lord (2007).

291 *World's largest known extinct rodent:* Rinderknecht and Blanco (2008).

293 *All three lineages reached the West Indies:* For monkey dispersal to West Indies, see notes to Chapter Nine. For sigmodontine dispersals to West Indies, McFarlane et al. (2002); McFarlane and Lundberg (2002). For caviomorph dispersal to the West Indies, Pregill (1981). For sigmodontine dispersal to the Galápagos, Steadman and Ray (1982); Key and Heredia (1994). For sigmodontine dispersal to Fernando de Noronha, Carleton and Olson (1999).

293 *Some that must indicate a massive impact:* The numbers of species are from Lord (2007).

293 *Tree of these nematodes largely mirrors:* The nematode and monkey study is Hugot (1998). For some of the many parasites that are likely restricted to South American overwater colonist mammals, Tantaleán and Gozalo (1994); Rossin et al. (2010).

294 *Snakes called the Xenodontinae:* The time period for colonization of South America by xenodontine snakes is inferred from the timetree in Hedges et al. (2009; nodes 1 and 3 in Table 3), in conjunction with the larger phylogeny presented in Vidal et al. (2010a). The number of South American xenodontine species is from Vidal et al. (2010a), with the Alsophiini and other non–South American taxa removed. For xenodontine diversity and diet, Greene (1997).

295 *The geographic origins of most South American plants:* Renner (2004a) has compiled a list of at least 110 plant genera, including species in both tropical America and

tropical Africa, most of which likely represent dispersal across the Atlantic either to or from the New World tropics. Molecular dating also has revealed many other cases of overwater dispersal by plants to South America (for instance, the tepui sundew and many of the bean-plant examples discussed in Chapter Seven). For recent reviews of the origins of Neotropical plants, see Pennington and Dick (2004); Christenhusz and Chase (2013).

295 *Plants must have ecological connections:* Interactions between animals and plants that are both probably descended from overseas colonists are given here. In each case, the first reference is for the ecological interaction and the second for overwater colonization of South America by the plant taxon in question. Monkeys and *Symphonia globulifera*, Riba-Hernández and Stoner (2005), Dick et al. (2003); *Oecomys* rodents and *Renealmia alpinia*, Bizerril and Gastal (1997), Särkinen et al. (2007); *Turdus* thrushes and *Miconia*, Marcondes-Machado (2002), Renner et al. (2001); *Turdus* thrushes and *Ocotea*, Francisco and Galetti (2002), Chanderbali et al. (2001).

297 *"What if?" scenarios:* Gaddis (2002).

298 *"Altered by an apparently insignificant jot or tittle . . . ":* Gould (1989), 289.

301 *The famous "butterfly effect":* Gleick (1987), describing the work of the meteorologist Edward Lorenz.

301 *Chaos theory—a research area:* For applications of chaos theory to various population phenomena, see Gleick (1987). An early classic paper in this area is May (1974).

REFERENCES

Abegg, C., and B. Thierry. 2002. Macaque evolution and dispersal in insular south-east Asia. *Biological Journal of the Linnean Society* 75, 555–576.

Adalsteinsson, Solny A., William R. Branch, Sébastien Trape, Laurie J. Vitt, and S. Blair Hedges. 2009. Molecular phylogeny, classification, and biogeography of snakes of the family Leptotyphlopidae (Reptilia, Squamata). *Zootaxa* 2244, 1–50.

Ali, Jason R., and Matthew Huber. 2010. Mammalian biodiversity on Madagascar controlled by ocean currents. *Nature* 463, 653–656.

Alroy, John. 2008. Dynamics of origination and extinction in the marine fossil record. *Proceedings of the National Academy of Sciences USA* 105, suppl. 1, 11536–11542.

Angier, Natalie. 2010. Reptile's pet-store looks belie its Triassic appeal. *New York Times* online edition, November 22, 2010, www.nytimes.com/2010/11/23/science/23angier.html?pagewanted=all.

Antoine, Pierre-Olivier, et al. 2012. Middle Eocene rodents from Peruvian Amazonia reveal the pattern and timing of caviomorph origins and biogeography. *Proceedings of the Royal Society of London B* 279, 1319–1326.

Arnason, Ulfur, Anette Gullberg, Alondra Schweizer Burguete, and Axel Janke. 2000. Molecular estimates of primate divergences and new hypotheses for primate dispersal and the origin of modern humans. *Hereditas* 133, 217–228.

Arnold, E. N. 2000. Using fossils and phylogenies to understand evolution of reptile communities on islands. Pp. 309–323 in G. Rheinwald, ed., *Isolated Vertebrate Communities in the Tropics*. Bonner Zoologische Monographien 46.

Associated Press. 2005. Five days at sea, now pregnant. *Sydney Morning Herald* online edition, January 6, 2005, www.smh.com.au/news/Asia-Tsunami/Five-days-at-sea-now-pregnant/2005/01/06/1104832227649.html.

Austin, J. J., and E. N. Arnold. 2006. Using ancient and recent DNA to explore relationships of extinct and endangered *Leiolopisma* skinks (Reptilia: Scincidae) in the Mascarene Islands. *Molecular Phylogenetics and Evolution* 39, 503–511.

Austin, J. J., E. N. Arnold, and C. G. Jones. 2004. Reconstructing an island radiation using ancient and recent DNA: the extinct and living day geckos (*Phelsuma*) of the Mascarene Islands. *Molecular Phylogenetics and Evolution* 31, 109–122.

Austin, Jeremy J., Julien Soubrier, Francisco J. Prevosti, Luciano Prates, Valentina Trejo, Francisco Mena, and Alan Cooper. 2013. The origins of the enigmatic Falkland Islands wolf. *Nature Communications* 4, no. 1552.

Azuma, Yoichiro, Yoshinori Kumazawa, Masaki Miya, Kohji Mabuchi, and Mutsumi Nishida. 2008. Mitogenomic evaluation of the historical biogeography of cichlids toward reliable dating of teleostean divergences. *BMC Evolutionary Biology* 8, 215.

Baker, Allan J., and Sergio L. Pereira. 2009. Ratites and tinamous (Paleognathae). Pp. 412–414 in S. B. Hedges and S. Kumar, eds., *The Timetree of Life*. Oxford University Press, Oxford, UK.

Baldo, Laura, Alan de Queiroz, Marshal Hedin, Cheryl Y. Hayashi, and John Gatesy. 2011. Nuclear-mitochondrial sequences as witnesses of past interbreeding and population diversity in the jumping bristletail *Mesomachilis*. *Molecular Biology and Evolution* 28, 195–210.

Baldwin, Bruce G., and Warren L. Wagner. 2010. Hawaiian angiosperm radiations of North American origin. *Annals of Botany* 105, 849–879.

Balke, Michael, Ignacio Ribera, Lars Hendrich, Michael A. Miller, Katayo Sagata, Aloysius

Posman, Alfried P. Vogler, and Rudolf Meier. 2009. New Guinea highland origin of a widespread arthropod supertramp. *Proceedings of the Royal Society of London B* 276, 2359–2367.

Ball, Ian R. 1975. Nature and formulation of biogeographical hypotheses. *Systematic Zoology* 24, 407–430.

Bandoni de Oliveira, Felipe, Eder Cassola Molina, and Gabriel Marroig. 2009. Paleogeography of the South Atlantic: a route for primates and rodents into the New World. Pp. 55–68 in P. A. Garber, A. Estrada, J. C. Bicca-Marques, E. W. Heymann, and K. B. Strier, eds., *South American Primates: Comparative Perspectives in the Study of Behavior, Ecology, and Conservation.* Springer Science+Business Media, New York.

Barker, Peter F., Gabriel M. Filippelli, Fabio Florindo, Ellen E. Martin, and Howard D. Scher. 2007. Onset and role of the Antarctic Circumpolar Current. *Deep-Sea Research II* 54, 2388–2398.

Barthlott, Wilhelm, Stefan Porembski, Eberhard Fischer, and Björn Gemmel. 1998. First protozoa-trapping plant found. *Nature* 392, 447.

Battistuzzi, Fabia U., Alan Filipski, S. Blair Hedges, and Sudhir Kumar. 2010. Performance of relaxed-clock methods in estimating evolutionary divergence times and their credibility intervals. *Molecular Biology and Evolution* 27, 1289–1300.

Baum, David A., Randall L. Small, and Jonathan F. Wendel. 1998. Biogeography and floral evolution of baobabs (*Adansonia*, Bombacaceae) as inferred from multiple data sets. *Systematic Biology* 47, 181–207.

Beard, K. Christopher. 1998. East of Eden: Asia as an important biogeographic center of taxonomic origination in mammalian evolution. *Bulletin of the Carnegie Museum of Natural History* 34, 5–39.

Bellemain, Eva, and Robert E. Ricklefs. 2008. Are islands the end of the colonization road? *Trends in Ecology and Evolution* 23, 461–468.

Benton, M., P. C. J. Donoghue, and R. J. Asher. 2009. Calibrating and constraining molecular clocks. Pp. 35–86 in S. B. Hedges and S. Kumar, eds., *The Timetree of Life.* Oxford University Press, Oxford, UK.

Biffin, Ed, Robert S. Hill, and Andrew J. Lowe. 2010. Did kauri (*Agathis*: Araucariaceae) really survive the Oligocene drowning of New Zealand? *Systematic Biology* 59, 594–602.

Biju, S. D., and Franky Bossuyt. 2003. New frog family from India reveals an ancient biogeographical link with the Seychelles. *Nature* 425, 711–714.

Bininda-Emonds, Olaf R. P., et al. 2007. The delayed rise of present-day mammals. *Nature* 446, 507–512.

Bizerril, M. X. A., and M. L. A. Gastal. 1997. Fruit phenology and mammal frugivory in *Renealmia alpinia* (Zingiberaceae) in a gallery forest of central Brazil. *Revista Brasileira de Biologia* 57, 305–309.

Blount, Zachary D., Christina Z. Borland, and Richard E. Lenski. 2008. Historical contingency and the evolution of a key innovation in an experimental population of *Escherichia coli*. *Proceedings of the National Academy of Sciences USA* 105, 7899–7906.

Bouchenak-Khelladi, Yanis, G. Anthony Verboom, Vincent Savolainen, and Trevor R. Hodkinson. 2010. Biogeography of the grasses (Poaceae): a phylogenetic approach to reveal evolutionary history in geographical space and geological time. *Botanical Journal of the Linnean Society* 162, 543–557.

Bousquet, Jean, Steven H. Strauss, Allan H. Doerksen, and Robert A. Price. 1992. Extensive variation in evolutionary rate of *rbcL* gene sequences among seed plants. *Proceedings of the National Academy of Sciences USA* 89, 7844–7848.

Bowler, Peter J. 1996. *Life's Splendid Drama.* University of Chicago Press, Chicago.

Boyer, Sarah L., Ronald M. Clouse, Ligia R. Benavides, Prashant Sharma, Peter J. Schwendinger, I. Karunarathna, and Gonzalo Giribet. 2007. Biogeography of the world: a case study from cyphophthalmid Opiliones, a globally distributed group of arachnids. *Journal of Biogeography* 34, 2070–2085.

Branner, John C. 1888. Notes on the fauna of the islands of Fernando de Noronha. *American Naturalist* 22, 861–871.

Bremer, Birgitta. 2009. Asterids. Pp. 177–187 in S. B. Hedges and S. Kumar, eds., *The Timetree of Life.* Oxford University Press, Oxford, UK.

Briceño, H., C. Schubert, and J. Paolini. 1990. Table-mountain geology and surficial geochemistry: Chimantá Massif, Venezuelan Guayana Shield. *Journal of South American Earth Sciences* 3, 179–184.
Briggs, John C. 1987. *Biogeography and Plate Tectonics.* Elsevier, Amsterdam.
———. 1995. *Global Biogeography.* Elsevier, Amsterdam.
Britten, Roy J. 1986. Rates of DNA sequence evolution differ between taxonomic groups. *Science* 231, 1393–1398.
Brochu, Christopher A. 2001. Congruence between physiology, phylogenetics, and the fossil record on crocodylian historical biogeography. Pp. 9–28 in G. Grigg, F. Seebacher, and C. Franklin, eds., *Crocodilian Biology and Evolution.* Surrey Beatty and Sons, Chipping Norton, New South Wales.
Bromham, L. 2009. Why do species vary in their rate of molecular evolution? *Biology Letters* 5, 401–404.
Brooks, Daniel R., and E. O. Wiley. 1988. *Evolution as Entropy: Toward a Unified Theory of Biology.* University of Chicago Press, Chicago.
Brown, Belinda, Carmen Gaina, and R. Dietmar Müller. 2006. Circum-Antarctic palaeobathymetry: illustrated examples from Cenozoic to recent times. *Palaeogeography, Palaeoclimatology, Palaeoecology* 231, 158–168.
Brown, James H. 1971. Mammals on mountaintops: nonequilibrium insular biogeography. *American Naturalist* 105, 467–478.
Browne, Janet. 1983. *The Secular Ark: Studies in the History of Biogeography.* Yale University Press, New Haven, CT.
———. 1995. *Charles Darwin: Voyaging.* Princeton University Press, Princeton, NJ.
Brundin, Lars. 1966. Transantarctic relationships and their significance as evidenced by chironomid midges with a monograph of the subfamilies Podonominae and Aphroteniinae and the austral Heptagyiae. *Kungliga Svenska Vetenskapsakademiens Handlingar*, Series 4, vol. 11, no. 1, 1–472.
Bryson, Bill. 2003. *A Short History of Nearly Everything.* Broadway Books, New York.
Buchanan, Mark. 2001. *Ubiquity: Why Catastrophes Happen.* Three Rivers Press, New York.
Burkhardt, Frederick, and Sydney Smith, eds. 1989. *The Correspondence of Charles Darwin*, vol. 5. Cambridge University Press, Cambridge, UK.
Butschi, Lorenz. 1989. Carnivorous plants of Auyantepui in Venezuela. *Carnivorous Plant Newsletter* 18, March, 15–18.
Caccone, Adalgisa, George Amato, Oliver C. Gratry, John Behler, and Jeffrey R. Powell. 1999. A molecular phylogeny of four endangered Madagascar tortoises based on mtDNA sequences. *Molecular Phylogenetics and Evolution* 12, 1–9.
Campbell, Hamish A., Matthew E. Watts, Scott Sullivan, Mark A. Read, Severine Choukroun, Steve R. Irwin, and Craig E. Franklin. 2010. Estuarine crocodiles ride surface currents to facilitate long-distance travel. *Journal of Animal Ecology* 79, 955–964.
Campbell, Hamish J., P. B. Andrews, A. G. Beu, A. R. Edwards, N. deB. Hornibrook, M. G. Laird, P. A. Maxwell, and W. A. Watters. 1988. Cretaceous-Cenozoic lithostratigraphy of the Chatham Islands. *Journal of the Royal Society of New Zealand* 18, 285–308.
Campbell, Hamish J., and Gerard D. Hutching. 2007. *In Search of Ancient New Zealand.* Penguin Books and GNS Science, Auckland, New Zealand.
Carleton, Michael D., and Storrs L. Olson. 1999. Amerigo Vespucci and the rat of Fernando de Noronha: a new genus and species of Rodentia (Muridae, Sigmodontinae) from a volcanic island off Brazil's continental shelf. *American Museum Novitates*, no. 3256, 1–59.
Carranza, S., and E. N. Arnold. 2003. Investigating the origin of transoceanic distributions: mtDNA shows *Mabuya* lizards (Reptilia, Scincidae) crossed the Atlantic twice. *Systematics and Biodiversity* 1, 275–282.
Carranza, S., E. N. Arnold, J. A. Mateo, and L. F. López-Jurado. 2000. Long-distance colonization and radiation in gekkonid lizards, *Tarentola* (Reptilia: Gekkonidae), revealed by mitochondrial DNA sequences. *Proceedings of the Royal Society of London B* 267, 637–649.
Carreño, Ana Luisa, and Javier Helenes. 2002. Geology and ages of the islands. Pp. 14–40 in T. J. Case, M. L. Cody, and E. Ezcurra, eds., *A New Island Biogeography of the Sea of Cortés.* Oxford University Press, New York.

Censky, Ellen J., Karim Hodge, and Judy Dudley. 1998. Overwater dispersal of lizards due to hurricanes. *Nature* 395, 556.

Chaimanee, Yaowalak, et al. 2012. Late Middle Eocene primate from Myanmar and the initial anthropoid colonization of Africa. *Proceedings of the National Academy of Sciences USA* 109, 10293–10297.

Chamberlin, Rollin T. 1928. Some of the objections to Wegener's theory. Pp. 83–87 in W. A. J. M. van Waterschoot van der Gracht, et al., eds., *Theory of Continental Drift: A Symposium on the Origin and Movement of Land Masses Both Inter-Continental and Intra-Continental, as Proposed by Alfred Wegener*. American Association of Petroleum Geologists, Tulsa, Oklahoma.

Chanderbali, Andre S., Henk van der Werff, and Susanne S. Renner. 2001. Phylogeny and historical biogeography of Lauraceae: evidence from the chloroplast and nuclear genomes. *Annals of the Missouri Botanical Garden* 88, 104–134.

Christenhusz, Maarten J. M., and Mark W. Chase. 2013. Biogeographical patterns of plants in the Neotropics—dispersal rather than plate tectonics is most explanatory. *Botanical Journal of the Linnean Society* 171, 277–286.

Clague, David A., Juan C. Braga, Davide Bassi, Paul D. Fullagar, Willem Renema, and Jody M. Webster. 2010. The maximum age of Hawaiian terrestrial lineages: geological constraints from Koko Seamount. *Journal of Biogeography* 37, 1022–1033.

Clarke, Julia A., Claudia P. Tambussi, Jorge I. Noriega, Gregory M. Erickson, and Richard A. Ketcham. 2005. Definitive fossil evidence for the extant avian radiation in the Cretaceous. *Nature* 433, 305–308.

Cleland, Carol E. 2002. Methodological and epistemic differences between historical science and experimental science. *Philosophy of Science* 69, 474–496.

Clement, Wendy L., Mark C. Tebbitt, Laura L. Forrest, Jaime E. Blair, Luc Brouillet, Torsten Eriksson, and Susan M. Swensen. 2004. Phylogenetic position and biogeography of *Hillebrandia sandwicensis* (Begoniaceae): a rare Hawaiian relict. *American Journal of Botany* 91, 905–917.

Colacino, Carmine, and John R. Grehan. 2003. Suppression at the frontiers of evolutionary biology: Léon Croizat's case. English translation, available at www.unibas.it/utenti/colacino/ccjrg-eng.pdf, of Colacino and Grehan, 2003, Ostracismo alle frontiere della biologia evoluzionistica: il caso Léon Croizat. Pp. 195–220 in M. Mamone Capria, ed., *Scienza e Democrazia*. Liguori Editore, Napoli.

Conan Doyle, Arthur. 2007 [1912]. *The Lost World*. Penguin, London.

Conway Morris, Simon. 1998. *The Crucible of Creation: The Burgess Shale and the Rise of Animals*. Oxford University Press, Oxford, UK.

Cook, Lyn G., and Michael D. Crisp. 2005. Not so ancient: the extant crown group of *Nothofagus* represents a post-Gondwanan radiation. *Proceedings of the Royal Society of London B* 272, 2535–2544.

Cooper, Alan, and Richard Fortey. 1999. Evolutionary explosions and the phylogenetic fuse. *Trends in Ecology and Evolution* 13, 151–156.

Cracraft, Joel. 1974. Continental drift and vertebrate distribution. *Annual Review of Ecology and Systematics* 5, 215–261.

———. 1975. Historical biogeography and Earth history: perspectives for a future synthesis. *Annals of the Missouri Botanical Garden* 62, 227–250.

Craw, Robin C. 1979. Generalized tracks and dispersal in biogeography: a response to R. M. McDowall. *Systematic Zoology* 28, 99–107.

Craw, Robin C., John R. Grehan, and Michael J. Heads. 1999. *Panbiogeography: Tracking the History of Life*. Oxford University Press, New York.

Crisp, Michael D., Mary T. K. Arroyo, Lyn G. Cook, Maria A. Gandolfo, Gregory J. Jordan, Matt S. McGlone, Peter H. Weston, Mark Westoby, Peter Wilf, and H. Peter Linder. 2009. Phylogenetic biome conservatism on a global scale. *Nature* 458, 754–756.

Crisp, Michael D., and Lyn G. Cook. 2013. How was the Australian flora assembled over the last 65 million years? A molecular phylogenetic perspective. *Annual Review of Ecology, Evolution, and Systematics* 44, in press.

Croizat, Léon. 1958. *Panbiogeography; or, An Introductory Synthesis of Zoogeography, Phytogeography, and Geology*. Vol. 1, *The New World*. Published by the author, Caracas, Venezuela.

———. 1962. *Space, Time, Form: The Biological Synthesis.* Published by the author, Caracas, Venezuela.
———. 1982. Vicariance/vicariism, panbiogeography, "vicariance biogeography," etc.: a clarification. *Systematic Zoology* 31, 291–304.
Croizat, Léon, Gareth Nelson, and Donn Eric Rosen. 1974. Centers of origin and related concepts. *Systematic Zoology* 23, 265–287.
Curtis, M. L., and D. M. Hyam. 1998. Late Palaeozoic to Mesozoic structural evolution of the Falkland Islands: a displaced segment of the Cape Fold Belt. *Journal of the Geological Society, London,* 155, 115–129.
Dalla Vecchia, Fabio M. 2009. *Tethyshadros insularis,* a new hadrosauroid dinosaur (Ornithischia) from the Upper Cretaceous of Italy. *Journal of Vertebrate Paleontology* 29, 1100–1116.
Daniels, Savel R. 2011. Reconstructing the colonisation and diversification history of the endemic freshwater crab (*Seychellum alluaudi*) in the granitic and volcanic Seychelles Archipelago. *Molecular Phylogenetics and Evolution* 61, 534–542.
Darlington, P. J., Jr. 1965. *Biogeography of the Southern End of the World.* Harvard University Press, Cambridge, MA.
———. 1970. A practical criticism of Hennig-Brundin "phylogenetic systematics" and Antarctic biogeography. *Systematic Zoology* 19, 1–18.
Darwin, Charles. 1989 [1839]. *Voyage of the Beagle.* Edited and abridged with an introduction by J. Browne and M. Neve. Penguin, London.
———. 1964 [1859]. *On the Origin of Species by Means of Natural Selection.* Facsimile of the First Edition. Harvard University Press, Cambridge, MA.
Darwin Correspondence Project, www.darwinproject.ac.uk/home.
Delfino, Massimo, Jeremy E. Martin, and Eric Buffetaut. 2008. A new species of *Acynodon* (Crocodylia) from the Upper Cretaceous (Santonian-Campanian) of Villaggio del Pescatore, Italy. *Palaeontology* 51, 1091–1106.
de Queiroz, Alan. 2005. The resurrection of oceanic dispersal in historical biogeography. *Trends in Ecology and Evolution* 20, 68–73.
de Queiroz, Alan, and Robin Lawson. 1994. Phylogenetic relationships of the garter snakes based on DNA sequence and allozyme variation. *Biological Journal of the Linnean Society* 53, 209–229.
———. 2008. A peninsula as an island: multiple forms of evidence for overwater colonization of Baja California by the garter snake *Thamnophis validus. Biological Journal of the Linnean Society* 95, 409–424.
de Queiroz, Alan, Robin Lawson, and Julio A. Lemos-Espinal. 2002. Phylogenetic relationships of North American garter snakes (*Thamnophis*) based on four mitochondrial genes: how much DNA sequence is enough? *Molecular Phylogenetics and Evolution* 22, 315–329.
Dick, Christopher W., Kobinah Abdul-Salim, and Eldredge Bermingham. 2003. Molecular systematic analysis reveals cryptic Tertiary diversification of a widespread tropical rain forest tree. *American Naturalist* 162, 691–703.
Donoghue, Michael J., and Brian R. Moore. 2003. Toward an integrative historical biogeography. *Integrative and Comparative Biology* 43, 261–270.
Donoghue, Michael J., and Stephen A. Smith. 2004. Patterns in the assembly of temperate forests around the Northern Hemisphere. *Philosophical Transactions of the Royal Society of London B* 359, 1633–1644.
Douzery, Emmanuel J., Frédéric Delsuc, Michael J. Stanhope, and Dorothée Huchon. 2003. Local molecular clocks in three nuclear genes: divergence times for rodents and other mammals and incompatibility among fossil calibrations. *Journal of Molecular Evolution* 57, suppl. 1, S201–S213.
Dray, William. 1957. *Laws and Explanation in History.* Oxford University Press, Oxford, UK.
Drummond, Alexei J., Simon Y. W. Ho, Matthew J. Phillips, and Andrew Rambaut. 2006. Relaxed phylogenetics and dating with confidence. *PLOS Biology* 4, e88.
Drummond, Alexei J., and Andrew Rambaut. 2007. BEAST: Bayesian evolutionary analysis by sampling trees. *BMC Evolutionary Biology* 7, 214.
du Toit, Alexander. 1937. *Our Wandering Continents.* Oliver and Boyd, London.
Ebach, Malte C. 2003. Area cladistics. *Biologist* 50, 169–172.
Ebach, Malte C., and Christopher J. Humphries. 2002. Cladistic biogeography and the art of

discovery. *Journal of Biogeography* 20, 427–444.

Enting, B., and L. Molloy. 1982. *The Ancient Islands*. Port Nicholson Press, Wellington, New Zealand.

Evans, Ben J., Rafe M. Brown, Jimmy A. McGuire, Jatna Supriatna, Noviar Andayani, Arvin Diesmos, Djoko Iskandar, Don J. Melnick, and David C. Cannatella. 2003. Phylogenetics of fanged frogs: testing biogeographical hypotheses at the interface of the Asian and Australian faunal zones. *Systematic Biology* 52, 794–819.

Evans, Susan E. 2003. At the feet of the dinosaurs: the early history and radiation of lizards. *Biological Reviews* 78, 513–551.

Fain, Matthew G., and Peter Houde. 2004. Parallel radiations in the primary clades of birds. *Evolution* 58, 2558–2573.

Felsenstein, Joseph. 2003. *Inferring Phylogenies*. Sinauer Associates, Sunderland, MA.

Fernandez, Frederick. 2005. Indonesian recalls 9-day sea ordeal. *The Star Online*, January 6, 2005, http://thestar.com.my/news/story.asp?file=/2005/1/6/nation/20050106070243&sec=nation.

Filardi, Christopher E., and Robert G. Moyle. 2005. Single origin of a pan-Pacific bird group and upstream colonization of Australasia. *Nature* 438, 216–219.

Flannery, Tim. 1994. *The Future Eaters*. Reed New Holland, Sydney.

Fleagle, John G. 1999. *Primate Adaptation and Evolution*, 2nd ed. Academic Press, San Diego.

Fleagle, John G., and Christopher C. Gilbert. 2006. The biogeography of primate evolution: the role of plate tectonics, climate and chance. Pp. 375–418 in S. M. Lehman and J. G. Fleagle, eds., *Primate Biogeography*. Springer, New York.

Fleming, Charles A. 1962. New Zealand biogeography: a paleontologist's approach. *Tuatara* 10, 53–108.

———. 1975. The geological history of New Zealand and its biota. Pp. 1–86 in G. Kuschel, ed., *Biogeography and Ecology in New Zealand*. Dr. W. Junk Publishers, The Hague.

Flora of New Zealand series webpage. Landcare Research. http://floraseries.landcareresearch.co.nz/pages/index.aspx.

Forest, Félix, and Mark W. Chase. 2009a. Magnoliids. Pp. 166–168 in S. B. Hedges and S. Kumar, eds., *The Timetree of Life*. Oxford University Press, Oxford, UK.

———. 2009b. Eudicots. Pp. 169–176 in S. B. Hedges and S. Kumar, eds., *The Timetree of Life*. Oxford University Press, Oxford, UK.

———. 2009c. Eurosid I. Pp. 188–196 in S. B. Hedges and S. Kumar, eds., *The Timetree of Life*. Oxford University Press, Oxford, UK.

———. 2009d. Eurosid II. Pp. 197–202 in S. B. Hedges and S. Kumar, eds., *The Timetree of Life*. Oxford University Press, Oxford, UK.

Fouquet, Antoine, Kévin Pineau, Miguel Trefaut Rodrigues, Julien Mailles, Jean-Baptiste Schneider, Raffael Ernst, and Maël Dewynter. 2013. Endemic or exotic: the phylogenetic position of the Martinique volcano frog *Allobates chalcopis* (Anura: Dendrobatidae) sheds light on its origin and challenges current conservation strategies. *Systematics and Biodiversity* 11, 87–101.

Francisco, Mercival, and Mauro Galetti. 2002. Aves como potenciais dispersoras de sementes de *Ocotea pulchella* Mart. (Lauraceae) numa área de vegetação de cerrado do sudeste brasileiro. *Revista Brasileira de Botanica* 25, 11–17.

Frankel, Henry. 1978. Arthur Holmes and continental drift. *British Journal for the History of Science* 11, 130–150.

———. 1981. The paleobiogeographical debate over the problem of disjunctively distributed life forms. *Studies in History and Philosophy of Science* 12, 211–259.

Frost, Darrel R. 2011. *Amphibian Species of the World: An Online Reference*. Version 5.5 (January 31, 2011). Electronic database accessible at http://research.amnh.org/vz/herpetology/amphibia/, American Museum of Natural History, New York.

Funk, Vicki A. 2004. Revolutions in historical biogeography. Pp. 647–657 in M. V. Lomolino, D. F. Sax, and J. H. Brown, eds., *Foundations of Biogeography*. University of Chicago Press, Chicago.

Funk, Vicki A., and Roy McDiarmid. 1988. An African connection? *Americas* 40, no. 5, September-October, 48–49.

Gaddis, John Lewis. 2002. *The Landscape of History*. Oxford University Press, Oxford, UK.
Gadow, Hans. 1913. *The Wanderings of Animals*. Cambridge University Press, London.
Gamble, T., A. M. Bauer, G. R. Colli, E. Greenbaum, T. R. Jackman, L. J. Vitt, and A. M. Simons. 2011. Coming to America: multiple origins of New World geckos. *Journal of Evolutionary Biology* 24, 231–244.
Gamerro, Juan Carlos, and Viviana Barreda. 2008. New fossil record of Lactoridaceae in southern South America: a palaeobiogeographical approach. *Botanical Journal of the Linnean Society* 158, 41–50.
Gans, Carl. 2005. Checklist and bibliography of the Amphisbaenia of the world. *Bulletin of the American Museum of Natural History*, no. 289, 1–130.
Gerlach, Justin, Catharine Muir, and Matthew D. Richmond. 2006. The first substantiated case of trans-oceanic tortoise dispersal. *Journal of Natural History* 40, 2403–2408.
Gibbs, George. 2006. *Ghosts of Gondwana: The History of Life in New Zealand*. Craig Potton Publishing, Nelson, New Zealand.
Gildart, Bert, and Jane Gildart. 2005. *Death Valley National Park: A Guide to Exploring the Great Outdoors*. Globe Pequot Press, Guilford, CT.
Giller, Paul S., Alan A. Myers, and Brett R. Riddle. 2004. Earth history, vicariance, and dispersal. Pp. 267–276 in M. V. Lomolino, D. F. Sax, and J. H. Brown, eds., *Foundations of Biogeography*. University of Chicago Press, Chicago.
Gillespie, Rosemary, Bruce G. Baldwin, Jonathan M. Waters, Ceridwen I. Fraser, Raisa Nikula, and George K. Roderick. 2012. Long-distance dispersal: a framework for hypothesis testing. *Trends in Ecology and Evolution* 27, 47–56.
Gillespie, Rosemary G., and George K. Roderick. 2002. Arthropods on islands: colonization, speciation, and conservation. *Annual Review of Entomology* 47, 595–632.
Giribet, Gonzalo, and Sarah L. Boyer. 2010. "Moa's Ark"; or, "Goodbye Gondwana": is the origin of New Zealand's terrestrial invertebrate fauna ancient, recent, or both? *Invertebrate Systematics* 24, 1–8.
Giribet, Gonzalo, Lars Vogt, Abel Pérez González, Prashant Sharma, and Adriano B. Kury. 2010. A multilocus approach to harvestmen (Arachnida: Opiliones) phylogeny with emphasis on biogeography and the systematics of Laniatores. *Cladistics* 26, 408–437.
Giribet, Gonzalo, et al. 2012. Evolutionary and biogeographical history of an ancient and global group of arachnids (Arachnida: Opiliones: Cyphophthalmi) with a new taxonomic arrangement. *Biological Journal of the Linnean Society* 105, 92–130.
Gladwell, Malcolm. 2000. *The Tipping Point: How Little Things Can Make a Big Difference*. Little, Brown, Boston.
Glazko, Galina V., and Masatoshi Nei. 2003. Estimation of divergence times for major lineages of primate species. *Molecular Biology and Evolution* 20, 424–434.
Gleick, James. 1987. *Chaos: Making a New Science*. Viking, New York.
Glor, Richard E., Jonathan B. Losos, and Allan Larson. 2005. Out of Cuba: overwater dispersal and speciation among lizards in the *Anolis carolinensis* subgroup. *Molecular Ecology* 14, 2419–2432.
Gohau, Gabriel. 1990. *A History of Geology*. Revised and translated by Albert V. Carozzi and Marguerite Carozzi. Rutgers University Press, New Brunswick, NJ.
Goldberg, Julia, Steven A. Trewick, and Adrian M. Paterson. 2008. Evolution of New Zealand's terrestrial fauna: a review of molecular evidence. *Philosophical Transactions of the Royal Society of London B* 363, 3319–3334.
Goldblatt, Peter, ed. 1993. *Biological Relationships Between Africa and South America*. Yale University Press, New Haven, CT.
Goodman, Morris, John Barnabas, Genji Matsuda, and William G. Moore. 1971. Molecular evolution in the descent of man. *Nature* 233, 604–613.
Goodman, Steven M., and Jonathan P. Benstead, eds. 2004. *The Natural History of Madagascar*. University of Chicago Press, Chicago.
Goswami, Anjali, and Paul Upchurch. 2010. The dating game: a reply to Heads (2010). *Zoologica Scripta* 39, 406–409.
Gould, Stephen Jay. 1977. *Ever Since Darwin: Reflections in Natural History*. W. W. Norton, New York.

———. 1986. The hardening of the Modern Synthesis. Pp. 71–93 in M. Grene, ed., *Dimensions of Darwinism*. Cambridge University Press, Cambridge, UK.

———. 1989. *Wonderful Life: The Burgess Shale and the Nature of History*. W. W. Norton, New York.

Grandcolas, Philippe, Jérôme Murienne, Tony Robillard, Laure Desutter-Grandcolas, Hervé Jourdan, Eric Guilbert, and Louis Deharveng. 2008. New Caledonia: a very old Darwinian island? *Philosophical Transactions of the Royal Society of London B* 363, 3309–3317.

Grayson, Donald K. 1993. *The Desert's Past: A Natural Prehistory of the Great Basin*. Smithsonian Institution Press, Washington, DC.

Greene, Brian. 1999. *The Elegant Universe: Superstrings, Hidden Dimensions, and the Quest for the Ultimate Theory*. W. W. Norton, New York.

Greene, Harry W. 1997. *Snakes: The Evolution of Mystery in Nature*. University of California Press, Berkeley.

Guo, Peng, Qin Liu, Yan Xu, Ke Jiang, Mian Hou, Li Ding, R. Alexander Pyron, and Frank T. Burbrink. 2012. Out of Asia: Natricine snakes support the Cenozoic Beringian dispersal hypothesis. *Molecular Phylogenetics and Evolution* 63, 825–833.

Guppy, Henry Brougham. 1906. *Observations of a Naturalist in the Pacific Between 1896 and 1899*. Vol. 2, *Plant Dispersal*. Macmillan, London.

Hallam, Anthony. 1973. *A Revolution in the Earth Sciences*. Clarendon Press, Oxford, UK.

———. 1994. *An Outline of Phanerozoic Biogeography*. Oxford University Press, Oxford, UK.

Handwerk, Brian. 2004. "Lost world" mesas showcase South America's evolution. *National Geographic News*, February 20, 2004, http://news.nationalgeographic.com/news/pf/20848173.html.

Harding, Paul, Carolyn Bain, and Neal Bedford. 2002. *New Zealand*, 11th ed. Lonely Planet Publications, Melbourne.

Harshman, John, et al. 2008. Phylogenomic evidence for multiple losses of flight in ratite birds. *Proceedings of the National Academy of Sciences USA* 105, 13462–13467.

Hawaii Audubon Society. 2005. *Hawaii's Birds*, 6th ed. Hawaii Audubon Society, Honolulu.

Hay, Jennifer M., Stephen D. Sarre, David M. Lambert, Fred W. Allendorf, and Charles H. Daugherty. 2010. Genetic diversity and taxonomy: a reassessment of species designation in tuatara (*Sphenodon*: Reptilia). *Conservation Genetics* 11, 1063–1081.

Heads, Michael. 2005. Dating nodes on molecular phylogenies: a critique of molecular biogeography. *Cladistics* 21, 62–78.

———. 2008. Panbiogeography of New Caledonia, south-west Pacific: basal angiosperms on basement terranes, ultramafic endemics inherited from volcanic island arcs and old taxa endemic to young islands. *Journal of Biogeography* 35, 2153–2175.

———. 2009. Inferring biogeographic history from molecular phylogenies. *Biological Journal of the Linnean Society* 98, 757–774.

———. 2010. Evolution and biogeography of primates: a new model based on molecular phylogenetics, vicariance and plate tectonics. *Zoologica Scripta* 39, 107–127.

———. 2011. Old taxa on young islands: a critique of the use of island age to date island-endemic clades and calibrate phylogenies. *Systematic Biology* 60, 204–218.

Hedges, S. Blair. 2006. Paleogeography of the Antilles and origin of West Indian terrestrial vertebrates. *Annals of the Missouri Botanical Garden* 93, 231–244.

Hedges, S. Blair, Arnaud Couloux, and Nicolas Vidal. 2009. Molecular phylogeny, classification, and biogeography of West Indian racer snakes of the tribe Alsophiini (Squamata, Dipsadidae, Xenodontinae). *Zootaxa* 2067, 1–28.

Hedges, S. Blair, and Sudhir Kumar, eds. 2009a. *The Timetree of Life*. Oxford University Press, Oxford UK.

———. 2009b. Discovering the timetree of life. Pp. 3–18 in S. B. Hedges and S. Kumar, eds., *The Timetree of Life*. Oxford University Press, Oxford UK.

Heenan, P. B., A. D. Mitchell, P. J. de Lange, J. Keeling, and A. M. Paterson. 2010. Late-Cenozoic origin and diversification of Chatham Islands endemic plant species revealed by analyses of DNA sequence data. *New Zealand Journal of Botany* 48, 83–136.

Heesy, Christopher P., Nancy J. Stevens, and Karen E. Samonds. 2006. Biogeographic origins of primate higher taxa. Pp. 419–437 in S. M. Lehman and J. G. Fleagle, eds., *Primate*

Biogeography. Springer, New York.

Heinicke, Matthew P., William E. Duellman, and S. Blair Hedges. 2007. Major Caribbean and Central American frog faunas originated by ancient oceanic dispersal. *Proceedings of the National Academy of Sciences USA* 104, 10092–10097.

Hennig, Willi. 1966. *Phylogenetic Systematics*. University of Illinois Press, Urbana.

Hess, H. H. 1946. Drowned ancient islands of the Pacific basin. *American Journal of Science* 244, 772–791.

———. 1962. History of ocean basins. Pp. 599–620 in A. E. J. Engel, H. L. James, and B. F. Leonard, eds., *Petrologic Studies: A Volume to Honor A. F. Buddington*. Geological Society of America, New York.

Higham, Thomas, Atholl Anderson, and Chris Jacomb. 1999. Dating the first New Zealanders: the chronology of Wairau Bar. *Antiquity* 73, 420–427.

Ho, Simon Y. W., and Matthew J. Phillips. 2009. Accounting for calibration uncertainty in phylogenetic estimation of evolutionary divergence times. *Systematic Biology* 58, 367–380.

Hoberg, Eric P., Nancy L. Alkire, Alan de Queiroz, and Arlene Jones. 2001. Out of Africa: origins of the *Taenia* tapeworms in humans. *Proceedings of the Royal Society of London B* 268, 781–787.

Holmes, Arthur. 1928. Continental drift. *Nature* 122, 431–433.

———. 1944. *Principles of Physical Geology*. Thomas Nelson and Sons, Edinburgh.

Houle, Alain. 1999. The origin of platyrrhines: an evaluation of the Antarctic scenario and the floating island model. *American Journal of Physical Anthropology* 109, 541–559.

Huelsenbeck, John P., Michael E. Alfaro, and Marc A. Suchard. 2011. Biologically inspired phylogenetic models strongly outperform the no common mechanism model. *Systematic Biology* 60, 225–232.

Hughes, Patrick. 1994. The meteorologist who started a revolution. *Weatherwise* 47, April 1, 1994, 29–35.

Hugot, Jean-Pierre. 1998. Phylogeny of Neotropical monkeys: the interplay of morphological, molecular, and parasitological data. *Molecular Phylogenetics and Evolution* 9, 408–413.

Hull, David L. 1988. *Science as a Process: An Evolutionary Account of the Social and Conceptual Development of Science*. University of Chicago Press, Chicago.

———. 2009. Leon Croizat: a radical biogeographer. Pp. 194–212 in O. Harman and M. R. Dietrich, eds., *Rebels, Mavericks, and Heretics in Biology*. Yale University Press, New Haven, CT.

Humphries, Christopher J., and Lynne R. Parenti. 1989. *Cladistic Biogeography*. Oxford University Press, Oxford, UK.

Inger, Robert F., and Harold K. Voris. 2001. The biogeographical relations of the frogs and snakes of Sundaland. *Journal of Biogeography* 28, 863–891.

Iturralde-Vinent, Manuel A. 2006. Meso-Cenozoic Caribbean paleogeography: implications for the historical biogeography of the region. *International Geology Review* 48, 791–827.

Iturralde-Vinent, Manuel A., and R. D. E. MacPhee. 1999. Paleogeography of the Caribbean region: implications for Cenozoic biogeography. *Bulletin of the American Museum of Natural History*, no. 238.

Jablonski, David. 1994. Extinctions in the fossil record [with discussion]. *Philosophical Transactions of the Royal Society of London B* 344, 11–17.

Jackson, Donald Dale. 1985. Scientists zero in on the strange "lost world" of Cerro de la Neblina. *Smithsonian* 16, no. 2, 51–63.

Jacobs, David S. 1994. Distribution and abundance of the endangered Hawaiian hoary bat, *Lasiurus cinereus semotus*, on the island of Hawai'i. *Pacific Science* 48, 193–200.

James, Harold L. 1973. Harry Hammond Hess, 1906–1969. *Biographical Memoirs, National Academy of Sciences*, 108–128.

Jockusch, Elizabeth L., and David B. Wake. 2002. Falling apart and merging: diversification of slender salamanders (Plethodontidae: *Batrachoseps*) in the American West. *Biological Journal of the Linnean Society* 76, 361–391.

Johnson, Donald Lee. 1980. Problems in the land vertebrate zoogeography of certain islands and the swimming powers of elephants. *Journal of Biogeography* 7, 383–398.

Jones, Marc E. H., Alan J. D. Tennyson, Jennifer P. Worthy, Susan E. Evans, and Trevor H.

Worthy. 2009. A sphenodontine (Rhynchocephalia) from the Miocene of New Zealand and palaeobiogeography of the tuatara (*Sphenodon*). *Proceedings of the Royal Society of London B* 276, 1385–1390.

Jones, Sam. 2005. Indonesian survivor recalls eight days at sea. *Guardian* online edition, January 6, 2005, www.guardian.co.uk/world/2005/jan/06/tsunami2004.samjones.

Jordan, Greg J. 2001. An investigation of long-distance dispersal based on species native to both Tasmania and New Zealand. *Australian Journal of Botany* 49, 333–340.

Joseph, Josy. 1980. The nutmeg—its botany, agronomy, production, composition, and uses. *Journal of Plantation Crops* 8, 61–72.

Kar, A., B. Weaver, J. Davidson, and M. Colucci. 1998. Origin of differentiated volcanic and plutonic rocks from Ascension Island, South Atlantic Ocean. *Journal of Petrology* 39, 1009–1024.

Kay, R. F., B. A. Williams, C. F. Ross, M. Takai, and N. Shigehara. 2004. Anthropoid origins: a phylogenetic analysis. Pp. 91–135 in C. F. Ross and R. F. Kay, eds., *Anthropoid Origins: New Visions*. Kluwer Academic Press, New York.

Kennedy, Liam, Paul S. Ell, E. M. Crawford, and L. A. Clarkson. 2000. *Mapping the Great Irish Famine: An Atlas of the Famine Years*. Four Courts Press, Dublin.

Key, Gillian, and Edgar Muñoz Heredia. 1994. Distribution and current status of rodents in the Galápagos. *Noticias de Galápagos*, no. 53, 21–25.

Kishino, Hirohisa, Jeffrey L. Thorne, and William J. Bruno. 2001. Performance of a divergence time estimation method under a probabilistic model of rate evolution. *Molecular Biology and Evolution* 18, 352–361.

Knapp, Michael, Ragini Mudaliar, David Havell, Steven J. Wagstaff, and Peter J. Lockhart. 2007. The drowning of New Zealand and the problem of *Agathis*. *Systematic Biology* 56, 826–870.

Knapp, Michael, Karen Stöckler, David Havell, Frédéric Delsuc, Federico Sebastiani, and Peter J. Lockhart. 2005. Relaxed molecular clock provides evidence for long-distance dispersal of *Nothofagus* (southern beech). *PLoS Biology* 3, no. 1, e14.

Koepfli, Klaus-Peter, Matthew E. Gompper, Eduardo Eizirik, Cheuk-Chung Ho, Leif Linden, Jesus E. Maldonado, and Robert K. Wayne. 2007. Phylogeny of the Procyonidae (Mammalia: Carnivora): molecules, morphology and the Great American Interchange. *Molecular Phylogenetics and Evolution* 43, 1076–1095.

Krause, David W. 2003. Late Cretaceous vertebrates from Madagascar: a window into Gondwanan biogeography at the end of the Age of Dinosaurs. Pp. 40–47 in S. M. Goodman and J. P. Benstead, eds., *The Natural History of Madagascar*. University of Chicago Press, Chicago.

Krause, David W., Patrick M. O'Connor, Kristina Curry Rogers, Scott D. Sampson, Gregory A. Buckley, and Raymond R. Rogers. 2006. Late Cretaceous terrestrial vertebrates from Madagascar: implications for Latin American biogeography. *Annals of the Missouri Botanical Garden* 93, 178–208.

Kuhn, Thomas S. 1970. *The Structure of Scientific Revolutions*, 2nd ed. University of Chicago Press, Chicago.

Kumar, Sudhir, Alan Filipski, Vinod Swarna, Alan Walker, and S. Blair Hedges. 2005. Placing confidence limits on the molecular age of the human-chimpanzee divergence. *Proceedings of the National Academy of Sciences USA* 102, 18842–18847.

Laguna, Marcia Maria, Renata Cecília Amaro, Tamí Mott, Yatiyo Yonenaga-Yassuda, and Miguel Trefaut Rodrigues. 2010. Karyological study of *Amphisbaena ridleyi* (Squamata, Amphisbaenidae), an endemic species of the archipelago of Fernando de Noronha, Pernambuco, Brazil. *Genetics and Molecular Biology* 33, 57–61.

Larink, O. 1972. Zur Struktur der Blastoderm-Cuticula von *Petrobius brevistylis* und *P. maritimus* (Thysanura, Insecta). *Cytobiologie* 5, 422–426.

Lavin, Matt, Brian P. Schrire, Gwilym Lewis, R. Toby Pennington, Alfonso Delgado-Salinas, Mats Thulin, Colin E. Hughes, Angela Beyra Matos, and Martin F. Wojciechowski. 2004. Metacommunity process rather than continental tectonic history better explains geographically structured phylogenies in legumes. *Philosophical Transactions of the Royal Society of London B* 359, 1509–1522.

Lavin, Matt, Mats Thulin, Jean-Noel Labat, and R. Toby Pennington. 2000. Africa, the odd man out: molecular biogeography of dalbergioid legumes (Fabaceae) suggests otherwise. *Systematic Botany* 25, 449–467.

Lawrence, David M. 2002. *Upheaval from the Abyss: Ocean Floor Mapping and the Earth Science Revolution*. Rutgers University Press, New Brunswick, NJ.

Lee, Daphne E., William G. Lee, and Nick Mortimer. 2001. Where and why have all the flowers gone? Depletion and turnover in the New Zealand Cenozoic angiosperm flora in relation to palaeogeography and climate. *Australian Journal of Botany* 49, 341–356.

Lessios, H. A. 2008. The Great American Schism: divergence of marine organisms after the rise of the Central American Isthmus. *Annual Review of Ecology, Evolution, and Systematics* 39, 63–91.

Levin, Donald A. 2006. Ancient dispersals, propagule pressure, and species selection in flowering plants. *Systematic Botany* 31, 443–448.

Lewis, Cherry L. E. 2002. Arthur Holmes: an ingenious geoscientist. *GSA Today*, March, 16–17.

Lord, Rexford D. 2007. *Mammals of South America*. Johns Hopkins University Press, Baltimore.

Lovejoy, N. P., S. P. Mullen, G. A. Sword, R. F. Chapman, and R. G. Harrison. 2006. Ancient trans-Atlantic flight explains locust biogeography: molecular phylogenetics of *Schistocerca*. *Proceedings of the Royal Society of London B* 273, 767–774.

Lundberg, John G. 1993. African–South American freshwater fish clades and continental drift: problems with a paradigm. Pp. 157–199 in P. Goldblatt, ed., *Biological Relationships Between Africa and South America*. Yale University Press, New Haven, CT.

Macdougall, J. D. 1996. *A Short History of Planet Earth: Mountains, Mammals, Fire, and Ice*. John Wiley and Sons, New York.

Macey, J. Robert, Theodore J. Papenfuss, Jennifer V. Kuehl, H. Mathew Fourcade, and Jeffrey L. Boore. 2004. Phylogenetic relationships among amphisbaenian reptiles based on complete mitochondrial genomic sequences. *Molecular Phylogenetics and Evolution* 33, 22–31.

MacPhee, R. D. E., M. A. Iturralde-Vinent, and Eugene S. Gaffney. 2003. Domo de Zaza, an early Miocene vertebrate locality in south-central Cuba, with notes on the tectonic evolution of Puerto Rico and the Mona Passage. *American Museum Novitates*, no. 3394, 1–42.

Malatesta, Parisina. 1996. Lost land of water and rock. *Americas* 48, November/December, 28–37.

Mann, Charles C. 2011. *1493: Uncovering the New World Columbus Created*. Vintage, New York.

Mann, Paul, Robert D. Rogers, and Lisa Gahagan. 2007. Overview of plate tectonic history and its unresolved problems. Pp. 201–238 in J. Bundschuh and G. E. Alvarado, eds., *Central America: Geology, Resources, Hazards*, vol. 1. Taylor and Francis, London.

Marcondes-Machado, Luiz Octavio. 2002. Comportamento alimentar de aves em *Miconia rubiginosa* (Melastomataceae) em fragmento de cerrado, São Paulo. *Iheringia, Série Zoologia, Porto Alegre* 92, 97–100.

Marks, Kathy. 2009. Henry the tuatara is a dad at 111. *The Independent*, January 26, 2009.

Marshall, Charles R., Elizabeth C. Raff, and Rudolf A. Raff. 1994. Dollo's law and the death and resurrection of genes. *Proceedings of the National Academy of Sciences USA* 91, 12283–12287.

Martin, Andrew P., and Stephen R. Palumbi. 1993. Body size, metabolic rate, generation time, and the molecular clock. *Proceedings of the National Academy of Sciences USA* 90, 4087–4091.

Martin, P. G., and J. M. Dowd. 1988. A molecular evolutionary clock for angiosperms. *Taxon* 37, 364–377.

Matthew, William Diller. 1915. Climate and evolution. *Annals of the New York Academy of Sciences* 24, 171–318.

Mausfeld, Patrick, Andreas Schmitz, Wolfgang Böhme, Bernhard Misof, Davor Vrcibradic, and Carlos Frederico Duarte Rocha. 2002. Phylogenetic affinities of *Mabuya atlantica* Schmidt, 1945, endemic to the Atlantic Ocean archipelago of Fernando de Noronha

(Brazil): necessity of partitioning the genus *Mabuya* Fitzinger, 1826 (Scincidae: Lygosominae). *Zoologischer Anzeiger* 241, 281–293.

May, Robert M. 1974. Biological populations with nonoverlapping generations: stable points, stable cycles, and chaos. *Science* 186, 645–647.

Mayr, Ernst. 1952. Conclusions. Pp. 255–258 in E. Mayr, ed., *The Problem of Land Connections Across the South Atlantic, with Special Reference to the Mesozoic. Bulletin of the American Museum of Natural History* 99, 79–258.

———. 1982. Review of *Vicariance Biogeography. The Auk* 99, 618–620.

———. 1983. *The Growth of Biological Thought: Diversity, Evolution, and Inheritance*. Harvard University Press, Cambridge, MA.

Mayr, Gerald. 2003. Phylogeny of Early Tertiary swifts and hummingbirds (Aves: Apodiformes). *The Auk* 120, 145–151.

McAllister, James W. 1998. Is beauty a sign of truth in scientific theories? *American Scientist* 86, 174–183.

McAtee, W. L. 1914. Birds transporting food supplies. *The Auk* 31, 404–405.

McCarthy, Dennis. 2005. Biogeography and scientific revolutions. *The Systematist*, no. 25, 3–12.

———. 2009. *Here Be Dragons*. Oxford University Press, New York.

McCoy, Roger M. 2006. *Ending in Ice: The Revolutionary Idea and Tragic Expedition of Alfred Wegener*. Oxford University Press, Oxford, UK.

McDowall, Robert M. 1978. Generalized tracks and dispersal in biogeography. *Systematic Zoology* 27, 88–104.

———. 2005. Falkland Islands biogeography: converging trajectories in the South Atlantic Ocean. *Journal of Biogeography* 32, 49–62.

McDowall, Robert M., R. M. Allibone, and W. L. Chadderton. 2005. *Falkland Islands Freshwater Fishes: A Natural History*. Falklands Conservation, London.

McFarlane, Donald A., and Joyce Lundberg. 2002. A Middle Pleistocene age and biogeography for the extinct rodent *Megalomys curazensis* from Curaçao, Netherlands, Antilles. *Caribbean Journal of Science* 38, 278–281.

McFarlane, D. A., J. Lundberg, and A. G. Fincham. 2002. A Late Quaternary paleoecological record from caves of southern Jamaica, West Indies. *Journal of Cave and Karst Studies* 64, 117–125.

McLoughlin, Stephen. 2001. The breakup history of Gondwana and its impact on pre-Cenozoic floristic provincialism. *Australian Journal of Botany* 49, 271–300.

McNeill, William H. 1999. How the potato changed the world's history. *Social Research* 66, 67–83.

McPhee, John. 1981. *Basin and Range*. Farrar, Straus and Giroux, New York.

Measey, G. John, Miguel Vences, Robert C. Drewes, Ylenia Chiari, Martim Melo, and Bernard Bourles. 2007. Freshwater paths across the ocean: molecular phylogeny of the frog *Ptychadena newtoni* gives insight into amphibian colonization of oceanic islands. *Journal of Biogeography* 34, 7–20.

Meredith, Robert, Evon Hekkala, George Amato, and John Gatesy. 2011a. A phylogenetic hypothesis for *Crocodylus* (Crocodylia) based on mitochondrial DNA: evidence for a trans-Atlantic voyage from Africa to the New World. *Molecular Phylogenetics and Evolution* 60, 183–191.

Meredith, Robert W., et al. 2011b. Impacts of the Cretaceous Terrestrial Revolution and the KPg Extinction on extant mammal diversification. *Science* 334, 521–524.

Metcalf, Lawrie. 2002. *A Photographic Guide to Trees of New Zealand*. New Holland Publishers, Auckland, New Zealand.

Mildenhall, Dallas C. 1980. New Zealand Late Cretaceous and Cenozoic plant biogeography: a contribution. *Palaeogeography, Palaeoclimatology, Palaeoecology* 31, 197–233.

Miller, Hilary C., and David M. Lambert. 2006. A molecular phylogeny of New Zealand's *Petroica* (Aves: Petroicidae) species based on mitochondrial DNA sequences. *Molecular Phylogenetics and Evolution* 40, 844–855.

Miller, Stanley L. 1953. A production of amino acids under possible primitive Earth conditions. *Science* 117, 528–529.

Miura, Osamu, Mark E. Torchin, Eldredge Bermingham, David K. Jacobs, and Ryan F. Hechinger. 2012. Flying shells: historical dispersal of marine snails across Central America. *Proceedings of the Royal Society of London B* 279, 1061–1067.

Moore, Peter D., Judith A. Webb, and Margaret E. Collinson. 1991. *Pollen Analysis*, 2nd ed. Blackwell Scientific, Oxford, UK.

Morgan, Gregory J. 1998. Emile Zuckerkandl, Linus Pauling, and the molecular evolutionary clock. *Journal of the History of Biology* 31, 155–178.

Morrone, Juan J., and Paula Posadas. 2005. Falklands: facts and fiction. *Journal of Biogeography* 32, 2183–2187.

Morwood, M. J., and W. L. Jungers. 2009. Conclusions: implications of the Liang Bua excavations for hominin evolution and biogeography. *Journal of Human Evolution* 57, 640–648.

Mullis, Kary. 1998. *Dancing Naked in the Mind Field*. Pantheon, New York.

Mullis, K. B., and F. A. Faloona. 1987. Specific synthesis of DNA *in vitro* via a polymerase catalyzed chain reaction. *Methods in Enzymology* 55, 335–350.

Murienne, Jerome, Gregory D. Edgecombe, and Gonzalo Giribet. 2010. Including secondary structure, fossils and molecular dating in the centipede tree of life. *Molecular Phylogenetics and Evolution* 57, 301–313.

Nathan, Ran, Frank M. Schurr, Orr Spiegel, Ofer Steinitz, Ana Trakhtenbrot, and Asaf Tsoar. 2008. Mechanisms of long-distance seed dispersal. *Trends in Ecology and Evolution* 23, 638–647.

Near, Thomas J., and Michael J. Sanderson. 2004. Assessing the quality of molecular divergence time estimates by fossil calibrations and fossil-based model selection. *Philosophical Transactions of the Royal Society of London B* 359, 1477–1483.

Nelson, Gareth. 1973. Comments on Leon Croizat's biogeography. *Systematic Zoology* 22, 312–320.

———. 1975. Review of *Biogeography and Ecology in New Zealand* by G. Kuschel, ed. *Systematic Zoology* 24, 494–495.

———. 1977. Review of *Biogeografia Analítica y Sintética ("Panbiogeografía") de las Américas* by L. Croizat-Chaley. *Systematic Zoology* 26, 449–452.

———. 1978a. From Candolle to Croizat: comments on the history of biogeography. *Journal of the History of Biology* 11, 269–305.

———. 1978b. Ontogeny, phylogeny, paleontology, and the biogenetic law. *Systematic Zoology* 27, 324–345.

———. 1978c. Refuges, humans, and vicariance (Review of *Biogeographie et Evolution en Amerique Tropicale* and *Human Biogeography*). *Systematic Zoology* 27, 484–487.

———. 2000. Ancient perspectives and influence in the theoretical systematics of a bold fisherman. Pp. 9–23 in P. L. Forey, B. G. Gardiner, and C. J. Humphries, eds., *Colin Patterson (1933–1998): A Celebration of His Life*. Special Issue no. 2, Linnean Society of London.

———. Unpublished manuscript. Cladistics at an earlier time.

Nelson, Gareth, and Pauline Y. Ladiges. 2001. Gondwana, vicariance biogeography and the New York School revisited. *Australian Journal of Botany* 49, 389–409.

———. 2009. Biogeography and the molecular dating game: a futile revival of phenetics? *Bulletin de la Société géologique de France* 180, 39–43.

Nelson, Gareth, and Norman Platnick. 1981. *Systematics and Biogeography: Cladistics and Vicariance*. Columbia University Press, New York.

Ni, Xijun, Daniel L. Gebo, Marian Dagosto, Jin Meng, Paul Tafforeau, John J. Flynn, and K. Christopher Beard. 2013. The oldest known primate skeleton and early haplorhine evolution. *Nature* 498, 60–64.

Nicholson, Kirsten E., Richard E. Glor, Jason J. Kolbe, Allan Larson, S. Blair Hedges, and Jonathan B. Losos. 2005. Mainland colonization by island lizards. *Journal of Biogeography* 32, 929–938.

Nielsen, Stuart V., Aaron M. Bauer, Todd R. Jackman, Rod A. Hitchmough, and Charles H. Daugherty. 2011. New Zealand geckos (Diplodactylidae): cryptic diversity in a post-Gondwanan lineage with trans-Tasman affinities. *Molecular Phylogenetics and Evolution* 59, 1–22.

Noonan, Brice P., and Paul T. Chippindale. 2006. Vicariant origin of Malagasy reptiles supports Late Cretaceous Antarctic land bridge. *American Naturalist* 168, 730–741.

Noonan, Brice P., and Jack W. Sites, Jr. 2010. Tracing the origins of iguanid lizards and boine snakes of the Pacific. *American Naturalist* 175, 61–72.

Norval, Gerrut, Jean-Jay Mao, Hsin-Pin Chu, and Lee-Chang Chen. 2002. A new record of an introduced species, the brown anole (*Anolis sagrei*) (Duméril & Bibron, 1837), in Taiwan. *Zoological Studies* 41, 332–336.

Nunn, Nathan, and Nancy Qian. 2011. The potato's contribution to population and urbanization: evidence from a historical experiment. *Quarterly Journal of Economics* 126, 593–650.

O'Grady, Patrick, and Rob DeSalle. 2008. Out of Hawaii: the origin and biogeography of the genus *Scaptomyza* (Diptera: Drosophilidae). *Biology Letters* 4, 195–199.

O'Grady, P. M., K. N. Magnacca, and R. T. Lapoint. 2010. Taxonomic relationships within the endemic Hawaiian Drosophilidae (Insecta: Diptera). In N. L. Evenhuis and L. G. Eldredge, eds., *Records of the Hawaii Biological Survey for 2008*. Bishop Museum Occasional Papers 108, 1–34.

O'Hara, Robert J. 1988. Homage to Clio; or, toward an historical philosophy for evolutionary biology. *Systematic Zoology* 37, 142–155.

O'Leary, Maureen A., et al. 2013. The placental mammal ancestor and the post-K-Pg radiation of placental mammals. *Science* 339, 662–667.

Olmstead, Richard G., and Jeffrey D. Palmer. 1997. Implications for the phylogeny, classification, and biogeography of *Solanum* from cpDNA restriction site variation. *Systematic Botany* 22, 19–29.

Oreskes, Naomi. 1988. The rejection of continental drift. *Historical Studies in the Physical and Biological Sciences* 18, 311–348.

Osi, Attila, Richard J. Butler, and David B. Weishampel. 2010. A Late Cretaceous ceratopsian dinosaur from Europe with Asian affinities. *Nature* 465, 466–468.

Overbye, Dennis. 2013. Universe as an infant: fatter than expected and kind of lumpy. *New York Times*, March 21, 2013.

Page, Roderic D. M. 1989. New Zealand and the new biogeography. *New Zealand Journal of Zoology* 16, 471–483.

Palmer, D. D. 2003. *Hawaiʻi's Ferns and Fern Allies*. University of Hawaii Press, Honolulu.

Parada, Andrés, Ulyses F. J. Pardiñas, Jorge Salazar-Bravo, Guillermo D'Elía, and R. Eduardo Palma. 2013. Dating an impressive Neotropical radiation: molecular time estimates for the Sigmodontinae (Rodentia) provide insights into its historical biogeography. *Molecular Phylogenetics and Evolution* 66, 960–968.

Parenti, Lynne R. 2006. Common cause and historical biogeography. Pp. 61–82 in M. Ebach and R. Tangney, eds., *Biogeography in a Changing World*. CRC Press, Boca Raton, FL.

Parenti, Lynne R., and Malte C. Ebach. 2009. *Comparative Biogeography: Discovering and Classifying Biogeographical Patterns of a Dynamic Earth*. University of California Press, Berkeley.

———. 2013. Evidence and hypothesis in biogeography. *Journal of Biogeography* 40, 813–820.

Paterson, Adrian, Steve Trewick, Karen Armstrong, Julia Goldberg, and Anthony Mitchell. 2006. Recent and emergent: molecular analysis of the biota supports a young Chatham Islands. Pp. 27–29 in S. Trewick and M. J. Phillips, eds., *Geology and Genes III* (extended abstracts for papers presented at the Geogenes III Conference, Wellington, July 14, 2006). Geological Society of New Zealand Miscellaneous Publication 121.

Patterson, Colin. 1997. Peter Humphry Greenwood. 21 April 1927–3 March 1995. *Biographical Memoirs of Fellows of the Royal Society* 43, 194–213.

Pelletier, Bernard. 2006. Geology of the New Caledonia region and its implications for the study of the New Caledonian biodiversity. Pp. 19–32 in C. Payri and B. Richer de Forges, eds., *Compendium of Marine Species from New Caledonia*. Documents Scientifiques et Techniques IRD, II 7, 2nd ed. Institut de Recherche pour le Développement, Nouméa, France.

Pennington, R. Toby, Quentin C. B. Cronk, and James A. Richardson. 2004. Introduction and synthesis: plant phylogeny and the origin of major biomes. *Philosophical Transactions of the Royal Society of London B* 359, 1455–1464.

Pennington, R. Toby, and Christopher W. Dick. 2004. The role of immigrants in the assembly of the South American rainforest tree flora. *Philosophical Transactions of the Royal Society of London B* 359, 1611–1622.

Phillips, Matthew J., Gillian C. Gibb, Elizabeth A. Crimp, and David Penny. 2010. Tinamous and moa flock together: mitochondrial genome sequence analysis reveals independent losses of flight among ratites. *Systematic Biology* 59, 90–107.

Pillon, Yohan. 2012. Time and tempo of diversification in the flora of New Caledonia. *Botanical Journal of the Linnean Society* 170, 288–298.

Pole, Mike. 1993. Keeping in touch: vegetation prehistory on both sides of the Tasman. *Australian Systematic Botany* 6, 387–397.

———. 1994. The New Zealand flora—entirely long-distance dispersal? *Journal of Biogeography* 21, 625–635.

Pole, Mike, C. A. Landis, H. J. Campbell, J. G. Begg, D. C. Mildenhall, A. M. Paterson, and S. A. Trewick. 2010. Discussion of "The Waipounamu Erosion Surface: questioning the antiquity of the New Zealand land surface and terrestrial fauna and flora." *Geological Magazine* 147, 151–155.

Pollan, Michael. 2001. *The Botany of Desire*. Random House, New York.

Popper, Karl. 2002 [1965]. *Conjectures and Refutations: The Growth of Scientific Knowledge*, 2nd ed. Routledge, London.

Pough, F. Harvey, Robin M. Andrews, John E. Cadle, Martha L. Crump, Alan H. Savitzky, and Kentwood D. Wells. 1998. *Herpetology*. Prentice Hall, Upper Saddle River, NJ.

Poux, Céline, Pascale Chevret, Dorothée Huchon, Wilfried W. de Jong, and Emmanuel J. P. Douzery. 2006. Arrival and diversification of caviomorph rodents and platyrrhine primates in South America. *Systematic Biology* 55, 228–244.

Powers, Sidney. 1911. Floating islands. *Popular Science Monthly* 79, 303–307.

Pramuk, Jennifer B., Tasia Robertson, Jack W. Sites Jr., and Brice P. Noonan. 2008. Around the world in 10 million years: biogeography of the nearly cosmopolitan true toads (Anura: Bufonidae). *Global Ecology and Biogeography* 17, 72–83.

Pregill, Gregory. 1981. An appraisal of the vicariance hypothesis of Caribbean biogeography and its application to West Indian terrestrial vertebrates. *Systematic Zoology* 30, 147–155.

Prevosti, Francisco J., and Ulyses F. J. Pardiñas. 2009. Comment on "The oldest South American Cricetidae (Rodentia) and Mustelidae (Carnivora): Late Miocene faunal turnover in central Argentina and the Great American Biotic Interchange," by D. H. Verzi and C. J. Montalvo [*Palaeogeography, Palaeoclimatology, Palaeoecology* 267 (2008), 284–291]. *Palaeogeography, Palaeoclimatology, Palaeoecology* 280, 543–547.

Price, Jonathan P., and David A. Clague. 2002. How old is the Hawaiian biota? Geology and phylogeny suggest recent divergence. *Proceedings of the Royal Society of London B* 269, 2429–2435.

Pyron, R. Alexander, and Frank T. Burbrink. 2011. Extinction, ecological opportunity, and the origins of global snake diversity. *Evolution* 66, 163–178.

Quammen, David. 1996. *The Song of the Dodo: Island Biogeography in an Age of Extinctions*. Simon and Schuster, New York.

———. 2006. *The Reluctant Mr. Darwin: An Intimate Portrait of Charles Darwin and the Making of His Theory of Evolution*. W. W. Norton, New York.

Raaum, Ryan L., Kirstin N. Sterner, Colleen M. Noviello, Caro-Beth Stewart, and Todd R. Disotell. 2005. Catarrhine primate divergence dates estimated from complete mitochondrial genomes: concordance with fossil and nuclear DNA evidence. *Journal of Human Evolution* 48, 237–257.

Raine, J. I., D. C. Mildenhall, and E. M. Kennedy. 2008. New Zealand fossil spores and pollen: an illustrated catalogue, 3rd ed. GNS Science Miscellaneous Series no. 4. Online at www.gns.cri.nz/what/earthhist/fossils/spore_pollen/catalog/index.htm.

Raven, Peter H., and Daniel I. Axelrod. 1972. Plate tectonics and Australasian paleobiogeography. *Science* 176, 1379–1386.

Reader, John. 2008. The fungus that conquered Europe. *New York Times* online edition, March 17, 2008, www.nytimes.com/2008/03/17/opinion/17reader.html?_r=1&.

———. 2009. *Potato: A History of the Propitious Esculent*. Yale University Press, New Haven, CT.

Renner, Susanne S. 2004a. Plant dispersal across the tropical Atlantic by wind and sea currents. *International Journal of Plant Sciences* 165, suppl. 4, S23–S33.

———. 2004b. Multiple Miocene Melastomataceae dispersal between Madagascar, Africa and India. *Philosophical Transactions of the Royal Society of London B* 359, 1485–1494.

———. 2005. Relaxed molecular clocks for dating historical plant dispersal events. *Trends in Ecology and Evolution* 10, 550–558.

Renner, Susanne S., G. Clausing, and K. Meyer. 2001. Historical biogeography of Melastomataceae: the roles of Tertiary migration and long-distance dispersal. *American Journal of Botany* 88, 1290–1300.

Riba-Hernández, Pablo, and Kathryn E. Stoner. 2005. Massive destruction of *Symphonia globulifera* (Clusiaceae) flowers by Central American spider monkeys (*Ateles geoffroyi*). *Biotropica* 37, 274–278.

Richardson, C. Howard, and David J. Nemeth. 1991. Hurricane-borne African locusts (*Schistocerca gregaria*) on the Windward Islands. *GeoJournal* 23, 349–357.

Rinderknecht, Andrés, and R. Ernesto Blanco. 2008. The largest fossil rodent. *Proceedings of the Royal Society of London B* 275, 923–928.

Rivadavia, Fernando, V. F. O. de Miranda, G. Hoogenstrijd, F. Pinheiro, G. Heubl, and A. Fleischmann. 2012. Is *Drosera meristocaulis* a pygmy sundew? Evidence of a long-distance dispersal between western Australia and northern South America. *Annals of Botany* 110, 11–21.

Rivadavia, Fernando, Katsuhiko Kondo, Masahiro Kato, and Mitsuyasu Hasebe. 2003. Phylogeny of the sundews, *Drosera* (Droseraceae), based on chloroplast *rbcL* and nuclear 18S ribosomal DNA sequences. *American Journal of Botany* 90, 123–130.

Robertson, Hugh, and Barrie Heather. 2005. *The Hand Guide to the Birds of New Zealand.* Penguin, Auckland, New Zealand.

Rocha, Sara, Miguel A. Carretero, Miguel Vences, Frank Glaw, and D. James Harris. 2006. Deciphering patterns of transoceanic dispersal: the evolutionary origin and biogeography of coastal lizards (*Cryptoblepharus*) in the Western Indian Ocean region. *Journal of Biogeography* 33, 13–22.

Rodríguez-Trelles, Francisco, Rosa Tarrío, and Francisco J. Ayala. 2004. Molecular clocks: whence and whither? Pp. 5–26 in P. C. J. Donoghue and M. Paul Smith, eds., *Telling the Evolutionary Time: Molecular Clocks and the Fossil Record.* CRC Press, Boca Raton, FL.

Romer, Alfred Sherwood. 1959. *The Vertebrate Story*, 4th ed. University of Chicago Press, Chicago.

Rosen, Donn E. 1978. Vicariant patterns and historical explanation in biogeography. *Systematic Zoology* 27, 159–188.

———. 1979. Fishes from the uplands and intermontane basins of Guatemala: revisionary studies and comparative geography. *Bulletin of the American Museum of Natural History* 162, article 5, 267–376.

Rosenzweig, Michael L. 2002. *Species Diversity in Space and Time.* Cambridge University Press, Cambridge, UK.

Rossie, James B., and Erik R. Seiffert. 2006. Continental paleobiogeography as phylogenetic evidence. Pp. 469–522 in S. M. Lehman and J. G. Fleagle, eds., *Primate Biogeography.* Springer, New York.

Rossin, Maria, Juan T. Timi, and Eric P. Hoberg. 2010. An endemic *Taenia* from South America: validation of *T. talicei* Dollfus, 1960 (Cestoda: Taeniidae) with characterization of metacestodes and adults. *Zootaxa* 2636, 49–58.

Rowe, Diane L., Katherine A. Dunn, Ronald M. Adkins, and Rodney L. Honeycutt. 2010. Molecular clocks keep dispersal hypotheses afloat: evidence for trans-Atlantic rafting by rodents. *Journal of Biogeography* 37, 305–324.

Rubinoff, Daniel, and William P. Haines. 2005. Web-spinning caterpillar stalks snails. *Science* 309, 575.

Rubinoff, Daniel, and Patrick Schmitz. 2010. Multiple aquatic invasions by an endemic, terrestrial Hawaiian moth radiation. *Proceedings of the National Academy of Sciences USA* 107, 5903–5906.

Saiki, R. K., S. J. Scharf, F. Faloona, K. B. Mullis, G. T. Horn, H. A. Erlich, and N. Arnheim. 1985. Enzymatic amplification of ß-globin genomic sequences and restriction site analysis for diagnosis of sickle cell anemia. *Science* 230, 1350–1354.

Salmon, John T. 1992. *A Field Guide to the Alpine Plants of New Zealand.* Random House New Zealand, Auckland.

Samonds, Karen E., Laurie R. Godfrey, Jason R. Ali, Steven M. Goodman, Miguel Vences, Michael R. Sutherland, Mitchell T. Irwin, and David W. Krause. 2012. Spatial and temporal arrival patterns of Madagascar's vertebrate fauna explained by distance, ocean currents, and ancestor type. *Proceedings of the National Academy of Sciences USA* 109, 5352–5357.

Samways, Michael J. 1999. Translocating fauna to foreign lands: here comes the Homogenocene. *Journal of Insect Conservation* 3, 65–66.

Sanmartín, Isabel, and Fredrik Ronquist. 2004. Southern Hemisphere biogeography inferred by event-based models: plant versus animal patterns. *Systematic Biology* 53, 216–243.

Sanmartín, Isabel, Livia Wanntorp, and Richard C. Winkworth. 2007. West Wind Drift revisited: testing for directional dispersal in the Southern Hemisphere using event-based tree fitting. *Journal of Biogeography* 34, 398–416.

Santos, Charles Morphy D. 2007. On basal clades and ancestral areas. *Journal of Biogeography* 34, 1470–1471.

Särkinen, Tiina E., Mark F. Newman, Paul J. M. Maas, Hiltje Maas, Axel D. Poulsen, David J. Harris, James E. Richardson, Alexandra Clark, Michelle Hollingsworth, and R. Toby Pennington. 2007. Recent oceanic long-distance dispersal and divergence in the amphi-Atlantic rain forest genus *Renealmia* L.f. (Zingiberaceae). *Molecular Phylogenetics and Evolution* 44, 968–980.

Schaefer, Hanno, Christoph Heibl, and Susanne Renner. 2009. Gourds afloat: a dated phylogeny reveals an Asian origin of the gourd family (Cucurbitaceae) and numerous oversea dispersal events. *Proceedings of the Royal Society of London B* 276, 843–851.

Schatz, G. E. 1996. Malagasy/Indo-australo-malesian phytogeographic connections. Pp. 73–84 in W. R. Lourenço, ed., *Biogéographie de Madagascar*. Editions ORSTOM, Paris.

Schmid, Rudolf. 1986. Léon Croizat's standing among biologists. *Cladistics* 2, 105–111.

Schoener, Amy, and Thomas W. Schoener. 1984. Experiments on dispersal: short-term floatation of insular anoles, with a review of similar abilities in other terrestrial animals. *Oecologia* 63, 289–294.

Schuchert, Charles. 1928. The hypothesis of continental displacement. Pp. 104–144 in W. A. J. M. van Waterschoot van der Gracht et al., eds., *Theory of Continental Drift: A Symposium on the Origin and Movement of Land Masses Both Inter-Continental and Intra-Continental, as Proposed by Alfred Wegener*. American Association of Petroleum Geologists, Tulsa, Oklahoma.

———. 1932. Gondwana land bridges. *Bulletin of the Geological Society of America* 43, 875–916.

Schweizer, Manuel, Ole Seehausen, Marcel Güntert, and Stefan T. Hertwig. 2010. The evolutionary diversification of parrots supports a taxon pulse model with multiple trans-oceanic dispersal events and local radiations. *Molecular Phylogenetics and Evolution* 54, 984–994.

Scotese, Christopher R. 2004. A continental drift flipbook. *Journal of Geology* 112, 729–741.

Scott, William B. 1932. Nature and origin of the Santa Cruz fauna. *Reports of the Princeton University Expeditions to Patagonia, 1896–1899*, vol. 7, *Palaeontology* 4, part 3, 193–238.

Seiffert, Erik R., Elwyn L. Simons, William C. Clyde, James B. Rossie, Yousry Attia, Thomas M. Bown, Prithijit Chatrath, and Mark E. Mathison. 2005. Basal anthropoids from Egypt and the antiquity of Africa's higher primate radiation. *Science* 310, 300–304.

Sharma, Prashant, and Gonzalo Giribet. 2009. A relict in New Caledonia: phylogenetic relationships of the family Troglosironidae (Opiliones: Cyphophthalmi). *Cladistics* 25, 279–294.

Sharp, Warren D., and David A. Clague. 2006. 50-Ma initiation of Hawaiian-Emperor bend records major change in Pacific Plate motion. *Science* 313, 1281–1284.

Shedlock, Andrew M., and Scott V. Edwards. 2009. Amniotes (Amniota). Pp. 375–379 in S. B. Hedges and S. Kumar, eds., *The Timetree of Life*. Oxford University Press, Oxford, UK.

Shermer, Michael. 2002. *In Darwin's Shadow: The Life and Science of Alfred Russel Wallace*. Oxford University Press, Oxford, UK.

Shultz, Susanne, Emma Nelson, and Robin I. M. Dunbar. 2012. Hominin cognitive evolution: identifying patterns and processes in the fossil and archaeological record. *Philosophical Transactions of the Royal Society of London B* 367, 2130–2140.

Sibley, C. G., and J. E. Ahlquist. 1987. Avian phylogeny reconstructed from comparisons of the genetic material, DNA. Pp. 95–121 in C. Patterson, ed., *Molecules and Morphology in Evolution: Conflict or Compromise?* Cambridge University Press, Cambridge, UK.

Simanek, Donald, and John Holden. 2001. *Science Askew: A Light-Hearted Look at the Scientific World.* Taylor and Francis, New York.

Simpson, George Gaylord. 1952. Probabilities of dispersal in geologic time. Pp. 163–176 in E. Mayr, ed., *The Problem of Land Connections Across the South Atlantic, with Special Reference to the Mesozoic. Bulletin of the American Museum of Natural History* 99, 79–258.

———. 1953. *The Major Features of Evolution.* Columbia University Press, New York.

———. 1980. *Splendid Isolation: The Curious History of South American Mammals.* Yale University Press, New Haven, CT.

Skottsberg, C. 1925. Juan Fernandez and Hawaii: a phytogeographical discussion. Bernice P. Bishop Museum, Bulletin 16, 1–47.

Skottsberg, C. 1941. The flora of the Hawaiian Islands and the history of the Pacific Basin. *Proceedings of the 6th Pacific Science Congress, California* 4, 685–701.

Smith, Albert C. 1973. Angiosperm evolution and the relationship of the floras of Africa and America. Pp. 49–62 in B. J. Meggers, E. S. Ayensu, and W. D. Duckworth, eds., *Tropical Forest Ecosystems in Africa and South America: A Comparative Review.* Smithsonian Institution Press, Washington, DC.

Smith, Andrew B., and Kevin J. Peterson. 2002. Dating the time of origin of major clades: molecular clocks and the fossil record. *Annual Review of Earth and Planetary Science* 30, 65–88.

Smith, Sarah A., Ross A. Sadlier, Aaron M. Bauer, Christopher C. Austin, and Todd Jackman. 2007. Molecular phylogeny of the scincid lizards of New Caledonia and adjacent areas: evidence for a single origin of the endemic skinks of Tasmantis. *Molecular Phylogenetics and Evolution* 43, 1151–1166.

Soltis, Douglas E., Charles D. Bell, Sangtae Kim, and Pamela S. Soltis. 2008. Origin and early evolution of angiosperms. *Annals of the New York Academy of Sciences,* no. 1133, 3–25.

Song, Hojun, Matthew J. Moulton, Kevin D. Hiatt, and Michael F. Whiting. 2013. Uncovering historical signature of mitochondrial DNA hidden in the nuclear genome: the biogeography of *Schistocerca* revisited. *Cladistics,* Early View (Online Version of Record published before inclusion in an issue).

Sousa, Wayne P. 1993. Size-dependent predation on the salt-marsh snail *Cerithidea californica* Haldeman. *Journal of Experimental Marine Biology and Ecology* 166, 19–37.

Spooner, David M., Karen McLean, Gavin Ramsay, Robbie Waugh, and Glenn J. Bryan. 2005. A single domestication for potato based on multilocus amplified fragment length polymorphism genotyping. *Proceedings of the National Academy of Sciences USA* 102, 14695–14699.

Springer, Mark S., et al. 2012. Macroevolutionary dynamics and historical biogeography of primate diversification inferred from a species supermatrix. *PloS One* 7, e49521.

Steadman, David W. 1995. Prehistoric extinctions of Pacific island birds: biodiversity meets zooarchaeology. *Science* 267, 1123–1131.

Steadman, David W., and Clayton E. Ray. 1982. The relationships of *Megaoryzomys curioi*, an extinct cricetine rodent (Muroidae: Muridae) from the Galápagos Islands, Ecuador. *Smithsonian Contributions to Paleobiology*, no. 51, 1–23.

Steiper, Michael E., and Nathan M. Young. 2009. Primates (Primates). Pp. 482–486 in S. B. Hedges and S. Kumar, eds., *The Timetree of Life.* Oxford University Press, Oxford, UK.

Steppan, Scott J., Ronald M. Adkins, and Joel Anderson. 2004. Phylogeny and divergence-date estimates of rapid radiations in muroid rodents based on multiple nuclear genes. *Systematic Biology* 53, 533–553.

Stillwell, Jeffrey D., and Christopher P. Consoli. 2012. Tectono-stratigraphic history of the Chatham Islands, SW Pacific—the emergence, flooding and reappearance of eastern "Zealandia." *Proceedings of the Geologists' Association* 123, 170–181.

Stillwell, Jeffrey D., Christopher P. Consoli, Rupert Sutherland, Steven Salisbury, Thomas H. Rich, Patricia A. Vickers-Rich, Philip J. Currie, and Graeme J. Wilson. 2006. Dinosaur sanctuary on the Chatham Islands, Southwest Pacific: first record of theropods from the K-T boundary Takatika Grit. *Palaeogeography, Palaeoclimatology, Palaeoecology* 230, 243–250.

Stoddart, D. R., ed. 1984. *Biogeography and Ecology of the Seychelles Islands.* Dr. W. Junk Publishers, The Hague.

Storey, B. C., M. L. Curtis, J. K. Ferris, M. A. Hunter, and R. A. Livermore. 1999. Reconstruction and break-out model for the Falkland Islands within Gondwana. *Journal of African Earth Sciences* 29, 153–163.

Strange, Ian J. 1972. *The Falkland Islands.* David and Charles, Newton Abbot, UK.

Sturm, Helmut. 1993. A new *Neomachilis* species from the Hawaiian Islands (Insecta: Archaeognatha: Machilidae). Bishop Museum Occasional Papers, no. 36, 1–16.

Sturm, Helmut, and Carmen Bach de Roca. 1988. Archaeognatha (Insecta) from the Krakatau Islands and the Sunda strait area, Indonesia. *Memoirs of Museum Victoria* 49, 367–383.

Sturm, Helmut, and Ryuichiro Machida, eds. 2001. *Archaeognatha: Handbook of Zoology.* Vol. 4, *Arthropoda: Insecta*, Part 37. Walter de Gruyter, Berlin.

Takai, Masanaru, Federico Anaya, Nobuo Shigehara, and Takeshi Setoguchi. 2000. New fossil materials of the earliest New World monkey, *Branisella boliviana*, and the problem of platyrrhine origins. *American Journal of Physical Anthropology* 111, 263–281.

Taleb, Nassim Nicholas. 2010. *The Black Swan: The Impact of the Highly Improbable*, 2nd ed. Random House, New York.

Tan, Heok Hui, and Kelvin K. P. Lim. 2012. Recent introduction of the brown anole *Norops sagrei* (Reptilia: Squamata: Dactyloidae) to Singapore. *Nature in Singapore* 5, 359–362.

Tantaleán, Manuel, and Alfonso Gozalo. 1994. Parasites of the *Aotus* monkey. Pp. 353–374 in J. Baer, R. E. Veller, and I. Kakoma, eds., *Aotus: The Owl Monkey.* Academic Press, San Diego.

Tennyson, Alan J. D. 2010. The origin and history of New Zealand's terrestrial vertebrates. *New Zealand Journal of Ecology* 34, 6–27.

Tennyson, Alan J. D., Trevor H. Worthy, Craig M. Jones, R. Paul Scofield, and Suzanne J. Hand. 2010. Moa's ark: Miocene fossils reveal the great antiquity of moa (Aves: Dinornithiformes) in Zealandia. *Records of the Australian Museum* 62, 105–114.

Thiel, Martin, and Lars Gutow. 2005. The ecology of rafting in the marine environment. Part 2, The rafting organisms and community. *Oceanography and Marine Biology: An Annual Review* 43, 279–418.

Thomas, Jessica A., John J. Welch, Megan Woolfit, and Lindell Bromham. 2006. There is no universal molecular clock for invertebrates, but rate variation does not scale with body size. *Proceedings of the National Academy of Sciences USA* 103, 7366–7371.

Thomson, K. 1998. When did the Falklands rotate? *Marine and Petroleum Geology* 15, 723–736.

Thorne, Robert F. 1973. Floristic relationships between tropical Africa and tropical America. Pp. 27–47 in B. J. Meggers, E. S. Ayensu, and W. D. Duckworth, eds., *Tropical Forest Ecosystems in Africa and South America: A Comparative Review.* Smithsonian Institution Press, Washington, DC.

Tilling, Robert I., Christina Heliker, and Donald A. Swanson. 2010. Eruptions of Hawaiian volcanoes—past, present, and future. United States Geological Survey General Information Product 117.

Tolley, Krystal A., Ted M. Townsend, and Miguel Vences. 2013. Large-scale phylogeny of chameleons suggests African origins and Eocene diversification. *Proceedings of the Royal Society of London B* 280, March 27, 2013, doi: 10.1098/rspb.2013.0184.

Towns, D. R., and Charles H. Daugherty. 1994. Patterns of range contractions and extinctions in the New Zealand herpetofauna following human colonisation. *New Zealand Journal of Zoology* 21, 325–339.

Trewick, Steven A., and Gillian C. Gibb. 2010. Vicars, tramps and assembly of the New Zealand avifauna: a review of molecular phylogenetic evidence. *Ibis* 152, 226–253.

Vawter, Lisa, and Wesley M. Brown. 1986. Nuclear and mitochondrial DNA comparisons reveal extreme rate variation in the molecular clock. *Science* 234, 194–196.

Vences, Miguel, David R. Vieites, Frank Glaw, Henner Brinkman, Joachim Kosuch, Michael

Veith, and Axel Meyer. 2003. Multiple overseas dispersal in amphibians. *Proceedings of the Royal Society of London B* 270, 2435–2442.

Verzi, Diego H., and Claudia I. Montalvo. 2008. The oldest South American Cricetidae (Rodentia) and Mustelidae (Carnivora): Late Miocene faunal turnover in central Argentina and the Great American Biotic Interchange. *Palaeogeography, Palaeoclimatology, Palaeoecology* 267, 284–291.

Vidal, Nicolas, Anna Azvolinsky, Corinne Cruaud, and S. Blair Hedges. 2008. Origin of tropical American burrowing reptiles by transatlantic rafting. *Biology Letters* 4, 115–118.

Vidal, Nicolas, Maël Dewynter, and David J. Gower. 2010a. Dissecting the major American snake radiation: a molecular phylogeny of the Dipsadidae Bonaparte (Serpentes, Caenophidia). *Comptes Rendus Biologies* 333, 48–55.

Vidal, Nicolas, and S. Blair Hedges. 2009. The molecular evolutionary tree of lizards, snakes, and amphisbaenians. *Comptes Rendus Biologies* 332, 129–139.

Vidal, Nicolas, Julie Marin, Marina Morini, Steve Donnellan, William R. Branch, Richard Thomas, Miguel Vences, Addison Wynn, Corinne Cruaud, and S. Blair Hedges. 2010b. Blindsnake evolutionary tree reveals long history on Gondwana. *Biology Letters* 6, 558–561.

Vidal, Nicolas, Jean-Claude Rage, Arnaud Couloux, and S. Blair Hedges. 2009. Snakes (Serpentes). Pp. 390–397 in S. B. Hedges and S. Kumar, eds., *The Timetree of Life*. Oxford University Press, Oxford, UK.

Vine, F. J., and D. H. Matthews. 1963. Magnetic anomalies over oceanic ridges. *Nature* 199, 947–949.

Voelker, Gary, Sievert Rohwer, Diana C. Outlaw, and Rauri C. K. Bowie. 2009. Repeated trans-Atlantic dispersal catalysed a global songbird radiation. *Global Ecology and Biogeography* 18, 41–49.

Wade, Nicholas. 1998. Scientist at work / Kary Mullis; after the "Eureka," a Nobelist drops out. *New York Times*, September 15, 1998.

Wagner, Warren L., Derral R. Herbst, and S. H. Sohmer. 1999. *Manual of the Flowering Plants of Hawai'i*, rev. ed. University of Hawaii Press, Honolulu.

Wallace, Alfred Russel. 1880. *Island Life*. Macmillan, London.

———. 1891. *Natural Selection and Tropical Nature*. Macmillan, London.

Wallis, Graham P., and Steven A. Trewick. 2009. New Zealand phylogeography: evolution on a small continent. *Molecular Ecology* 18, 3548–3580.

Webb, S. David. 2006. The Great American Biotic Interchange: patterns and processes. *Annals of the Missouri Botanical Garden* 93, 245–257.

Wegener, Alfred. 1915. *Die Entstehung der Kontinente und Ozeane*. Friedr. Vieweg und Sohn, Braunschweig, Germany.

———. 1924. *The Origin of Continents and Oceans*. Translated from the 3rd German ed. by J. G. A. Skerl. E. P. Dutton, New York.

Welch, John J., and Lindell Bromham. 2005. Molecular dating when rates vary. *Trends in Ecology and Evolution* 20, 320–327.

Wendell, Jonathan F., Curt L. Brubaker, and Tosak Seelanan. 2010. The origin and evolution of *Gossypium*. Pp. 1–18 in J. McD. Stewart, Derrick M. Oosterhuis, James J. Heitholt, and Jack R. Mauney, eds., *Physiology of Cotton*. Springer, Dordrecht, Germany.

Whiting, Alison S., Jack W. Sites Jr., Katia C. M. Pellegrino, and Miguel T. Rodrigues. 2006. Comparing alignment methods for inferring the history of the new world lizard genus *Mabuya* (Squamata: Scincidae). *Molecular Phylogenetics and Evolution* 38, 719–730.

Wickens, G. E. 1982. The baobab—Africa's upside-down tree. *Kew Bulletin* 37, 173–209.

Wiley, E. O. 1988. Vicariance biogeography. *Annual Review of Ecology and Systematics* 19, 513–542.

Williams, Ernest E. 1969. The ecology of colonization as seen in the zoogeography of anoline lizards on small islands. *Quarterly Review of Biology* 44, 345–389.

Wilson, Don E., and DeeAnn M. Reeder, eds. 2005. *Mammal Species of the World: A Taxonomic and Geographic Reference*, 3rd ed. Johns Hopkins University Press, Baltimore.

Winkworth, Richard C., Steven J. Wagstaff, David Glenny, and Peter J. Lockhart. 2002. Plant dispersal N.E.W.S. from New Zealand. *Trends in Ecology and Evolution* 17, 514–520.

Yanoviak, Stephen P., Michael Kaspari, and Robert Dudley. 2009. Gliding hexapods and the origins of insect aerial behaviour. *Biology Letters* 5, 510–512.

Yoder, Anne D., Melissa M. Burns, Sarah Zehr, Thomas Delefosse, Geraldine Veron, Steven M. Goodman, and John J. Flynn. 2003. Single origin of Malagasy Carnivora from an African ancestor. *Nature* 421, 734–737.

Yoder, Anne D., and Michael D. Nowak. 2006. Has vicariance or dispersal been the predominant biogeographic force in Madagascar? Only time will tell. *Annual Review of Ecology, Evolution, and Systematics* 37, 405–431.

Yoon, Carol Kaesuk. 2009. *Naming Nature: The Clash Between Instinct and Science*. W. W. Norton, New York.

Young, J. Z. 1962. *The Life of Vertebrates*, 2nd ed. Oxford University Press, New York.

Yount, Lisa. 2009. *Alfred Wegener: Creator of the Continental Drift Theory*. Chelsea House, New York.

Zhang, Peng, and Marvalee H. Wake. 2009. A mitogenomic perspective on the phylogeny and biogeography of living caecilians. *Molecular Phylogenetics and Evolution* 53, 479–491.

Ziegler, Alan C. 2002. *Hawaiian Natural History, Ecology, and Evolution*. University of Hawaii Press, Honolulu.

Zuckerkandl, Emile, and Linus Pauling. 1962. Molecular disease, evolution and genic heterogeneity. Pp. 189–225 in M. Kasha and B. Pullman, eds., *Horizons in Biochemistry: Albert Szent Györgyi Dedicatory Volume*. Academic Press, New York.

INDEX

Aciphylla speargrass, 105
Adriatic-Dinaric Carbonate Platform (ADCP), 221–222 (box)
Aegean Sea, 31
Ae'o, 75
Aerial animals, generalization about, 262
Africa, 116, 118, 181, 186, 208
 and the Cameroon Line, 178
 Carboniferous and Permian period, 37
 contingency and unpredictability scenario involving, 300
 and the Cretaceous, 221 (box)
 early assumptions about, 33
 and the Eocene, 223
 and the Falklands, 39n, 229, 230, 232, 233, 234, 238, 250–251, 250–251
 final separation of, 214 (fig.)
 and the fit with South America, 34, 36, 37, 40
 and Gondwanan breakup, 4, 11 (box), 61 (fig.), 142 (box), 143 (box), 229, 248, 288
 and the Gondwanan relict idea, 97, 152, 232–233
 island-hopping routes from, 217 (fig.)
 and the Jurassic, 229, 230 (fig.)
 and land bridges, 42–43, 43 (fig.), 266
 Madagascar compared to, 247
 and the Mesozoic, 97
 and the Miocene, 248, 296
 and ocean currents, 182, 219
 and the Pliocene, 296
 rainfall in, 193
 taxa involving, 2, 3, 14, 15, 44, 54, 79, 87, 97, 135, 142 (box), 143 (box), 148, 153, 158, 161 (fig.), 165, 166, 169, 173, 179, 187, 190–191, 192, 195, 197, 198, 199, 200, 204, 205–206, 209, 211–212, 213, 215, 218, 231, 248, 249, 261, 265,

267, 279, 280 (fig.), 286, 287–288, 290, 295 (fig.), 300
 tracks and, 80 (fig.), 81
 and the Triassic, 272
Agapornis lovebird, 286
Age estimates
 of Earth, 28, 30, 54, 116n
 skepticism about, 88, 89
 and the timing of relevant events, addressing, 115–118
 of the universe, 115–116
 See also Fossil records; Molecular clock analyses
Airline route map, shift toward scenario resembling, 15
Ajkaceratops kozmai dinosaur, 222 (box)
Alaska, 74, 75n
Aldabra giant tortoise (*Dipsochelys dussumieri*), 19, 168–169
Aleutians, 74
Alpine, 105
Amalgamation of landmasses, 82, 270n
Amazon River, 193–194
Amazonia, 303
Amborella trichopoda shrub, 242, 244
American Museum of Natural History, 45
 See also "New York School" dispersalists
Americas. *See* New World; North America; South America
Amphibian skin, 179–180
Amphisbaenians. *See* Worm lizards
Anacondas, 294
Anasazi, 287
Anatomical evidence, 69, 97, 137, 147, 153, 184, 185, 187, 190, 198, 209, 210, 216, 220, 262, 291
Anguilla, 253
Annals & Magazine of Natural History (journal), 29
Anolis lizards, 72, 265
Antarctica, 2, 4, 29, 49, 54, 61

(fig.), 63, 64 (fig.), 152, 230 (fig.), 248, 265, 288
Antarctica/Australia, 4, 97, 99, 102, 104, 105, 109, 160, 162, 211, 246
Anthropoids, 287–288
Antigua, 253, 279
Apes, 90, 136, 209, 287–288
Aporostylis bifolia orchid, 105
Appalachians, 37
Araucariaceae (araucarians) trees, 158–159, 163, 240, 242, 299
Archaeognatha. *See* Jumping bristletails
Arctic Ocean, 56
Arctic region, 1, 9
Area cladograms, defined, 68
Argentina, 97, 156, 290, 303
Arizonasaurus fossil, 137, 138 (fig.)
Ark ideas, 4, 27, 163, 178, 234, 246
 See also Landmasses-as-life-rafts story
Arnason, Ulfur, 212–213, 214
Arthropod eggs, 169
Arthur's Pass, 104–105, 162
Ascension, 76, 266, 268
Asia, 2, 4, 33, 37, 58, 80 (fig.), 179, 200, *211*, 212, 213, 214, 215, 222 (box), 262, 283, 287
 See also specific parts of Asia
Asteraceae (sunflower), 18, 162–163
Atlantic Ocean, 3, 28, 33, 158
 and continental fit, 60–61
 continued deepening of, 217
 continued spreading of, 216
 crossings involving, 79, 148, 154, 166, 191, 205–206, 207, 209, 215, 216–217, 218, 219, 222 (box), 223, 261, 266, 279, 281, 287, 290, 293, 303
 currents in, 219
 and the Eocene, 216–217
 formation of, 4, 97, 229
 and frogs, 200
 opening of, 116, 118, 135, 142 (box), 210, 215, 267, 288

possible island-hopping across, 217–218
ridges in, 56–57, 58, 59, 216
tracks across, 80 (fig.)
width of, 216
See also specific islands; specific landmasses bordering the ocean
Atlantis, 27
Attrition, 235, 236, 238, 239
See also Extinctions
Australia, 248
  Carboniferous and Permian period, 37
  and Gondwanan breakup, 4, 11 (box), 61 (fig.), 170
  and the Jurassic, 230 (fig.)
  and land bridges, 42
  and the Mesozoic, 63
  taxa involving, 2, 3, 15, 25, 49, 54, 63, 64 (fig.), 65, 69 (fig.), 86, 87 (fig.), 107–108, 109, 111, 117, 135, 153–154, 161 (fig.), 162, 165, 166, 170–171, 200, 243 (fig.), 265
  tracks and, 80 (fig.), 81, 82
  Wallace's Line and, 33
  See also Antarctica/Australia
Australian Plate, 99
Aye-ayes (*Daubentonia madagascariensis*), 249 (fig.)
Azores, 76, 261
Aztecs, 287

Baboons, 210, 214 (fig.), 287
Baconian view, 80
Bacteria, 121, 122, 125, 302n
Bahamas, 72
Bair, Janet, 185
Baja California, 4–8, 9 (fig.), 14, 188, 190
Bald-headed uakaris, 223
Banda Islands, 286
Bandoni de Oliveira, Felipe, 217–218
Baobab trees, 3, 15, 80–81, 135, 248
Barbuda, 253
Bats, 75, 219–220
Beach strawberry, 75
Beagle voyage, 27, 46, 226, 231n, 293
Bealor, Matthew, 7
Beans/legumes, 156–158, 159, 162, 286
BEAST program, 134, 138n
Belgium, 37, 283–284
Bering Land Bridge, 2, 214
Beyer, G. E., 93
Bible/God. *See* Religion, influence of
Big Bang Theory, 115, 116
Big Five extinctions, 237, 238

Biogeography
  beginnings of, 27
  conundrums of, 3, 161, 281
  critical studies for, 130
  defined, 2
  definitions of common terms and concepts, 10–12 (box)
  father of, 33
  focal point of, 100, 110
  fundamental fact of, 2
  history of debate within, 273–274
  mindset essential for, 146, 147
  stagnant period in, 118
  unified theory of, 23–24
  See also Dispersalism; Historical biogeography; Vicariance biogeography
Biogeography and Plate Tectonics (Briggs), 118
Biogeography of the Southern End of the World (Darlington), 67–68
Bioko, 178
Biological Relationships Between Africa and South America (Goldblatt), 170
Bipes worm lizard, 142–143 (box)
Bird-croc branching point, 52 (fig.), 137, 138 (fig.), 144, 299
Black Robin, 240, 241
"Black swan" events, 220, 222–223
Black Swan, The (Taleb), 220
Black-and-white thinking, addressing, 90–92, 146
Black-necked garter snake, 128
Black-Necked Stilt, 75
Blindsnakes, 206
Boas, 244, 247, 251, 294
Bolivia, 213
Bolyeriidae boas, 244
Boophis frogs, 184–185
Borneo, 12, 28, 109, 110, 180, 189, 216
Boston swordfern, 75
Bourles, Bernard, 192, 193, 194
Brachylophus iguanas, 215n
Brazil, 37, 151, 203, 266, 303
Brazilian porcupine (*Coendou prehensilis*), 292 (fig.)
Briggs, Derek, 298
Briggs, John, 70, 117–118, 268, 274–275
Bristletails. *See* Jumping bristletails
British colonization of the Chatham Islands, 239
British Isles, 37
Brooks, Dan, 91–92

Browne, Janet, 27
Brundin, Lars, 48–49, 51–52, 53, 54, 62, 63, 64 (fig.), 65–67, 68, 70, 77, 83–84, 85, 91, 164–165, 171, 269
Burgess Shale Formation, 297, 298, 299n, 301
Buttercups, 18, 105, 162
"Butterfly effect," 301

Caecilians, 176, 177, 180, 182, 189, 195, 197, 198, 199, 200, 251
Calibration points
  defined, 135–136
  Gondwanan breakup events used as, problem with, 142–143 (box)
  and incorporating uncertainty, 137–138, 141
  multiple, use of, 138–139, 141
  See also Fossil calibrations; Molecular clock analyses
California, 6, 52, 53, 208
California Channel Islands, 200
California Quail, 185, 186
Camalotes (Amazon giant rafts), 193–194, 198
Cambrian, 282, 297
Cambrian Explosion, 298
Cameroon Line, 178
Campbell, Doug, 107
Campbell Island, 97, 108
Canadian Rockies, 297
Capuchin monkey, 3, 223
Capybara (*Hydrochoerus hydrochaeris*), 291, 292 (fig.), 303
Carboniferous period, 37
Caribbean islands, 72, 93, 157, 200, 279
  See also specific Caribbean islands
Caribbean Sea, 13, 158, 214, 224
Carl G. Washburne Memorial State Park, 305–306
Catarrhini monkeys. *See* Old World monkeys
Cause and effect, 116, 119, 297
  See also Contingency
Caviomorph rodent, 219, 290, 291, 292 (fig.), 293, 294, 301, 302–303
Celmisia daisies, 105, 162
Cenozoic, 44, 102n, 156, 157, 240
  See also Eocene; Miocene; Oligocene; Paleocene; Pleistocene; Pliocene
"Centers of Origin and Related Concepts" (Croizat, Nelson, & Rosen), 84–85

Central America, 2, 265, 290
  See also Panamanian Isthmus
Centrifugal force, 38, 40, 54
Cerithideopsis snails, 224
Cerro de la Neblina, 152–154
Cetus, 120, 122, 124
Chamberlin, R. Thomas, 41
Chameleons, 247, 286
Chance dispersal, defined, 10 (box)
Changuu Island, 19
Chatham Island Oystercatcher, 241
Chatham Islands, 97, 225, 228 (fig.), 239–243, 246n, 251, 271, 281
Chatham Island Warbler, 241
Chatham Rise, 239–240
Checkered garter snake, 128
Chile, 25, 69 (fig.)
Chimpanzees, 12 (box), 88, 90, 116, 135, 136, 210, 288
China, 97, 284
Chionochloa pallens grass, 105
Chironomid midges, 48–49, 53–54, 63–65, 83, 89, 165, 171, 269
Chordates, 298, 299
Circular reasoning, 142 (box)
Clades, defined and described, 49–50
Cladistics, 50, 51–52, 53, 63, 64 (fig.), 66–67, 69, 70–71, 76, 83, 84, 87, 91, 165–166, 216, 267, 275, 276, 277
Cladistics (journal), 71
Cladograms, 50, 51, 52 (fig.), 53, 63, 65, 68, 69, 71, 84, 91, 100, 117, 118, 157, 170, 267, 270, 276
"Climate and Evolution" (Matthew), 43, 45
Climate change, 10 (box), 13, 99, 156, 157n, 158, 301n
  See also Ice ages
Cobra bobo (*Schistometopum thomense*), 175, 176, 177, 181, 189, 191, 195, 199
Colobus monkey, 287
Colombia, 151
Colorado River, 28
Columbia University, 45
  See also "New York School" dispersalists
Columbus, 115, 285
Comoros Archipelago, 182, 184, 199
Complexity vs. simplicity, addressing, 90–92, 146, 269–270
Congo Current, 191 (fig.), 194, 197
Congo, Ituri Province, 199
Congo River & Basin, 181, 191, 192–193, 194, 195, 197, 198
Continental crust, 36, 38, 55, 58, 108, 242, 266, 268
  See also Plate tectonics
Continental drift, 4, 33–34, 35, 36–41, 42, 43, 45, 54–60, 61–62, 63, 65, 67–68, 71, 76, 106, 118, 142 (box), 163, 169–170, 172, 200, 226, 267, 268
  See also Amalgamation of landmasses; Gondwanan breakup; Landmasses-as-life-rafts story; Plate tectonics
Continental fit, 36, 37, 60–61
Continental island, defined, 12 (box)
Continental position, early assumptions about, 33
  See also Fixed continents and ocean basins, belief in
Contingency
  defined, 296–297
  and unpredictability, 296, 297–299
Convection currents, theory involving, 55, 57, 62
Conway Morris, Simon, 298, 299n
Cook, Lyn, 166
Copernicus, 272, 276
Coprosma plant, 105
Corn, 286, 287
Cotton (*Gossypium hirsutum*), 285, 286–287
Coursetia legumes, 156
Cracraft, Joel, 69, 269
Creationism, 25, 31, 32, 179, 186, 197n, 207, 219
  See also Religion, influence of
Cretaceous, 34, 57, 97, 100, 101, 102n, 104, 109, 214, 221–222 (box), 237, 238, 239, 240, 244, 248, 249–250
Crick, Francis, 130
Crisp, Michael, 162n, 166
Crocodiles, 3, 4, 137, 138 (fig.), 144, 148, 219, 222 (box), 242, 249, 281, 286, 287, 299
Croizat, Léon, 23–24, 32, 45, 77–79, 80–81, 82, 83–84, 84–85, 85–86, 87, 89–90, 91, 92, 165, 227, 233, 252, 269, 270, 274–275, 276, 303
Cuba, 72
Cucurbitaceae (cucumbers/squash), 160, 161 (fig.)
Cyprus, 173

Daisies, 105, 162
Dalbergioid legumes, 157
Darlington, Philip J., 45, 65, 67–68, 70, 100
Darwin, Charles, 3, 38, 72, 108, 170, 268, 288
  addressing how-possibly questions, 197–198
  on batrachians (amphibians), 178, 179, 180, 185
  and the *Beagle* voyage, 27, 46, 226, 231n, 293
  and construction of the argument for evolution, 147
  on continents, 33, 43, 272
  conversion to belief in evolution of, 120
  Croizat's criticism of, 23, 24, 77 (fig.), 78, 79, 80, 81, 82, 85, 86, 91, 274
  and disjunct distributions, 27–28, 32, 33
  on the Falklands, 226–227, 231n, 233
  on fossil records, 88
  and how he constructed his argument for evolution, 147
  hypothetical scenario involving question posed to, 266, 267–268
  and land bridges, 42, 43, 275
  on long-distance dispersal, 24–26, 32, 43–44, 45, 79, 81–82, 109, 154, 168, 169, 186, 219–220, 261, 273, 290
  major theme of, 134
  on New Zealand, 100
  stock phrase of, 154
  view of historical biogeography, 23, 24, 28, 30, 31, 32, 273n
Darwinian dispersalism, 42, 86, 272
Darwinian Revolution, 30, 52, 273n
De Queiroz, Eiji (author's son), 1, 305–306
De Queiroz, Hana (author's daughter), 1, 207–208, 305
De Queiroz, Kevin (author's brother), 52, 179
De Queiroz, Sean (author's brother), 179
De Queiroz, Tara (author's wife), 4–5, 7, 16, 95, 104, 162, 185, 187, 207–208, 240, 305
Desert locusts, 279, 280 (fig.)
Devil's Hole pupfish, 235, 236
Devonian, 39n, 229, 260
Diamond, Jared, 16, 18
Dietz, R. S., 57n
Dinosaurs, 16, 50, 90, 97, 102, 143 (box), 152, 221–222

INDEX   353

(box), 234, 236, 237, 240, 242, 249, 250, 252, 281
Diplodactylid geckos, 242–244
Dirac, Paul, 270n
Disjunct distribution, defined, 10–11 (box)
Dispersal, normal versus long-distance, defining, 10 (box)
Dispersalism
  assumptions in, 227
  back-and-forth debate involving, 273–274
  height of, 45
  long history of, 24
  messiness of, 91, 270
  pendulum swings involving, 268, 272–273
  persistence of, 168, 169–170, 273
  rejection and ridicule of, history of, 62, 85, 86, 192, 209
  see also Oceanic dispersal
  shift back toward, 118, 166, 270–272, 277–278
  See also Historical biogeography
DNA sequencing, 27
  availability of, 127–128, 140, 146, 209–210
  changes to, 128–129, 139
  database of, 127, 133
  Heads's rejection of, 76
  and how-possibly arguments, 198
  and an ideal clock, 135
  as indirect evidence, 147
  invention of methods for, 119, 121–126, 130, 271
  and modeling of changes, 128–130
  precision of, 128
  revelations from, 153, 182, 187, 220, 271
  and the shift away from vicariance, 118, 271
  usage of, 7, 9 (fig.), 184–185, 189–190, 198, 205–206, 220, 224, 241, 262, 263, 265, 279
  See also Molecular clock analyses
DNA-DNA hybridization, 128
Dominica, 279
Donoghue, Michael, 70, 71, 92, 118, 134–135, 277
Doyle, Arthur Conan, 152
Dragonflies, 172n, 201
Dray, William, 196
Drewes, Bob, 175, 176, 177 (fig.), 189–191, 193–195, 199
Drift signal, 172

Driftwood, 262, 305–306
Drosera (sundews), 105, 152–154, 155, 299
Drosophila melanogaster fly, 126
Drunkard's walk (random fluctuations), 235–236, 238, 265
du Toit, Alexander, 39n, 42
Duck-feet experiments, 26, 27, 32, 198
Duke Lemur Center, 248
Dusicyon australis wolf, 231, 232 (fig.)

Earthquakes, 38
Easter Island, 76
Ecological effects, 293–294
Ediacaran fauna, 299
Eggeling, Tom, 229
Eggplant, 285, 286, 287
Egypt, 190
Elephant birds, 245, 247
Elephants, 173
Elk, 2
Emus, 3, 244–245
England, 97
Eocene, 104, 105, 136, 215, 216–217, 223, 238
Establishment requirement, 12 (box)
Estuarine crocodile (*Crocodylus porosus*), 148
Eurasia, 2, 172, 221 (box), 245, 265
  See also Asia; Europe
Europe, 34, 37, 39, 42, 61, 80 (fig.), 156, 283
  See also specific parts of Europe
Event cost in biogeographic analysis, 165–166n
Evidence, following the, importance of, 268–269, 271–272, 277
  See also Fossil records; Molecular clock analyses
Evolution
  construction of the argument for, 147
  versus creationism, 31–32, 178–179, 197n
  great metaphor of, 235
  and the landmasses-as-life-rafts story, 4
  overarching theory of, 2
  paradigm for, 273n
  powerful argument for, 179
  reality of, acceptance of, 40n
  unpredictability of, 282
Evolutionary biology, as a science, approach of, 92
  See also Historical biogeography

Evolutionary dead ends, islands as, rule of, addressing, 263–264
Evolutionary diagrams
  clarity of, 51, 52 (fig.), 71, 276
  muddiness of, 50–51, 71, 276
  See also Cladograms
Evolutionary mirrors, 293–294
Extinctions, 44, 99, 104, 106, 164n, 172, 197, 203, 211 (fig.), 226, 231, 232 (fig.), 234, 235–236, 237, 238–239, 240, 242, 246n, 250, 251, 252, 291, 298, 302
Extraterrestrial event, 222 (box), 237, 238

Fabaceae plants, 156
  See also Beans/legumes
Falkland Islands, 39n, 225, 226–227, 228–233, 234, 235, 238–239, 250–251, 252
Falkland Islands Freshwater Fishes (McDowall et al.), 229
False water cobras, 294
Falsification, 87, 91, 192, 215, 274
Fast-running clocks, 126, 139
Fernando de Noronha, 203–205, 206n, 223, 293
Fiji, 215n
Fixed continents and ocean basins, belief in, 33, 43, 54, 59, 65, 272, 276
Fleming, Charles, 106, 107, 109, 110
Flightless birds. *See* Ratites
Floating islands, 112, 193–195, 197, 198, 206–207, 216, 222 (box)
Flores, 173, 209
Forbes, Edward, 31
Fork-Tailed Flycatcher, 154–155
Fossil calibrations, 135–139, 141, 142–143 (box), 157, 163, 212–213
  See also Molecular clock analyses
Fossil records
  comparing molecular age estimates with estimates based on, 144–145, 146
  criticism of, 89, 136, 273, 277
  data collection for building up, 236–237
  and how-possibly arguments, 198
  incompleteness of, 88, 89, 90, 117, 136, 137, 142 (box), 197, 213, 215, 250, 267, 277

Fossil records *(continued)*
  oldest of, 88
  strong interest in, 117–118
  use of, 29, 34, 37, 39, 44, 57, 91, 101–102, 102–104, 105–106, 109, 110, 143 (box), 147, 148, 153, 160, 169, 170, 173, 198, 211–212, 218, 220, 222 (box), 236, 240, 243, 244, 245, 246, 249–250, 268, 272, 286, 287, 290, 297–298
Fossils, problem with, 135, 136, 146
France, 283
Frank, Phil, 7
Frogs, 12, 13, 32, 79, 101, 172, 177, 178, 179–180, 181, 182, 183 (fig.), 184–185, 186–187, 188, 189–194, 195, 197, 198, 199–200, 246n, 251, 271
Fruit flies, 52, 264–265
Fundamental tracks, 81, 82–83, 84, 85
Funk, Vicki, 152

Galápagos, 12 (box), 19, 25, 28, 44, 76, 175, 176, 226, 236, 293
Galaxiod fishes, 227, 228, 229, 231, 244
Ganges River, 173
Garter snakes (*Thamnophis*), 5–6, 7–8, 9 (fig.), 29, 124, 125–126, 128, 188, 190, 198
Gatesy, John, 92, 141, 257, 258, 260, 266
Geckos, 101, 200, 206, 219, 225, 242–244, 281, 295 (fig.)
GenBank, 127, 133
Generalizations, issue of, addressing, 14, 92, 165, 261, 262
Genetic evidence. *See* DNA sequencing; Molecular clock analyses; Phylogenetics
Geographic structure, defined, 153
Geological Society of America, 41
Germany, 9, 284
Gibbons, 88, 288
Glaciation, 9, 38, 104, 231n
  See also Ice ages
Gladwell, Malcolm, 47
Glossopteris flora, 37, 272
God/Bible. *See* Religion, influence of
Gondwana, 3, 229, 230 (fig.)
Gondwanaland, 33n, 67, 102
Gondwanan breakup, 3–4, 6, 7, 11 (box), 13, 14, 17,
18, 61 (fig.), 63, 65, 69, 80 (fig.), 82, 86, 97, 99, 102, 108, 109, 110, 117, 118, 142–143 (box), 151, 158, 164, 165, 166, 169, 170, 171, 210, 226, 229, 234, 241, 245, 247, 250, 252, 269, 270, 278, 288, 294
Gondwanan islands. *See* Chatham Islands; Falkland Islands; Madagascar; New Caledonia; New Zealand; Seychelles
Gondwanan landmasses. *See* Africa; Antarctica; Australia; India; South America; Zealandia
Gondwanan relicts, 96–97, 99, 100, 101–102, 103, 106, 107, 108, 109, 110, 152, 154, 160, 161–163, 166, 182, 225–226, 232–233, 240, 242, 244, 246, 247, 248, 250, 251, 281
Goose barnacles, 19
Gorillas, 210, 288
"Gossamer Spider," 46
Gould, Stephen Jay, 85, 147, 223, 237, 297–299
Grand Canyon, 28
Gravity measurements, 36
Great American Interchange, 289–290
"Great Dying," 237
Great Famine, 284
Greater Antilles, 156, 209
Green iguanas (*Iguana iguana*), 253
Green web, 155, 167
Greene, Brian, 270n
Greenland, 35, 41, 54
Greenwood, Humphry, 66–67
Grenada, 279
Grey Warbler, 241
Griqualand mountains, 37
Guadeloupe, 253, 279
Guinea pigs, 3, 127, 219, 286, 287, 290, 291
Gulf Coast, 93
Gulf of Guinea, 175, 178, 181, 190, 192, 192–193, 194–195, 197
Gulf of Guinea islands, 175, 178, 181, 189, 192, 194, 196, 197, 198, 199, 200
  See also Príncipe; São Tomé
Gulf of Mexico, 93
Gulf Stream, 112
Guppy, H. B., 201
Guyana, 151, 279
Guyot, Arnold, 56
Guyots, 56, 58

Hallam, Anthony, 117–118
Harrison, Rick, 124
Harvey, Janice (author's mother-in-law), 16, 95
Hawaii (Big Island), 73, 261
Hawaii (Hawaiian Islands), 12 (box), 28, 42, 52, 53, 60, 73, 74–76, 79, 82, 89, 117, 169, 180, 189, 208, 236, 244, 257–260, 260–261, 262–263, 264–265, 266, 268, 277, 303, 306
Hawaiian Hotspot, 73, 74
Hayashi, Cheryl, 257, 258, 260, 266
Heads, Michael, 74, 75, 76, 89, 107, 119, 135, 143 (box), 168, 211, 274, 275
Hebe shrubs, 104–105, 162
Heibl, Christoph, 160
Heliocentrism, 272, 276
Hennig Society, 71
Hennig, Willi, 49–50, 51, 52, 53, 66, 67, 71, 87
Hess, Harry, 56, 57, 58, 59, 62
Hillebrandia herb, 244
Himalayas, 4, 58
Historical biogeography
  beginning of, as a new field, 28
  core understanding needed for, 53
  current direction of, 277
  defined, 4
  general paradigm established for, 273
  iconic tale of, 4
  importance of timing information for, 115–116
  importance of tree diagrams to, 276
  as part of evolutionary biology, 92
  pendulum swings in, 268, 272–273
  possibility of achieving a paradigm for, 278
  pre-paradigm period in, 273–275
  steps toward maturation of, 275–278
  See also specific scientists and aspects of historical biogeography
Historical evidence, nature of, 146–147
"History of Ocean Basins" (Hess), 57, 59, 62
Hoary bat, 75
Hoberg, Eric, 133, 134
Hog-nosed snake, 294, 295
Holmes, Arthur, 54–55, 57, 58, 62
Holmes, Sherlock, 215
Hominids, 90
Homo floresiensis ("hobbit"), 209, 216

Homogenocene, 285
Honeycreepers, 75
Hooker, Joseph, 25, 26, 42, 45, 100
Houle, Alain, 216–217
Howler monkeys, 210, 223, 291
How-possibly arguments, 196–199, 216–217
Hull, David, 47–48
Human drifting incidents, 131
Human introductions, 27, 72, 179, 180, 200, 258, 291
  impact of, 99, 282–285, 288
  ruling out, 19, 75, 181, 184, 186
Humans (*Homo sapiens*), 12 (box), 89–90, 116, 117, 126, 131, 135, 136, 185, 208, 210, 237, 288, 299
Hummingbirds, 136–137
Hutton, James, 28
Huxley, Thomas Henry, 31
Hyperoliidae frogs, 186–187
Hyposmocoma moth, 257
Hystricognath rodents, 3

Ice ages, 2, 9, 10 (box), 12–13, 37, 40n, 99, 135, 178, 188, 238, 239
  See also Glaciation
Icebergs, 25, 26, 32, 82, 168, 227, 231n
Iguanas, 215n, 226, 253
Immobilism, 81
Inca, 283
India, 4, 33n, 37, 42, 43 (fig.), 58, 61 (fig.), 97, 161 (fig.), 165, 173, 180, 200, 221 (box), 230 (fig.), 247, 248, 251n, 284, 286
Indian Ocean, 19, 59, 80 (fig.), 81, 154, 165, 181, 186, 187, 206, 207
  See also specific islands; specific landmasses bordering the ocean
Inferences, convergence of, persuasion through, 147
Ireland, 37, 283, 284
Isla Clarión, 76
Islas Malvinas. *See* Falkland Islands
"Island rule," 173n
Islands-as-dead-ends rule, addressing, 263–264
Islands-as-stepping-stones explanations, 27, 74, 75n, 76, 212, 214, 217–218, 222 (box), 262, 290
Île Amsterdam, 76

Jamaica, 279
Japan, 200
Java, 12
Johnson, Donald, 173

Josephoartigasia monesi rodent, 291
Juan Fernandez Islands, 244
Jumping bristletails (Archaeognatha), 52, 258, 259, 260–261, 262–263, 264, 266, 306
Jurassic, 117, 229, 230 (fig.), 234, 250, 252, 301

Kagu (*Rhynochetos jubatus*), 242
Kakapo (*Strigops habroptilus*), 100, 101 (fig.)
Kalanikupule, 258
Kamehameha I, 258
Kauai, 74, 261
Kauri trees, 16, 102, 109, 163
Keeling Atoll, 201
Kenya, 190
Kilauea, 73
Kirtland's Warbler, 236
Kiwis, 3, 69, 80 (fig.), 100, 101, 102, 111, 225
Koolau Mountains, 257
Krause, David, 249–250
Kubitzki, Klaus, 170
Kuhn, Thomas, 42, 196, 272, 273, 275, 276
Kupukupu, 75

Lactoris shrub, 244
Lanai, 261
Land bridges, 2, 12 (box), 23, 25, 27, 28, 33, 42–43, 45, 76, 100, 156, 158, 169, 172, 173, 214, 266, 267, 268, 273, 274, 275, 277, 290
  See also Panamanian Isthmus
Landmasses-as-life-rafts story, 4, 7, 9, 14, 15, 18, 234, 251
  See also Continental drift; Gondwanan breakup
Lavin, Matt, 155–159, 166–167, 187, 265, 271
Laws and Explanation in History (Dray), 196
Lawson, Robin, 7, 188, 190, 198
Legumes/beans, 156–158, 159, 162, 286
Leiopelmatid frogs, 246n, 251
Lemurs, 44, 198, 209, 214 (fig.), 215, 225, 247, 248, 249 (fig.)
Lesser Antilles, 56, 253
Lesser Sundas, 200, 216
Lifespan of species, 235
Line Islands, 74n
Lizard floating experiments, 72
Lizards, 79, 143 (box), 206–207, 219, 247, 251, 265, 299

See also specific type
Locust swarms, 279, 280 (fig.)
Lo'ihi, 73
Long-distance dispersal, defined, 10 (box)
  See also Dispersalism; Oceanic dispersal
Lord Howe Island, 97, 108, 110, 161, 261
Lost World, The (Doyle), 152
Lyell, Charles, 28, 30

Mabuya skinks, 204–206, 293, 295 (fig.)
Macaques, 209, 216, 287
Macrofossil records, 103
Madagascar, 200
  as a continental island, 12 (box)
  and the Cretaceous, 248n, 249, 249–250
  dinosaurs of, 221 (box)
  evidence of dispersal origins for, 277
  and extinction risk, 236
  and Gondwanan breakup, 7, 143 (box), 247, 248
  and the Jurassic, 230 (fig.)
  and land bridges, 42, 43 (fig.), 45
  and the Miocene, 248
  myth applied to, 252
  as pivotal point in the vicariance-dispersal debate, 225
  and the Pleistocene, 250
  taxa involving, 3, 14, 44, 135, 143 (box), 157, 158, 161 (fig.), 165, 180, 182, 183 (fig.), 184, 186–187, 198, 199, 209, 225, 245, 247–250, 271, 286
  tracks and, 80 (fig.)
Magnetic field, 59, 60
Mahajanga Basin, 249–250
Majungasaurus dinosaur, 250
Malagasy. *See* Madagascar
Malay Archipelago, 29, 33
Malta, 173
Maluku Islands (Moluccas), 198, 200
Mantidactylus frogs, 184–185
Mantle convection theory, 55, 57, 62
  See also Seafloor spreading
Manual of Phytogeography (Croizat), 78
Maori, 239
Marmosets, 210, 289, 291
Marquesas, 76, 264
Mascarene Plateau, 251
Mass extinctions, 237, 238, 239, 240, 242, 246n, 250, 251, 252

INDEX

Matthew, William Diller, 43–45, 67, 78–79, 100, 118, 267, 274
Matthews, Drummond, 39, 59–60, 62
Maui, 261
Maui Nui, 261n
Mauna Loa, 73
Mauritius, 76
Mayans, 287
Mayotte, 182, 184–185, 186
Mayr, Ernst, 45, 48, 50, 65, 67, 78, 79, 83, 87, 118, 147
McDiarmid, Roy, 152
McDowall, Robert, 86, 227–233, 238, 244, 252
McPhee, John, 28
Measey, John, 175, 176–178, 180, 181–182, 188, 189–191, 192, 193, 194, 196, 198, 199, 261
Mendel, Gregor, 39
Meredith, Robert, 140–141
Merrill, E. D., 78
Mesoamerica, 287
Mesosaurus reptile, 37
Mesozoic, 63, 88, 90, 96, 97, 215, 242, 243, 250, 268
   See also Cretaceous; Jurassic; Triassic
Mexican black-bellied garter snake, 125–126
Mexico, 6, 7, 68, 76, 93, 97, 155, 156, 157
   See also Baja California
Michigan, northern, 236
Miconia (Melastomataceae) plants, 296
Microfossil records, 103
   See also Pollen
Mid-Atlantic Ridge, 34, 56–57, 58, 216
Middle America, 68, 158, 269
   See also Caribbean; Central America; Mexico
Midges. *See* Chironomid midges
Mildenhall, Dallas, 106–107, 109, 110, 117, 118
Miller, Stanley, 196–197
Miocene, 99, 102, 104, 106, 107, 161 (fig.), 238, 244, 248, 289 (fig.), 296
Mirroring, 293–294
Mississippi drainage, 287
Mite harvestmen, 246, 251
Mitochondrial DNA, 7, 124–125, 184, 189, 212–213, 224
Moas, 95, 101, 102, 245, 246
"Moa's Ark" analogy, 163, 234, 246
Mobilism, 81
Mojave Desert, 235
Molecular clock analyses
   bias in, addressing, 144, 145, 146
   calibration alternatives for, extremism involving, 142–143 (box)
   comeback of, 127–130
   comparing fossil-based age estimates with estimates from, 144–145, 146
   as a discarded idea, 126–127
   early basis for, 119n
   general agreement in massive survey of results from, 141
   importance and potential of, 276–278
   making the case in support of using, 146–147
   and the pace of genetic change, 126, 139–140
   problems with, addressing, 135–141
   providing evidence of animal oceanic dispersal, 7–8, 9 (fig.), 148, 186–187, 206, 212, 213, 215, 219, 231n, 241–242, 243, 244, 245, 248–249, 286, 290, 295 (fig.)
   providing evidence of plant oceanic dispersal, 75, 153–154, 157–160, 161–163, 166, 167–168, 169, 170, 242, 244
   reason for use of, 15
   refining, 128
   and the shift away from vicariance, 118, 271, 277–278
   skepticism about, 88–89, 90, 91, 111, 130, 134–135, 146
   strict, 88
   and suitability for modeling, 128–129
   upward trend in using, 118–119
   use of, 133–134
   See also DNA sequencing; Relaxed clock methods
Molecular dating. *See* Molecular clock analyses
Molecules, problem with, 135, 139
Molokai, 261
Monkeys, 3, 12, 13, 33, 44, 90, 209, 210–215, 216, 217–220, 223, 266, 267–268, 281, 287–288, 290, 291, 293, 294, 296, 299, 302–303, 304
Montgomery, Steve, 257–260, 264
Moose, 2
Moriori, 239
Morley, Lawrence, 60n
Morlière, Alain, 181, 182, 191
Morocco, 97
Morrone, Juan, 233n
Mount Ararat, 27, 28, 178, 234
Mozambique Channel, 44, 158, 165, 200, 248, 303
"Muddy" diagrams, 50, 51 (fig.), 71, 276
Muir, John, 198
Mullis, Kary, 120–122, 123, 124, 126, 130
"Multiple Overseas Dispersal in Amphibians" (Vences et al.), 187
Musicians Seamounts, 74n
Mutations, 129, 139, 302
Myths
   power of, and attempt at breaking, 252
   science and, Popper on, 1
   See also Falsification

Nachlinger, Jan (author's friend), 16, 95
National Geographic Society, 228
Natural crossings, influence of, 285, 286–288, 293–296
Natural selection, 23, 29, 30, 31, 40n, 85n, 103, 139, 169, 197, 208, 263
Nature (journal), 42, 59
Nazca Plate, 59
Nelson, Gareth (Gary), 47–48, 49, 50, 51–52, 53–54, 65–67, 68, 69, 70, 71, 77, 83, 84–85, 86, 87, 89–90, 91, 92, 100, 106, 110, 117, 118, 119, 135, 165, 192, 209, 212, 226, 267, 268, 269, 274, 275
Nematodes, 293–294
Nene Goose, 75
Neomachilellus bristletail, 261
Neomachilis bristletails, 52, 259 (fig.), 262–263, 264, 306
Neotropical Sunbittern, 242
New Caledonia, 4, 7, 69 (fig.), 82, 97, 98, 108, 152, 166, 171, 225, 242–244, 246n, 251, 252
New Guinea, 3, 69 (fig.), 80 (fig.), 82, 265, 286
New World
   discovery of, 115, 285
   and land bridges, 2n, 156
   taxa involving, 27, 148, 156, 157, 206, 212, 213, 214–215, 216, 219, 223, 261, 262, 279, 286–287, 295 (fig.), 302–303
   tracks and, 82
   unnatural crossings involving, 282, 283–285
   See also North America; South America

INDEX 357

New World monkeys, 210–215, 216, 217–218, 223, 291, 293–294, 300
New York, 155
"New York School" dispersalists, 45, 65, 66, 71, 78, 85, 87, 118, 169, 273
  See also specific scientists
New Zealand
  ark analogy involving, 163, 234, 246
  and biota like an oceanic island, 251
  and the Chatham Rise, 240
  comparing past floras of Australia with, 107–108
  as a continental island, 12 (box)
  and the Cretaceous, 97, 98, 100, 101, 102n, 103–104, 109
  and the Eocene, 104, 105
  fading prominence of panbiogeographers in, 277
  as a focal point of biogeography, 100–101, 110
  fossil records of, 101–102, 102–104, 105–106, 109, 160, 246
  and Gondwanan breakup, 4, 7, 17, 18, 99, 170
  and the Gondwanan relict idea, 96–97, 99, 100–102, 110, 152, 160, 161
  humans crossing to the Chathams from, 239
  and the Jurassic, 230 (fig.)
  and the Mesozoic, 63
  and the Miocene, 104, 106, 107
  myth applied to, 252
  and the Oligocene, 98–99, 104, 106
  and the Paleocene, 102n, 104
  as pivotal point in the vicariance-dispersal debate, 225
  and the Pleistocene, 104
  and the Pliocene, 104
  taxa involving, 3, 14, 16, 17–18, 25, 49, 54, 63, 64 (fig.), 65, 69 (fig.), 86, 87 (fig.), 95–97, 99, 100, 103–109, 110, 111, 117, 160, 161, 162–163, 164, 165, 166, 170, 171, 225, 226, 227, 241–242, 244–246
  tracks and, 80 (fig.), 82
  See also Chatham Islands; Zealandia
"New Zealand Biogeography—A Paleontologist's Approach" (Fleming), 107
New Zealand Geological Survey, 106

Newfoundland, 37
Newtonian physics, 91–92
Nicobar Islands, 209
Niger River, 192, 197
Noah's Ark, 4, 27, 178, 234
Nonaerial animals, generalization about dispersal and, 262
Norfolk Island, 108, 110, 161
Norfolk Island pine (Araucaria heterophylla), 108, 109
Normal dispersal, defined, 10 (box)
Normark, Ben, 124
Norops sagrei (brown anole), 72
North America, 34, 39
  and Australia, 200
  Carboniferous and Permian period, 37
  distance of Hawaiian Islands from, 73, 260
  eastern seaboard of, 287
  and the Eocene, 215
  and the Great American Interchange, 289–290
  Great Lakes of, 9, 287
  and ice ages, 2, 238
  and land bridges, 2, 42, 156, 172
  and the Panamanian Isthmus, 288, 289–290
  rock formations in, 61
  taxa involving, 2, 29, 179, 200, 214, 215, 238, 245, 259 (fig.), 260–261, 262, 265, 267, 286, 289–290, 294, 295 (fig.)
  treeline in mountains of, 17
North American Plate, 6, 60
North Equatorial Counter Current, 219
North Island, 99
North Island Robin, 241 (fig.)
Northern Channel Islands, 173
Northern Hemisphere, 79, 172
  See also specific northern landmasses, islands, and bodies of water
Nothofagus trees/shrubs. See Southern beeches
Nowak, Mike, 248, 249, 250
Nuclear DNA, 125, 213
Nutmeg, 285, 286, 287
Nuuanu Pali Wayside Park, 258

Oahu, 257–261
Observations of a Naturalist in the Pacific, vol. 2 (Guppy), 201
Ocean basins, early assumptions about, 33
  See also Fixed continents and ocean basins, belief in

Ocean crust, 36, 38, 55, 58, 242
Ocean floor, studies of, 55, 56–58, 59, 60
  See also Plate tectonics; Seafloor spreading
Oceanic dispersal
  envisioning, 18, 306
  inevitability of, 304
  invoking, facing constant disbelief when, 198
  long-distance dispersal usually as, 11 (box)
  molecular dating providing evidence of, 7–8, 9 (fig.), 75, 148, 153–154, 157–160, 161–163, 166, 167–168, 169, 170, 186–187, 206, 212, 213, 215, 219, 231n, 241–242, 243, 244, 245, 248–249, 286, 290, 295 (fig.)
  natural, influence of, 285, 286–288, 293–296
  number of studies supporting, epiphany resulting from, 14–15
  observed cases of, 19, 46, 93, 112, 173, 201, 253, 279
  probability of, awareness of, 32
  ubiquity of, awareness of, 281
  underestimating, 146
  unnatural, impact of, examples of, 99, 282–285
  See also specific taxa, landmasses, islands, and oceans
Oceanic island, defined, 12 (box)
Ocotea (Lauraceae) plants, 296
Oecomys rodents, 296
O'Grady, Patrick, 264, 265, 266
O'Hara, Robert, 197–198
O'helo papa, 75
Okinawa, 173
Old World
  and land bridges, 156
  rise in population of people in, 284
  taxa involving, 27, 156, 157, 206, 212, 214–215, 219, 279, 283, 284
  unnatural crossings involving, 282, 283–285
  See also Africa; Asia; Europe
Old World monkeys, 210–215, 216, 217–218
Oligocene, 98, 99, 104, 106, 213, 246n, 270
"On the Law Which Has Regulated the Introduction of New Species" (Wallace), 29, 110, 153n

Ope'ape'a, 75
Oregon coast, 305–306
Oreskes, Naomi, 40
Origin of Continents and Oceans, The (Wegener), 38
Origin of Species, The (Darwin), 24, 30, 32, 38, 88, 100, 168, 178, 179, 197, 219–220, 267, 276
Original continent. *See* Pangea
Ortelius, Abraham, 34
Ostriches, 3, 69, 80 (fig.), 244–245
Oval-leaf clustervine, 75
Overall picture, seeing the, importance of, 141, 146, 147
Owl-eyed night monkeys, 223

Pacific Ocean
  crossings involving, 75, 154, 162, 165, 215n, 224, 293
  and generalizations about Hawaiian crossings, 262, 263, 303
  and the Panamanian Isthmus, 13
  tectonic plates and rift involving, 6–7
  unnatural crossings involving, 99, 283
  volcanoes encircling, 58
  See also specific islands; specific landmasses bordering the ocean
Pacific Plate, 6, 60, 73, 99
Paleocene, 102n, 104
Paleontological data, reams of, 236–237
  See also Fossil records
Paleozoic, 268
Panamanian Isthmus, 13, 224, 288, 291n
Panbiogeography, 23–24, 77 (fig.), 80 (fig.), 81, 270n
Panbiogeography (Croizat), 78–79, 83
Pangea, 36–37, 38, 45, 61, 67, 145, 151
  See also Gondwana
Paradigm shift, process of, 272, 273–275
Parasites, 133–134, 139, 242, 293–294
Parlor game interest, 281
Patagonia, 232, 233
Paterson, Adrian, 241
Patterson, Colin, 66–67, 77, 269
Pa'u o Hi'iaka, 75
Pauling, Linus, 119, 126, 146
Pennsylvanian period, 34
Permian, 37, 229, 237
Peru, 283
Philippines, 198, 200

Phyllodactylid geckos, 295 (fig.)
Phylogenetics, 63, 65, 67, 69, 71, 89, 124, 153, 199–200, 199–200, 245, 262
Physics, theories in, 270n
Phytophthora infestans oomycete, 284
Pied Oystercatcher, 241
Pikaia (worm-like chordate), 298–299, 301
Pilgrims, 287
Pisonia trees, 259
Pittosporum shrubs/trees, 162
Pizarro, Francisco, 283
Plains viscacha (*Lagostomus maximus*), 291, 292 (fig.)
Plate tectonics, 4, 6, 15, 58–59, 60, 62, 63, 67, 69, 97, 117–118, 170, 180, 217, 229, 269–270, 275–276, 281
  See also Continental drift
Platnick, Norm, 68, 87, 89–90, 118
Platyrrhini monkeys. *See* New World monkeys
Pleistocene, 104, 135, 238, 239, 250
Pliocene, 68, 104, 291, 296
Podocarp conifers, 162, 163, 240, 242
Poland, 9
Pole, Mike, 106–111, 117, 118, 160–161, 163, 164, 246n
Pollen, 103, 106, 109, 153
Polymerase chain reaction (PCR), 122–123, 124–126, 129, 133, 271
Polynesia, 75, 262
Popper, Karl, 1, 87, 91, 192, 196, 215
Posadas, Paula, 233n
Potatoes (*Solanum tuberosum*), 283–285, 297
Pough, Harvey, 179
Power law, 237n
Preconceptions, blinded by, 185–186, 261, 268
Pre-paradigm period, 273–275
Primates, 131, 209, 220, 287–288
  See also specific type
Príncipe, 175–177, 178, 179, 180–181, 189, 190 (fig.), 191 (fig.), 192, 193, 194, 195, 196, 199, 200
Principles of Geology (Lyell), 28
Principles of Physical Geology (Holmes), 55
Proceedings of the Royal Society of London (journal), 187, 188
Procyonids, 290, 291n
Proficient dispersers, 101, 160, 168–169, 172, 219

Pseudogenes, 260
Pterosaurs, 102, 222 (box), 234
Ptychadena newtoni frog, 177, 189–192, 195
Pueo, 75
Pygmy hippos, 44
Pygmy sundew (*Drosera meristocaulis*), 152–154, 155
Pyramid Lake, 207–208

Radiation, impact of, 223, 291–296
Radiometric dating, 54–55, 61, 136
Ranunculus lyalli buttercup, 18, 105, 162
Ratites, 3, 4, 11 (box), 13, 33, 69, 80–81, 89, 111, 244–246, 244–246, 269
Rats, 44, 79, 99, 204, 291
Red Sea, 97
Refutation. *See* Falsification
Relationships of groups, importance of, 52–53
Relativity theory, 270n, 276
Relaxed clock methods, 140–141, 157, 186, 268
Religion, influence of, 25, 27, 28, 31, 32, 81, 178–179, 197n, 207, 219, 234
Renealmia (Zingiberaceae) gingers, 296, 301
Renner, Susanne, 160, 170, 171 (fig.)
Retrograde motion, 272
Rhantus beetles, 265
Rheas, 3, 69, 80 (fig.), 244–245
Rhinella toads, 295 (fig.)
Ribbon snakes, 125
Rivas, Jesús, 152
Rocky Mountains, 17, 297
Rodent clock, 126, 139
Rodents, 15, 44, 141, 219, 299
  See also specific type
Rodrigues, 76
Ronquist, Fred, 164–166, 167 (fig.), 170–171
Rosen, Donn, 48, 66–67, 68–69, 77, 84, 85, 91, 227, 269, 275
Round Island, 244
Russia, 284, 297

Salamanders, 179–180, 200
Salinity reduction, 192–193, 198
Samoa, 42, 268
Sampson, Scott, 249
San Andreas Fault, 6, 60
Sanger, Fred, 130
Sanmartín, Isabel, 164–166, 167 (fig.), 170–171
São Tomé, 175–177, 178, 179, 180–181, 189, 190, 191, 192, 193, 194, 195, 196, 199, 261

"Sarawak paper" (Wallace), 29, 110, 153n
Scaptomyza fly, 264–265
Schaefer, Hanno, 160
Schell Creek Range, 187, 188
Schistocerca gregaria locust, 279, 280 (fig.)
Schoener, Amy, 72
Schoener, Tom, 72
Schuchert, Charles, 266, 267, 268
Science
  Popper's philosophy of, 1, 87
  textbook version of, 62
Scientific revolutions, structure of, view on, 272, 273
Scotland, 37
Sea of Cortés, 7, 8, 9, 190, 198, 200
Seafloor spreading, 57, 58, 59–60, 60n, 61, 90, 163, 216, 267, 270
  See also Plate tectonics
Seed experiments, 26, 27, 32, 82, 168, 186, 198
Seeds, properties of, 79, 153, 169
Seychelles, 19, 180, 182, 186–187, 199, 225, 251, 252
Shahputra, Rizal, 131
Shared derived traits, defined, 49
Shark clock, 126, 139
Shattered-glass analogy, 70
Short-Eared Owl, 75
Sierra de la Laguna, 5, 7
Sierra Nevada, 17, 156, 187–188
Sigmodontine rodents, 290, 291, 293, 294, 296
Silversword plant, 75
Simplicity vs. complexity, addressing, 90–92, 146, 269–270
Simpson, George Gaylord, 45, 50, 65, 67, 78, 79, 83, 118, 147, 164, 266–267, 268, 290
Sister groups, defined, 12 (box)
Skinks, 203, 204–206, 223, 242, 281, 293, 295 (fig.)
Skottsberg, Carl, 277
"Slacker" species, notion of, 263, 264–265
Slow-running clocks, 126, 139
Smithsonian Institution, 170
"Smoking gun" concept, 218n
Snails, 26, 44, 93, 198, 219, 224, 281
Snake Range, 187, 188
Snakes, 97, 100, 141, 198, 206–207, 219, 247, 249, 294, 295
  See also specific type

Snider-Pellegrini, Antonio, 34, 36
Somali tectonic plate, 182
Sonoran Desert, 156
Sooglossid frogs, 251
Sophora bean trees, 162
South America, 116, 118, 285
  Carboniferous and Permian period, 37
  continental shelf of, 229
  contingency and unpredictability scenario involving, 300
  and the Cretaceous, 248n
  early assumptions about, 33
  and ease of Atlantic crossings, 79
  and the Eocene, 223
  and the Falklands, 229, 230–231, 232, 233
  final separation of, timetree depicting, 214 (fig.)
  and the fit with Africa, 34, 36, 37, 40
  and the Gondwana relict idea, 152
  and Gondwanan breakup, 4, 11 (box), 61 (fig.), 142 (box), 170, 288
  and the Great American Interchange, 289–290
  influence of natural crossings on, 288–296
  island-hopping routes to, 217 (fig.)
  as an isolated island, 288–289
  and the Isthmus of Panama, 288, 289–290
  and the Jurassic, 230 (fig.)
  and land bridges, 42–43, 290
  and the Mesozoic, 63
  ocean currents heading east from, 219
  taxa involving, 3, 15, 21, 25, 49, 54, 63, 64 (fig.), 69 (fig.), 79, 87, 108, 109, 135, 142 (box), 153–154, 155, 161 (fig.), 165, 166, 170–171, 200, 204, 205, 206, 207 (fig.), 209, 212, 213, 214, 215, 218, 223, 231, 233, 265, 267, 279, 283, 286, 289–296, 294, 295–296, 300, 301
  tracks and, 80 (fig.), 82
  and the Triassic, 272
  unnatural crossing from, example of, 283–284
South American Plate, 58–59
South Equatorial Current, 205
South Island, 17–18, 99, 104–106, 162, 240
Southeast Asia, 9, 12–13, 82, 161 (fig.), 180, 209, 248

Southern Alps, 104
Southern beeches (*Nothofagus*), 3, 4, 14, 17, 18, 69, 89, 102, 103, 106, 109, 111, 117, 162, 163, 166, 240, 242, 244, 246, 269, 281, 299
Southern Hemisphere, 3, 4, 14–15, 69, 272
  See also specific southern landmasses, islands, and bodies of water
Space, Time, Form (Croizat), 79
Spanish colonization, 283, 284, 297
Speciation event, defined, 50
Speciation, importance of, for islands, 236, 251
Spider monkeys, 210, 286, 291
Spiders, 46, 172n, 219, 262, 281
Sporopollenin, 103
Squirrel monkeys, 223, 303
Sri Lanka, 173
St. Helena, 266, 268
Steamer-ducks, 225, 226
Stephens Island, 99
Stewart Island, 163
Stochastic extinction, 235, 238
Stone runs, 226
String theory, 270n
Structure of Scientific Revolutions, The (Kuhn), 272
Sturm, Helmut, 262–263, 306
Suess, Eduard, 33n
Sulawesi, 44, 173, 200, 209, 216, 286
Sumatra, 12, 131, 189
Sunda Shelf, 189
Sundews (*Drosera*), 105, 152–154, 155, 299
Sunflowers (Asteraceae), 18, 162–163
Supercontinents. *See* Gondwana; Pangea
Surface tension, 72
Suriname, 151, 279
Sweepstakes dispersal, defined, 10 (box)
Sweetwater Mountains, 156
Swordtail fishes and relatives, 68, 269
Symphonia globulifera (Clusiaceae) trees, 296
Systematic Zoology (journal), 84
Systematics and Biogeography (Nelson & Platnick), 68, 87, 89, 118

Taleb, Nassim Nicholas, 220, 222
Tamarins, 214 (fig.)
Tanzania, 19, 190
Tapeworms, 133–134, 139
Tarsiers, 214 (fig.), 215

Tasman Sea, 65, 97, 101, 108, 111, 162, 165, 166
Tasmania, 14, 82
Tasmantis. *See* Zealandia
Taxon, defined, 12 (box)
Taylor, Frank Bursley, 34, 36n
Technological "advances," 237
Tectonic plates. *See* Plate tectonics
Tenrecs, 44, 143 (box), 247, 248
Tepuis, 151–154, 155
Tethys Sea, 221 (box), 287
Tethyshadros insularis dinosaur, 221–222 (box)
Thamnophis snakes. *See* Garter snakes
Thamnophis validus snake, 5–6, 7–8, 9 (fig.), 188, 190, 198
Thermus aquaticus (*Taq*) bacterium, 122, 125
Threadsnakes, 206, 281
Tidal force, 34, 38, 40, 54
Tierra del Fuego, 233
Timetree, defined, 12 (box)
Timetree of Life project, 141
Timor, 173
Tinamous, 245–246
Tipping point, 277
Tipping Point, The (Gladwell), 47
Toads, 178, 180, 200, 295
Toltecs, 287
Tomtit, 241
Tonga, 215n
Tortoises, 14, 19, 168–169, 219
Tracks, 80–81, 82, 82–83, 84, 85, 91, 100, 117
Transantarctic relationships, 49
Tree-of-life metaphor, 234
Trewick, Steve, 161–163, 241, 271, 277
Triassic, 117, 137, 272
Trinidad, 279
Tristan da Cunha, 76, 266, 268
Troglosironid harvestmen, 243
Tsunami, 131
Tuataras (*Sphenodon punctatus*), 16, 95, 96–97, 99, 100, 102, 246
Tuco-tucos, 291, 293, 303
Turdus thrushes, 295 (fig.), 296
Turtle, 68, 102, 242, 249, 250
Tussock grass, 17, 18, 104, 105
Tyrannosaurus dinosaur, 236, 240

Uganda, 190
Uniformitarianism, 28
United States, 153, 284, 287
Unnatural crossings, impact of, examples of, 99, 282–285
See also Human introductions
Unpredictability
archetypes of, 299–303
and contingency, 296, 297–299
Upland Sandpiper (*Bartramia longicauda*), 93

Vegetable sheep, 18, 244
Vences, Miguel, 182, 183 (fig.), 184–185, 186–187, 188–189, 189–191, 199, 271
Venezuela, 151, 152–154, 156
Vicariance biogeography
appeal of, 269–270
back-and-forth debate involving, 273–274
drivers of, 70, 269–270
emphasis in, 13, 14
evidence versus preconceived theories in, addressing, 268–269, 270, 271
fading of extreme, 278
and the focus on New Zealand, 100–101
hardening of, 85–86
lack of universal adoption, 276
lens of, 234
near dominance of, 100
pendulum swings involving, 268, 273
rise of, 13–14, 170, 270
shift away from, back toward dispersal, 118, 166, 270–272, 277–278
spread of, 47, 48
See also Historical biogeography
Vicariance, defined, 11 (box)
Vicariance event, defined, 13
See also Climate change; Continental drift; Gondwanan breakup; Ice ages
Villumsen, Rasmus, 41
Vine, Fred, 39, 59–60, 62
Vine-Matthews-Morley hypothesis, 60n
Volcanic mountains/islands.
See specific mountains/islands

Volcanic rock, 37, 58, 59
Volcanoes, 12 (box), 38, 56, 58, 60, 108, 178
Voyage of the Beagle, The (Darwin), 46, 293n

Wallace, Alfred Russel, 28–30, 31, 32, 33, 38, 78, 79, 81–82, 92, 96, 110, 153n, 198, 273n, 288
Wallace's Line, 33
Wallis, Graham, 161–163
Watson, James D., 130
Weather unpredictability, 301
Wegener, Alfred, 34–41, 42, 45, 54, 55, 58, 59, 60, 61, 62, 229, 267
Wegenerians, 67, 68
West Falkland Island, 250
West Indies, 206, 208, 223, 265, 279, 293
"What if" scenarios, 297
Whipsnake, 76
White-Faced Herons, 86, 87 (fig.)
Whittington, Harry, 298
Wiley, Ed, 91–92
Willets, 224
Wolves, 2, 226, 227, 231, 232 (fig.), 233
Wonderful Life (Gould), 297–298, 299
Woody legumes, 156–158, 159
"World of Wild Animals" map, 1–3
Worm lizards (Amphisbaenians), 142–143 (box), 203, 204, 206, 207 (fig.), 223, 281, 295 (fig.)

Xenodontine snakes, 294–295, 301

Yellowstone National Park, 122
Yoder, Anne, 198, 248, 249, 250, 277

Zealandia, 4, 97–98, 99, 101, 102, 103, 104, 105, 108, 109, 117, 160, 162, 171, 221 (box), 240, 246
See also Lord Howe Island; New Caledonia; New Zealand; Norfolk Island
Zucchini, 285, 286, 287
Zuckerkandl, Emile, 119, 126, 146